OCR
A LEVEL

1

# CHEMISTRY

John Older
Mike Smith

**HODDER**
EDUCATION
AN HACHETTE UK COMPANY

## Acknowledgements

### Photo credits

p 1 © Ryan McVay/Photodisc/Getty Images; *p. 10 t © koya979 – Fotolia, b* © Harold Cunningham/Getty Images; p. 14 © INSADCO Photography / Alamy; p. 28 © Alfred Pasieka/ Science Photo Library; p. 39 © Nneirda – Fotolia; p. 42 © Martyn Chillmaid; p. 62 © Science Photo Library; p. 75 © Charles D. Winters/Science Photo Library; p. 89 © Rita Jayaraman/ iStockphoto.com; p. 109 © DWW334 – Fotolia; p. 111 © Charles D. Winters/Science Photo Library; p. 115 © RTimages – Fotolia; p. 132 © Martin Bond/Science Photo Library; p. 134 *l* © Andrew Lambert Photography/Science Photo Library, *c* © sciencephotos / Alamy, *r* © Andrew Lambert Photography/Science Photo Library; p. 136 © Andrew Lambert Photography/Science Photo Library; p. 138 © Randy Faris/Corbis; p. 139 © Andrew Lambert Photography/Science Photo Library; p. 143 © NASA History Office and Kennedy Space Center; p. 144 *t* © Hodder, *b* © Phil Degginger / Alamy; p. 164 © NASA History Office and Kennedy Space Center; p. 165 © Philip Evans/Getty Images; p. 173 © Danny Lehman/Corbis; p. 184 © vovan – Fotolia; p. 202 © Paul Glendell / Alamy; p. 205 © Paul Glendell / Alamy; p. 219 *l* © Kevin Dyer/Getty Images, *r* © Olesia Bilkei – Fotolia; p. 225 © picsfive – Fotolia; p. 228 © Andrew Lambert Photography/ Science Photo Library; p. 236 *t* © Robert Rozbora – Fotolia, *b* © Science Photo Library; p. 244 © NASA/Phoenix Mission/University of Arizona/Corby Waste; p. 253 © Amy Sinisterra/AP/Press Association Images; p. 260 © NASA/Phoenix Mission/University of Arizona/Corby Waste; p. 266 © vinzstudio – Fotolia; p. 275 © littleny – Fotolia

Although every effort has been made to ensure that website addresses are correct at time of going to press, Hodder Education cannot be held responsible for the content of any website mentioned in this book. It is sometimes possible to find a relocated web page by typing in the address of the home page for a website in the URL window of your browser.

Hachette UK's policy is to use papers that are natural, renewable and recyclable products and made from wood grown in sustainable forests. The logging and manufacturing processes are expected to conform to the environmental regulations of the country of origin.

Orders: please contact Bookpoint Ltd, 130 Milton Park, Abingdon, Oxon OX14 4SB. Telephone: +44 (0)1235 827720. Fax: +44 (0)1235 400454. Lines are open 9.00a.m.–5.00p.m., Monday to Saturday, with a 24-hour message answering service. Visit our website at www.hoddereducation.co.uk

© John Older, Mike Smith 2015

First published in 2015 by

Hodder Education,

An Hachette UK Company

338 Euston Road

London NW1 3BH

| Impression number | 5 | 4 | 3 | 2 | 1 |
|---|---|---|---|---|---|
| Year | 2019 | 2018 | 2017 | 2016 | 2015 |

Cover photo © marcel – Fotolia

Illustrations by [design to complete]

Typeset in 11/13 pt Bliss Light by Integra Software Services Pvt. Ltd., Pondicherry, India

Printed in Italy

A catalogue record for this title is available from the British Library.

ISBN 9781471827068

# Contents

# Get the most from this book

Welcome to the **OCR A-level Chemistry 1 Student's Book**! This book covers Year 1 of the OCR A-level Chemistry specification and all content for the OCR AS Chemistry specification.

The following features have been included to help you get the most from this book.

## Prior knowledge

This is a short list of topics that you should be familiar with before starting a chapter. The questions will help to test your understanding.

## Key terms and formulae

These are highlighted in the text and definitions are given in the margin to help you pick out and learn these important concepts.

## Tips

These highlight important facts, common misconceptions and signpost you towards other relevant chapters.

## Test yourself questions

These short questions, found throughout each chapter, are useful for checking your understanding as you progress.

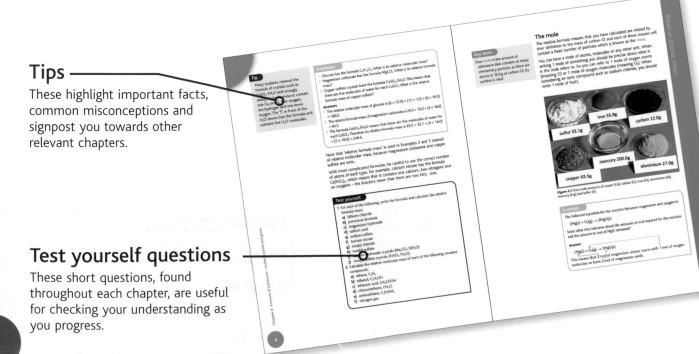

## Activities

These practical-based activities will help consolidate your learning and test your practical skills.

## Examples

Examples of questions or calculations are included to illustrate chapters and feature full workings and answers.

## Practice questions

You will find Practice questions, including multiple-choice questions, at the end of every chapter. These follow the style of the different types of questions with short and longer answers that you might see in your examination, and they are colour coded to highlight the level of difficulty. Challenge questions are also provided.

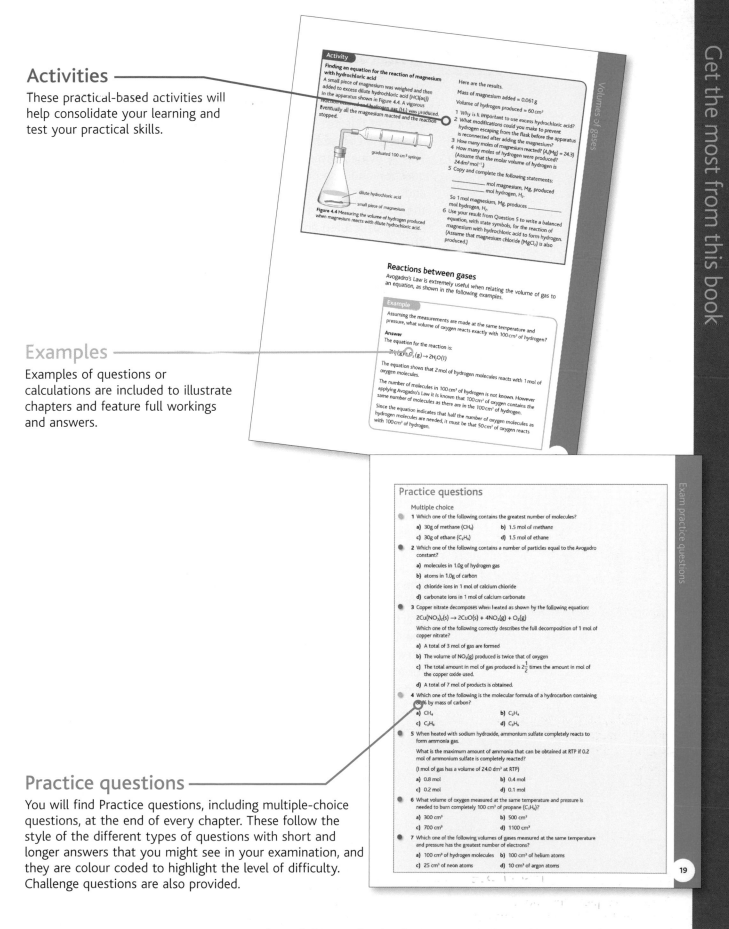

Dedicated chapters for developing your **Maths** and **Preparing for the exam** can be found at the back of this book.

# The questions in this book

The 'Test yourself' questions within the text of each chapter are designed to check that you have understood the short section of work that has just been covered. They are intended to be used during the course but can also be helpful while you are revising. The 'Activities' provide a chance to tackle a more extended question and also provide a check that you are able to understand and interpret experiments. The 'Practice' questions at the end of the chapters cover a broader section of work. They include multiple choice, structured and a smaller number of free response questions. They also sometimes focus on the evaluation of some experimental results. They are graded as follows:

- Basic questions that everyone must be able to answer without difficulty within the exam.

- Questions which cover work that are a regular feature of exams and that all competent candidates should be able to handle.

- More demanding questions that sometimes go beyond the normal requirements of the exam but which the best candidates should be able to do.

Challenge    These are questions for the most able candidates to test their full understanding and sometimes their ability to use ideas in a novel situation.

It must be emphasised that although these questions cover the skills and knowledge required to be successful in the examination, they are not exam questions as such. Many only cover components of questions which on the exam paper may come from more than one section of the specification. When you are confident that your knowledge base and your understanding is sufficient you must start practising the past examination questions which are available from the OCR website. Mark schemes are also available and these are very helpful in making it clear what points examiners are looking for and the depth that they expect. The best practice for the exam is doing past exams.

# Chapter 1

# Practical skills

As part of the full A Level course you must carry out a number of experiments which are separately assessed and lead to a practical endorsement. Contained in the chapters of this book you will find sections to help you understand the correct use of apparatus and the methods required to carry out these tasks. However, within both the AS and A Level courses, the written papers may also include questions that require you to understand the limits of reliability of the techniques that you have studied. This chapter gives some guidance to help support your understanding of laboratory work.

## The precision of apparatus

Each piece of laboratory apparatus used to measure a quantity has a limit to its precision. For example, a basic balance may give a measurement to the nearest gram; greater precision may be achieved by using a balance capable of reading to one, two, three or even four decimal places. The volume of a liquid can be measured using a graduated beaker, a measuring cylinder or volumetric apparatus. Each of these has a different level of precision. For most measuring equipment, the manufacturer will give the maximum error that is inherent in using that piece of apparatus; this is sometimes etched onto the apparatus but, in other cases, will need to be looked up. Not all experiments require the quantities measured to be exact, but if a quantitative result is needed, you should be aware of the limitations of the apparatus used.

The maximum error is an inevitable part of using that piece of equipment and is distinct from the competence with which the experiment is carried out. Important decisions are sometimes based on the results of experiments. For example, titrations are used in health care, in the food industry and in forensic science. It is crucial that the people making decisions based on the results obtained understand the extent to which they can rely on the data from their analysis.

## Measuring volumes of liquids

### Beakers and conical flasks

Markings on the sides of beakers and conical flasks serve only as a very rough guide to the volume taken and it is impossible to assess the maximum error involved in their use; however, it is likely to be large. Calibrated beakers and conical flasks are not suitable for measuring volumes reliably.

### Measuring cylinders

Unless otherwise indicated, a useful rule of thumb for measuring cylinders is that they have a maximum error of one-half of the difference between the graduation marks. If a $25\,cm^3$ measuring cylinder has graduations every $0.5\,cm^3$, the error is $0.25\,cm^3$. If it is used

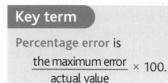

**Key term**

Percentage error is

$$\frac{\text{the maximum error}}{\text{actual value}} \times 100.$$

to measure a volume of 20 cm³, then the true volume lies between 19.75 cm³ and 20.25 cm³. This can usefully be expressed as a percentage error because it can then be carried forward and applied to other results that are based on the measurement. In this case the percentage error is:

$$\frac{0.25}{20.0} \times 100 = 1.25\%$$

A 100 cm³ measuring cylinder may have graduations every 1 cm³. The maximum error is, therefore, 0.5 cm³. So, 20 cm³ measured in a 100 cm³ cylinder is 20 ± 0.5 cm³. The percentage error is:

$$\frac{0.5}{20.0} \times 100 = 2.5\%$$

A 250 cm³ measuring cylinder usually has graduations every 2 cm³. The maximum error is, therefore, 1 cm³. So, 20 cm³ measured in a 250 cm³ cylinder is 20 ± 1 cm³. This has a large percentage error of 5%.

The difference in the error made in taking these measurements is considerable.

An increased error arises if both the start and end graduations are used to measure a volume. For example, if 20 cm³ of water is measured by filling a 100 cm³ measuring cylinder to the 100 cm³ mark and then pouring off the water until the 80 cm³ mark is reached, both the start and the end gradations are subject to a maximum error of 0.5 cm³. It is possible that the starting volume could be 100.5 cm³ and the end volume 79.5 cm³. The possible error is therefore 1 cm³ in the 20 cm³ measured, which is 5% (although this is probably greater than the experimental reality).

# Volumetric equipment

Volumetric apparatus includes pipettes, burettes and volumetric flasks.

- A **pipette** is usually selected when a small fixed volume of liquid or solution is required.

- A **burette** is used to provide more variable volumes.

- A **volumetric flask** is used when a larger volume of liquid or solution is required.

## Using a pipette

A pipette is used to deliver a precise volume of solution (see Figure 1.1). The maximum error will vary according to the quality of its manufacture. For routine work a 25.0 cm³ pipette with a maximum error of 0.06 cm³ is often used.

The percentage error is:

$$\frac{0.06}{25.0} \times 100 = 0.24\%$$

This is a much smaller error than would occur if a measuring cylinder had been used. The percentage error using a 25 cm³ measuring cylinder was 1.25% (see above), hence the pipette is over 5 times more reliable.

Some pipettes are graduated to allow a volume less than the full capacity to be dispensed. These are useful but usually have a greater maximum error than a standard pipette.

When the pipette is filled, the meniscus of the liquid should sit on the volume mark on the neck of the pipette. To do this reliably requires practice.

The solution in the pipette should be allowed to run out freely. When this is done a small amount will remain in the bottom of the pipette. Touch the tip of the pipette on the surface of the liquid that has been run out and then ignore any further solution that remains in the pipette.

**Figure 1.1** Using a pipette.

**Tip**

Since the solution that is sucked into the pipette may be poisonous, a pipette filler must always be used.

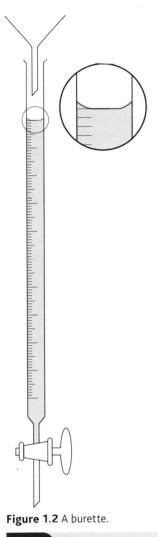

**Figure 1.2** A burette.

## Using a burette

When used correctly a volume measured by a burette has only a small error.

> **Tip**
>
> A funnel should be used to fill a burette so that solution is not spilled down the outside (Figure 1.2). Care must be taken not to overfill it. The tap should, of course, be closed during this procedure. However, before recording a volume, the funnel should be removed and some solution should be run out so that the space below the tap is filled.

The required volume is measured as the difference between final and initial readings. It is usual to read from the bottom of the meniscus but, because two readings are being subtracted, the top of the solution could be used.

Burettes usually have graduations every $0.1\,cm^3$ and should, therefore, have a maximum error of $0.05\,cm^3$. The volume delivered is the difference between two measurements, so the volume obtained has a potential uncertainty of $0.1\,cm^3$. The percentage error will vary with the volume that has been measured.

> **Test yourself**
>
> 1 Figure 1.3 shows four burettes.
>
>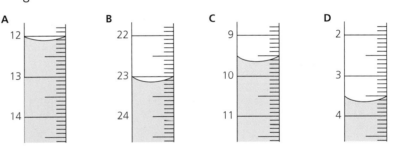
>
> **Figure 1.3**
>
> a) Record the readings for each of the four burettes.
> b) Calculate the percentage error in each of the burette readings. Assume the maximum error of each reading to be $0.05\,cm^3$ and give your answer to 2 decimal places.
> c) If the burette reading in A is the initial reading and burette reading in B is the final reading
>    i) calculate the titre volume
>    ii) calculate the percentage error in the titre volume.

> **Tip**
>
> There are different grades of burette, so the maximum error in any reading will vary. In an exam the maximum error will be stated if you are required to consider it.

> **Tip**
>
> If $250\,cm^3$ were measured using a $50\,cm^3$ burette filled five times, the exercise would not only be more tedious, but it would also have a greater error because each time the burette was used there would be an error in the volume measured.

## Using a volumetric flask

The mark on the neck of a volumetric flask indicates a specific volume. The maximum error is almost certain to be etched on the flask and is usually of the order of 0.125%. If a volumetric flask is used to measure $250.0\,cm^3$, the volume obtained has a maximum error of:

$$\frac{0.125}{100} \times 250 = 0.313\,cm^3$$

This means that the true volume lies between $249.7\,cm^3$ and $250.3\,cm^3$.

A volume measured in a volumetric flask is very precise. It holds the specified volume when the bottom of the meniscus of the solution sits on the engraved marking on the neck of the flask.

If it is necessary to dissolve a solid in water to make the solution to place in a volumetric flask, care is needed. It is usual to dissolve the solid first in a container such as a beaker using a volume of distilled water that is less than that required for the solution. A beaker is used so that, if necessary, the contents can be warmed to encourage the solid to dissolve. Once dissolved, the solution must be allowed to cool to room temperature before being transferred to the volumetric flask, using a funnel. Any solution remaining in the beaker is washed into the flask. Distilled water is added carefully to the flask until the meniscus sits on the engraved mark on the neck. With the stopper firmly in place, the flask is inverted a few times to ensure that the solution is completely mixed. The procedure is illustrated in Figure 1.4.

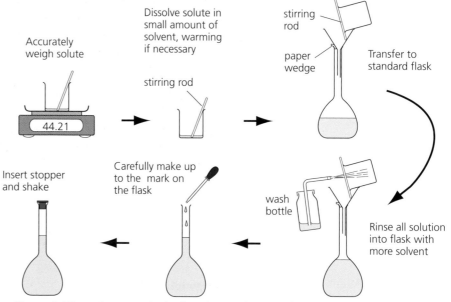

**Figure 1.4** Preparing a standard solution in a volumetric flask.

Solutions made up in a volumetric flask are sometimes referred to as standard solutions.

## Measurement of mass

If a **balance** measures to one decimal place, the mass measured will normally have a maximum error of 0.1 g (sometimes 0.05 g). For a balance measuring to two decimal places, the mass will be to within 0.01 g (sometimes 0.005 g). If carefully maintained and cleaned, a four decimal place balance will give a reading within 0.0001 g.

So if 1 g is weighed using a balance measuring to two decimal places, the mass may have an inevitable maximum error of 0.01 g. This means that the true mass lies anywhere between 0.99 g and 1.01 g.

The percentage error is:

$$\frac{0.01}{1} \times 100 = 1\%$$

Any result based on this measurement would also have a 1% error. A 1% error might be acceptable for most experimental work but weighing

1 g using a balance which can measure to four decimal places would give a percentage error of only

$$\frac{0.0001}{1} \times 100 = 0.01\%$$

If the mass measured is much smaller, more thought needs to be given. A two decimal place balance used to measure a mass of 0.10 g would have a large percentage error:

$$\frac{0.01}{0.10} \times 100 = 10\%$$

Using a balance accurate to four decimal places to weigh a mass of 0.10 g would give a more acceptable percentage error of 0.10%.

## Measurement of temperature

As with equipment used to measure volumes, different thermometers have different maximum errors in the readings taken. To obtain the percentage error in the readings, it is best to check the graduations. As with measuring cylinders, a useful rule of thumb for thermometers is that they have a maximum error of one-half the difference between their graduation marks.

In some experiments, thermometers are used to measure small differences in temperature. Care is needed in assessing the error resulting from the measurement of temperature change. In order to determine a temperature difference, both the initial and final temperatures are measured, which doubles the error.

Table 1.1 shows how the percentage error varies depending on the actual temperature change.

**Table 1.1** Errors in temperature readings. All readings are made using a thermometer graduated in units of 1 °C.

| Initial temperature/°C | Final temperature/°C | Change in temperature/°C | Error/°C | % error |
|---|---|---|---|---|
| 20 | 22 | 2 | 1 | $\left(\frac{1}{2}\right) \times 100 = 50$ |
| 20 | 60 | 40 | 1 | $\left(\frac{1}{40}\right) \times 100 = 2.5$ |

A digital thermometer should be more precise than a standard thermometer. It is normal practice with digital equipment to record all the digits shown, but there is still some inherent error.

**Test yourself**

2 Calculate the percentage error in measuring a mass of 2.5 g using a balance with a maximum error of:
   a) 0.1 g          b) 0.05 g          c) 0.001 g.
3 Calculate the percentage error in measuring 50 cm³ of a solution using:
   a) a 100 cm³ measuring cylinder with a maximum error of 0.5 cm³
   b) a 50 cm³ pipette with a maximum error of 0.1 cm³
   c) a 25 cm³ pipette with a maximum error of 0.06 cm³, used twice
   d) a 50 cm³ burette where each reading has a maximum error of 0.05 cm³.
▶▶▶

4 Calculate the percentage error in measuring a temperature rise of 6°C using a thermometer with a maximum error of:
   a) 0.5°C
   b) 0.1°C.
5 0.25 g of calcium carbonate is weighed out with a maximum error of 0.05 g to collect the carbon dioxide that it will produce when reacted with excess hydrochloric acid. The volume of carbon dioxide formed is 60 cm³ and is collected in a syringe with a maximum error of 1 cm³.
   a) Why does it matter that the hydrochloric acid is in excess?
   b) Calculate whether the balance or the syringe contributes the more significant error in this experiment.
   c) Write the equation for the reaction.
   d) Assuming that the volume of 1 mol of gas in the experiment is 24 dm³, show that the volume of gas collected should be 60 cm³.

# Recording results

When recording data, the precision should be indicated appropriately and the units must always be given. For example, if you use a balance that reads to two decimal places, the masses recorded should indicate this. This may seem obvious for a mass of, for example, 2.48 g. However, you must remember that this applies equally for a mass of, for example, 2.50 g. Here the '0' should be included after the '5' to indicate that this mass is also precise to two decimal places. Recording the mass as 2.5 g is incorrect and could be penalised in an exam.

When using a measuring cylinder with a maximum error of 1 cm³ to measure a volume, the result would normally be recorded without any decimal places. If you recorded a volume as 224.5 cm³ this would be considered incorrect.

Burette readings are normally recorded to 0.05 cm³ as this represents the appropriate maximum error. So a reading of 23.45 cm³ is acceptable, but 23.47 cm³ is not. The reading should not be recorded as 23.4 cm³ because this implies a maximum error of 0.1 cm³. If the initial reading is zero, it should be written as 0.00 cm³ because this reading has the same precision as any other.

## Identifying anomalies

If measurements are repeated in an experiment the results should be examined carefully to identify any that are anomalous. Any that clearly differ from the general trend should be rejected and excluded if an average is taken. For example, if titres obtained had volumes of 23.40 cm³, 23.90 cm³ and 23.50 cm³, the 23.90 cm³ stands out as being inconsistent with the other readings and should be excluded from the mean titre which should be given as 23.45 cm³. If a graph is drawn and one result is clearly inconsistent with the general shape of a line being drawn it should also be rejected and the line should not deviate to pass through that point.

## Overall accuracy of experimental results

When performing a quantitative experiment it is likely that you will use several pieces of apparatus whose maximum errors will vary. For example, a mass might have a percentage error of 0.5%, but a

**Tip**

You will not be required to carry out a detailed analysis. However, you should be able to identify the effect of the measurement with the largest error on the final answer.

temperature might have an percentage error of 4%. It is important to consider the appropriate precision of a value calculated using these data.

## Significant figures

Errors in measurements and results have been discussed, as has the importance of using a suitable number of figures when results are recorded, i.e. the correct number of **significant figures**. Applied properly, this indicates the level of precision. Significant figures are explained in more detail in Chapter 16.

- 0.12 has two digits beyond the decimal point. It has two significant figures and implies that the accurate number is between 0.11 and 0.13. In other words, the *final digit* is considered to have an accuracy of ± 1.

- 0.120 has three digits beyond the decimal point. Therefore, it has three significant figures and the number lies between 0.119 and 0.121.

- 0.004 has just one significant figure, because the two zeros beyond the decimal point are not counted. The number is between 0.003 and 0.005.

- 0.0042 has two significant figures. The number lies between 0.0041 and 0.0043.

- 0.004 00 has three significant figures. The two zeros after the '4' do count, so the number lies between 0.003 99 and 0.004 01.

A number such as 151.25 suggests a five-figure accuracy; it has five significant figures. The true number lies between 151.24 and 151.26.

If 151.25 is rounded to 151, then it has three significant figures. This indicates that the number lies between 150 and 152.

Writing 150 does not make the meaning clear. This is because when the final digit is zero it implies that the number is either between 149 and 151 or between 140 and 160. To get around this difficulty, scientific notation is used:

- $1.50 \times 10^2$ indicates that the number lies between 149 and 151 (remember the zero is a significant figure)

- $1.5 \times 10^2$ indicates that the number lies between 140 and 160.

### Test yourself

6 Using the appropriate number of significant figures, write down the precision possible of a mass of exactly 14 g measured using:
   a) a two decimal place balance
   b) a four decimal place balance.
7 A student performed a titration using a burette with a maximum error of 0.05 cm³. The student's results are recorded in Table 1.2, some of which are recorded incorrectly.

**Table 1.2**

| Titration | Rough | 1 | 2 | 3 |
|---|---|---|---|---|
| Final volume/cm³ | 24.0 | 23.57 | 23.9 | 23.62 |
| Initial volume/cm³ | 0 | 0.1 | 0 | 0 |
| Volume used/cm³ | 24 | 23.47 | 23.9 | 23.62 |

   a) Copy this table, but write the results appropriately and to the correct number of significant figures.
   b) State the mean value for the titre that should be used in any subsequent calculation.

# Errors in experimental procedure

Apart from the limitations imposed by apparatus, experiments also have errors caused by the procedure adopted. Such errors are difficult to quantify, but the following checklist might help you to assess an experiment.

### Handling substances

- Can you be sure that the substances you are using are pure? Solids may be damp and if a damp solid is weighed, the absorbed moisture is included in the mass.

- Does the experiment involve a reactant or a product that could react with the air? Remember that air contains carbon dioxide, which is acidic and reacts with alkalis. Some substances react with the oxygen present. Air is also always damp.

### Heating substances

- If you need to heat a substance until it decomposes, can you be sure that the decomposition is complete?

- Is it possible that heating is too strong and the product has decomposed further?

### Solutions

- If an experiment is quantitative, can you guarantee that any solution used is exactly at the stated concentration?

### Gases

- If a gas is collected during the experiment, can you be sure that none has escaped?

- If you collect a gas over water, are you sure it is not soluble?

- The volume of a gas is temperature-dependent.

### Timing

- If an experiment involves timing, are you sure that you can start and stop the timing exactly when required?

### Enthalpy experiments

- Heat loss is always a problem in enthalpy experiments, particularly if the reaction is slow.

## Improving experiments

As with limitations resulting from procedure, it is not possible to provide a comprehensive list of ways to improve experiments. However, it is worth emphasising that simple changes are often effective.

Using more precise measuring equipment is only useful if it focuses on a significant weakness. For example, if the volume of a solution does not need to be measured precisely, there is no point in using volumetric equipment.

If a change in the apparatus seems necessary, look for a straightforward alternative — for example:

- If the problem is that a gas to be collected over water is slightly soluble, using a syringe might be an effective remedy.

- If a gas might escape from the apparatus when a reagent is added, the procedure could be improved by placing one reagent in a container inside the flask containing the other reagent and mixing them when the apparatus is sealed.

- If the problem with an enthalpy reaction is that it is too slow, using powders would help.

It is a mistake to imagine that perfection can be achieved by using more complicated apparatus if the fault lies in the method that is being employed.

# Chapter 2

# Atoms and electrons

> **Prior knowledge**
>
> *In this chapter it is assumed that you know that:*
> - elements consist of atoms
> - compounds are formed when the atoms of elements combine together
> - matter is made up of elements and compounds.

> **Test yourself on prior knowledge**
>
> Explain the difference between:
> a) an element and a compound
> b) a compound and a mixture
> c) an atom and a molecule.

## Atoms

Elements are made up of atoms but the ways in which atoms combine together is very varied and forms the basis of a study of chemistry. By the end of the nineteenth century scientists began to appreciate that atoms were not indivisible as had originally been believed but had an intricate structure made up of different kinds of subatomic particles. Chemists have largely focused on the three main particles – protons, neutrons and electrons – although experiments done using the underground Large Hadron Collider at CERN on the borders of France and Switzerland have shown that protons and neutrons are made up of even smaller units.

**Figure 2.1** In the Hadron Large Collider at CERN two beams of high-energy particles travelling at speeds close to that of light are made to collide and the fragments created are analysed. In 2013 an analysis of the results of such an experiment tentatively confirmed the presence of a fundamental particle called the Higgs boson, whose existence had been predicted by Peter Higgs in 1964.

## Protons, neutrons and electrons

- An atom consists of a nucleus that is made up of positively charged particles called **protons** and uncharged particles called **neutrons**.

- Negatively charged particles called **electrons** orbit the nucleus.

- Protons and neutrons have almost identical masses. Electrons are much lighter, being approximately 1/2000 of the mass of a proton or neutron.

- All atoms are neutral. Therefore, the number of electrons is equal to the number of protons.

It is worth appreciating that the diameter of the nucleus is only about 1/100 000 of the total diameter of the atom. So, if the nucleus were the size of an average bedroom in a house in Birmingham (Figure 2.2), the electron would follow roughly an orbit through Blackpool, Swansea, London and Skegness. An atom has a large amount of space between its nucleus and the outer boundary of electrons.

The actual size of the charge on a proton or an electron is very small ($1.60 \times 10^{-19}$ coulombs) and the masses of protons, neutrons and electrons are equally tiny. A proton has a mass of only $1.67 \times 10^{-24}$ g.

You are not expected to recall these values but it is important that you remember the relative charges and sizes as shown in Table 2.1.

**Table 2.1** Relative charges and masses of the components of the atom.

| Particle | Relative charge | Relative mass |
|---|---|---|
| Proton (p) | +1 | 1 |
| Neutron (n) | 0 | 1 |
| Electron (e) | −1 | 1/1836 (approx. 1/2000) |

Some points to note are:

- The atoms of any one element are different from the atoms of all other elements.
- All atoms of the same element contain the same number of protons; atoms of different elements contain a different number of protons.
- The number of protons contained in the atom of an element is called its atomic number.
- The sum of the numbers of protons and neutrons in an atom is called its mass number. This implies that the protons and neutrons supply the total mass of an atom. This is not completely true, because there are also electrons present but, because electrons are so much lighter than protons and neutrons, it is a reasonable approximation.

It is easy to work out the numbers of particles present in an atom using its atomic number and mass number.

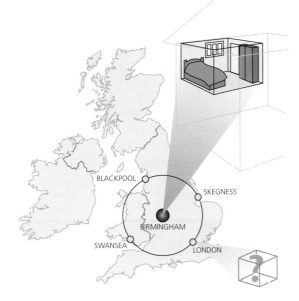

**Figure 2.2** A geographical illustration of the scale of an atom.

**Tip**

No one knows the size of an individual electron. As explained later, it cannot even be fully described as a particle.

**Key terms**

**Atomic number** is the number of protons in an atom of an element.
**Mass number** is the number of protons + neutrons in the nucleus of an atom.
An **ion** is an electrically charged particle formed by the loss or gain of one or more electrons from an atom or a group of atoms.

**Tip**

The atomic number is sometimes referred to as the **proton number**. The mass number is sometimes referred to as the **nucleon number**.

**Tip**

Mass
number $—_{13}^{27}$**Al**
Atomic
number

Subtract the bottom number from the top number to get the number of neutrons.

**Example 1**

An atom has atomic number 13 and mass number 27. How many protons, neutrons and electrons does it contain?

**Answer**
The atomic number is the number of protons, which is, therefore, 13.

Since an atom is neutral, the number of electrons is also 13.

The number of protons plus the number of neutrons is 27. Therefore, the number of neutrons is 14.

(The element is aluminium and it is usually written as $_{13}^{27}$Al.)

## Formation of ions

Losing or gaining electrons is the process by which many chemical reactions take place. When atoms do this they form charged particles called **ions**. An ion has the same atomic number and mass number as the atom of the element from which it was made.

If an atom loses one electron, the number of electrons in the atom becomes one fewer than the number of protons. Therefore, it has an electrical charge of +1. For example, a sodium atom loses an electron, forming an ion that is represented by $Na^+$. A calcium atom can lose two electrons and, as it then has two fewer electrons than it does protons, it is written as $Ca^{2+}$.

If one electron is gained, the atom of the element contains one more electron than it does protons and, therefore, has an electrical charge of −1. For example, a chlorine atom gains an electron, forming a chloride ion, which is represented by $Cl^-$. An oxygen atom gains two electrons, forming an oxide ion. This has two more electrons than protons and is written as $O^{2-}$.

Ions have different properties from those of the atoms of the element from which they came. For example, sodium metal is extremely reactive but sodium ions are not.

**Tip**

For non-metals, the element and its ion do not have the same name. A non-metallic ion has the ending '-ide'. For example, the ion from oxygen is oxide, from bromine it is bromide and from sulfur it is sulfide.

**Test yourself**

1 Use the atomic numbers and mass numbers given to deduce the number of protons, neutrons and electrons present in the following atoms and the ion formed from them.
   a) i) oxygen, O (atomic number 8, mass number 16)
      ii) $O^{2-}$
   b) i) sodium, Na (atomic number 11, mass number 23)
      ii) $Na^+$
   c) i) fluorine, F (atomic number 9, mass number 19)
      ii) $F^-$
   d) i) calcium, Ca (atomic number 20, mass number 40)
      ii) $Ca^{2+}$

**Example 2**

How many protons, electrons and neutrons are present in a sulfide ion, $S^{2-}$? Sulfur has atomic number 16 and mass number 32.

**Answer**

Sulfur has atomic number 16, so a sulfur atom contains 16 protons.

Since the electrical charge of a sulfide ion is '2−', it contains two more electrons than it does protons. Therefore, the number of electrons is 16 + 2 = 18.

The atomic mass is 32. The number of neutrons is the mass number minus the number of protons (32 − 16), which is 16.

Therefore, a sulfide ion contains 16 protons, 16 neutrons and 18 electrons.

It is written as $^{32}_{16}S^{2-}$.

**Example 3**

How many protons, neutrons and electrons are present in a potassium ion, $K^+$? Potassium has atomic number 19 and mass number 39.

**Answer**

Potassium has atomic number 19, so a potassium atom contains 19 protons.

Since the electrical charge of a potassium ion is '1+', it contains one more proton than it does electrons. Therefore, the number of electrons is 18.

The atomic mass is 39. The number of neutrons is 39 − 19, which is 20.

Therefore, a potassium ion contains 19 protons, 20 neutrons and 18 electrons.

It is written as $^{39}_{19}K^+$.

## Isotopes and mass spectra

Most elements are made up of more than one type of atom. Although these atoms contain the same number of protons they differ in the number of neutrons that are present. Potassium, for example, consists

Isotopes are atoms of the same element that have different masses. The isotopes of an element have the same number of protons (and electrons), but different numbers of neutrons.

of three different types of atom. All have 19 protons and most have 20 neutrons but a few have 22 neutrons and a very small number have 21. These variations are called isotopes of the element.

Evidence for the existence of isotopes was obtained by using an instrument called a mass spectrometer. Although you will not be expected to have knowledge of a mass spectrometer in an exam it is useful for you to understand the general principle by which it works. A gaseous sample of the substance is bombarded with high-energy electrons, which knock off electrons from the edge of an atom to create positive ions. These are separated according to their charge and those with a 1+ charge are selected. After being accelerated by an electric field, a focused stream of these ions is then passed at constant speed through a magnetic field. Here, the ions are deflected and the degree of the deflection depends on the mass of the ion. A detector is calibrated to record the degree of deflection and interpret this in terms of mass. The key features are summarised in Figure 2.3. In order to avoid interference from gases in the air, the spectrometer is evacuated using a pump.

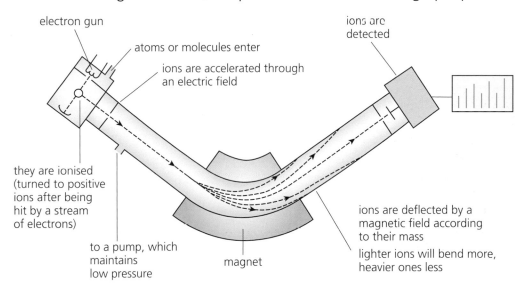

**Figure 2.3** A representation of a mass spectrometer.

A mass spectrometer analyses positive ions only. Do not confuse these ions with the ions that are formed under normal laboratory conditions. The conditions employed in the apparatus are such that it is impossible to create negative ions. An alternative version is the 'time of flight (TOF)' mass spectrometer. In this spectrometer ions are accelerated so that those of the same charge have the same kinetic energy. The velocity of the ion then depends only on its mass with those of greater mass moving more slowly. The time taken for ions to move across a fixed distance to a detector is then measured and this is interpreted to obtain an accurate mass of the particle.

## Mass spectra of elements

The mass spectrometer provides information about the masses of every particle that has been formed inside the machine and how often this particle has been detected. For an atom, a typical print-out from a mass spectrometer looks like a 'stick-diagram'. Each 'stick' shows the presence of an ion formed from an isotope and the height of the 'stick' represents its abundance.

Although you will not be asked to describe a mass spectrometer in an exam, you need to understand how the results obtained from a mass spectrometer are used.

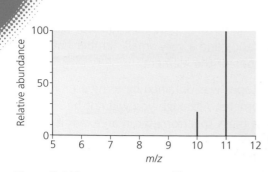

Figure 2.4 The mass spectrum of boron.

The mass spectrum for boron is shown in Figure 2.4.

The *x*-axis is labelled '*m/z*', which means mass/charge. However, since the charge is 1+ it is effectively the relative mass of the ions that is recorded. A mass spectrometer is calibrated so that the isotope carbon-12 is given a value of exactly 12. This is the standard used universally for the comparison of the masses of atoms (see page 40). The *y*-axis records relative abundance, which is the number of times an ion of that mass has been detected.

Figure 2.5 A mass spectrometer.

**Tip**

The word 'relative' is used because the figures are the masses of the isotopes based on a comparison with the mass of carbon, which is given a mass of 12 (see Chapter 4 page 40).

**Tip**

Strictly, it is better to use the expression 'weighted mean relative atomic mass' rather than 'relative atomic mass'. The average of 10 and 11 is 10.5. However, here the numbers of atoms of each isotope are considered and, in these circumstances, the word 'weighted' is included which takes account of the fact that there are more atoms of boron-11 than there are of boron-10.

The mass spectrum of boron has two lines. They are both a result of the formation of B$^+$ ions. One line is equivalent to a mass of 10 and the other to a mass of 11. This occurs because boron consists of two isotopes, the atoms of one isotope containing five neutrons and the other containing six neutrons. The relative abundance of these two isotopes is shown by the heights of the lines – they are in the ratio of 23 (boron-10) to 100 (boron-11). Knowing this ratio allows the mean relative atomic mass to be determined.

Consider a total of 123 boron particles. 23 will be boron-10 and 100 will be boron-11. Therefore, the atomic mass has a contribution of (23/123) from boron-10, which is (23/123) × 10 = 1.87, and a contribution of (100/123) from boron-11, which is (100/123) × 11 = 8.94.

The relative atomic mass of boron is 1.87 + 8.94 = 10.81 (or 10.8 to 1 decimal place).

In some problems the percentage abundances obtained from a mass spectrum may be supplied. The following example shows how the atomic mass is then obtained.

**Example 4**

Silver has two isotopes: 51.35% of the atoms are silver-107 and 48.65% of the atoms are silver-109. Calculate the relative atomic mass of silver.

**Answer**

The fraction that is $^{107}Ag^+$ is 51.35/100; the fraction that is $^{109}Ag^+$ is 48.65/100.

$$\text{weighted mean relative atomic mass} = \frac{51.35}{100} \times 107 + \frac{48.65}{100} \times 109 = 107.97$$

or 108.0 to 1 decimal place

**Activity**

**Finding the relative atomic mass of neon**

Look carefully at the mass spectrum of neon in Figure 2.6.

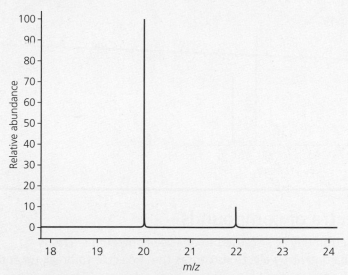

**Figure 2.6** The mass spectrum of neon.

1 How many isotopes does neon contain?
2 What are the relative masses of the isotopes of neon?
3 What are the relative amounts of the isotopes of neon?
4 Start a spreadsheet program (e.g. Excel) on your computer and open up a new spreadsheet for your results.
 a) Enter the relative masses of the isotopes of neon in column 1 and the relative amounts of these isotopes in column 2.
 b) Enter a formula in column 3 to work out the contribution of each isotope to the overall relative atomic mass of neon. (Hint: Look at the calculation of the relative atomic mass of boron or silver in the section above.)
 c) Finally, enter a formula in column 4 to calculate the relative atomic mass of neon.
 (If you do not have access to a computer with a spreadsheet program, calculate the relative atomic mass of neon yourself.)
5 Pure samples of the isotopes of neon have different densities.
 a) Why is this?
 b) Which isotope of neon will have the highest density?
6 Isotopes of the same element have different physical properties but the same chemical properties. Why is this?
7 Samples of neon collected in different parts of the world have slightly different relative atomic masses. Why is this?

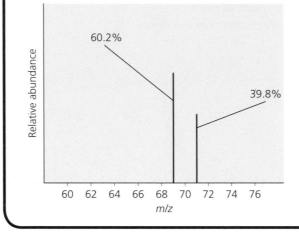
**Key term**

A molecular ion peak is the peak corresponding to the relative molecular mass of the compound. (It is sometimes referred to as the parent ion peak.)

# Mass spectra of compounds

The mass spectrometer is also used to analyse compounds. This is because bombardment by electrons causes the molecule to break up and each of the fragments obtained registers on the detector. This can be an advantage, as it is sometimes possible to obtain details of the molecule's structure as well as its overall molecular mass (see Chapter 15).

## Mass spectrum of propane

The mass spectrum of propane, $C_3H_8$, is shown in Figure 2.7.

The relative molecular mass of propane is 44. The peak furthest to the right of the spectrum represents the ion $[C_3H_8]^+$. This is called the molecular ion peak. The molecular ion is produced by the removal of an electron from the molecule within the mass spectrometer:

$$H_3C{-}CH_2{-}CH_3(g) \rightarrow [H_3C{-}CH_2{-}CH_3]^+(g) + e^-$$

However, under the bombardment of electrons, the molecular ion is unstable and breaks down to ion fragments of the molecule, which are also detected. Some examples are labelled on Figure 2.7. The peak at $m/z$ 29 occurs because a C–C bond in the $CH_3CH_2CH_3$ chain of the molecular ion is broken, producing two fragments. One of the fragments has a positive charge and the other is neutral:

$$[H_3C{-}CH_2{-}CH_3]^+(g) \rightarrow [CH_3{-}CH_2]^+(g) + CH_3(g)$$

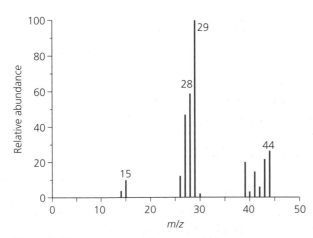

**Figure 2.7** Mass spectrum of propane, $C_3H_8$.

In this case, the ion $[CH_3CH_2]^+$ is then detected with a peak at $m/z = 29$.

The peak at $m/z$ 15 represents a $[CH_3]^+$ ion. Again, a C–C bond in the $CH_3CH_2CH_3$ chain is broken, producing one fragment with a positive charge and one fragment that is neutral:

$$[H_3C–CH_2–CH_3]^+(g) \rightarrow [CH_3]^+(g) + CH_3–CH_2(g)$$

A fragment can be identified for all other peaks in the spectrum. It is because of this fragmentation that the set of peaks is often described as a **fingerprint** of the molecule.

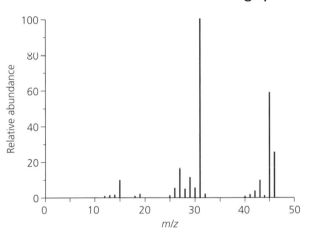

**Figure 2.8** Mass spectrum of ethanol, $CH_3CH_2OH$

## Mass spectrum of ethanol, $CH_3CH_2OH$

The mass spectrum of ethanol is shown in Figure 2.8.

The molecular ion is represented by the line at $m/z$ 46. This is the line furthest to the right, but it is not the highest peak. The height of the line represents the relative amount of that ion formed from the sample. It is often the case that a molecule breaks down readily, rather than surviving as the molecular ion. The peak at $m/z$ 45 is due to the loss of H from the –OH group, resulting in the ion $[CH_3CH_2O]^+(g)$. The most abundant peak is at $m/z$ 31, which corresponds to the ion $[CH_2OH]^+(g)$, formed by the fragmentation of ethanol by breaking the C–C bond.

### Tip

Remember that it is positive ions that are detected by a mass spectrometer and you must include the positive charge if you are asked to identify any of the fragments.

## Lines due to isomers

The precision of a mass spectrometer is such that lines formed from different isotopes can be observed (though these have been excluded from Figures 2.7 and 2.8). These lines have small peak heights. One characteristic is the presence of such small peaks beyond the molecular ion peak. These are usually the result of small numbers of ions containing the isotope $^{13}C$, although other isotopes may contribute. These peaks do not usually need to be considered when using a mass spectrum to identify a compound.

## The accuracy of mass spectrometers

Whereas the original machine designed by Francis Aston was large and only moderately accurate, today's instruments are much smaller and the results are accurate to 4 decimal places. This precision means that, for example, a mass spectrometer can distinguish between carbon monoxide, nitrogen and ethene even though to a first approximation each has a mass of 28. The accurate figures are: carbon monoxide, CO, 28.0101; nitrogen, $N_2$, 28.0134; ethene, $C_2H_4$, 28.0530.

A mass spectrometer uses very small samples of gaseous molecules and, today, can be operated remotely. This makes it a useful piece of equipment for analysing samples in remote locations. For example, the equipment on board the Mars space probe included a mass spectrometer that analysed samples of gas and returned the results to Earth. The mass spectrometer also plays an important role in forensic science. Chemicals are separated and a mass spectrum is obtained and the fragmentation pattern is matched by computer to a database. This type of analysis is used extensively in sport for detecting the use of illegal substances.

**Test yourself**

**6** The mass spectrum below is of a compound containing only hydrogen and carbon.

a) State which peak is the molecular ion peak.

b) Give the molecular formula of the hydrocarbon.

c) Suggest the formulae of each of the labelled fragments.

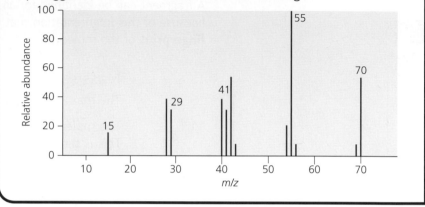

# Electrons

**Key term**

The ground state of an atom shows how it naturally exists with its electrons in their lowest energy position.

**Tip**

If you are asked to give the electron configuration of an element you should give the ground state for that element.

You may be familiar with a picture of atoms based on a model suggested by the Danish scientist, Niels Bohr. Each atom has electrons in orbits that are outside the nucleus and are of increasing size. The first shell can accommodate two electrons, the second eight electrons, and so on. Such diagrams show what is known as the ground state of the atom. This means that its electrons are in positions that are as close as allowable to the nucleus of the atom. These diagrams are useful as a basic representation of atomic structure and help us to understand the nature of chemical bonding (see Chapter 7). However, their use is limited to atoms with relatively few electrons. Electrons can however move from one orbit to another within an atom. If the electrons are in a position of higher energy they are said to be **excited**.

## Ground states of some elements

The simplest atom is hydrogen. With atomic number 1, an atom of hydrogen contains one proton and hence one electron. The next atom, helium, has atomic number 2. Therefore, there are two protons in its nucleus. The first shell can accommodate two electrons only and it is therefore completely filled. An atom of lithium has three protons in its nucleus. The third electron is in a new, larger shell.

Table 2.2 shows the arrangement of the electrons for the first three elements.

**Table 2.2** Electron arrangement for the ground state of the first three elements.

| Atomic number/ number of protons | 1 | 2 | 3 |
|---|---|---|---|
| Symbol | H | He | Li |
| Arrangement of electrons | | | |

The second shell can accommodate eight electrons. As the atomic number increases, the electrons progressively fill this shell. This is illustrated by beryllium and carbon. Beryllium, $_4$Be, has four electrons arranged 2, 2; carbon, $_6$C, has six electrons arranged 2, 4.

**Tip**

The shells are also referred to as electron orbits.

Neon, element 10, contains the maximum eight electrons in the second shell. The next element, sodium, $_{11}$Na, utilises a third shell. Its electrons are arranged 2, 8, 1.

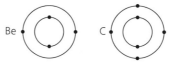

Chlorine and argon are further examples of elements with electrons in the third shell.

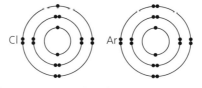

When an element forms an ion, the electrons are also accommodated in the appropriate shells.

So, for example, an oxide ion, $_8$O$^{2-}$ contains 10 electrons which between them will fill both the first and second shell.

**Key terms**

The **1st ionisation energy** is the energy required to remove 1 electron from each atom in 1 mole of gaseous atoms to form 1 mole of gaseous 1+ ions.

The **2nd ionisation energy** is the energy required to remove 1 electron from each 1+ ion in 1 mole of gaseous 1+ ions to form 1 mole of gaseous 2+ ions.

The **$n$th ionisation energy** is the energy required to remove 1 electron from each $(n-1)+$ ion in 1 mole of gaseous $(n-1)+$ ions to form 1 mole of gaseous $n+$ ions.

**Test yourself**

7 Draw the electrons in their shells for each of the following elements:
  a) nitrogen (atomic number 7)
  b) magnesium (atomic number 12)
  c) chloride ion, Cl$^-$ (atomic number 17)
  d) calcium ion, Ca$^{2+}$(atomic number 20).

## Ionisation energies

Some evidence for the existence of electron shells can be obtained by plotting the energy required to remove electrons, one by one, from a gaseous atom. These energies can be measured and are called ionisation energies.

For example, when successive ionisation energies are plotted for chlorine the graph has the shape shown in Figure 2.9.

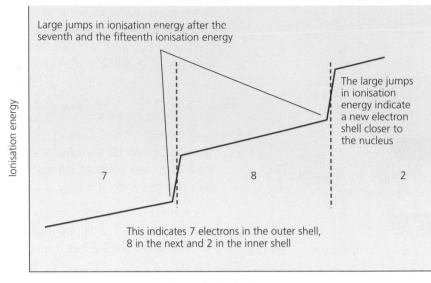

Large jumps in ionisation energy after the seventh and the fifteenth ionisation energy

The large jumps in ionisation energy indicate a new electron shell closer to the nucleus

This indicates 7 electrons in the outer shell, 8 in the next and 2 in the inner shell

Successive ionisation energy

**Figure 2.9** Successive ionisation energies of chlorine, $^{17}Cl$.

Chlorine has 17 electrons, arranged 2, 8, 7. There is a steady increase in the energy required to remove the electrons. This is to be expected because, as each electron is removed, the nucleus exerts an increasing pull on the remaining electrons (see page 120 for more details on ionisation energies). There is a noticeable jump in the energy required after the seventh and the fifteenth electrons have been removed. This is because at these two stages the electron is now being removed from a shell nearer to the nucleus.

## Evidence for the existence of sub-shells

Knowing the basic structure of the atom is helpful in understanding the way that atoms behave and you will use these ideas as your knowledge of chemistry develops. However, there are some aspects of the behaviour of electrons that cannot be explained by assuming they are in shells. Further research during the twentieth century revealed that electrons are more complicated than just small particles revolving around a central nucleus. There is experimental evidence to suggest that each shell is made up of sub-shells.

By studying the first ionisation energy (the energy required to remove the first electron from an atom), it is found that the first 20 elements show a more complex pattern than might be expected (Figure 2.10). Trends in the ionisation energies of the elements are looked at in detail in Chapter 8. For now, notice that, although there is a general increase in ionisation energy until each shell is filled, the trend is not quite uniform. There are several small peaks and troughs. These peaks and troughs are repeated in each shell and provide some evidence that the arrangement of electrons involves sub-shells.

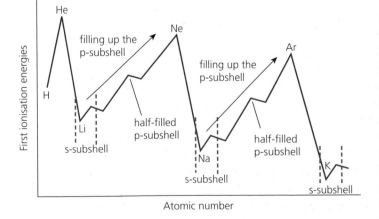

**Figure 2.10** The pattern of ionisation energies gives evidence for the existence of sub-shells.

# Electron energy levels and orbitals

## Energy levels

Apart from the evidence obtained from the patterns of ionisation energies, other research, particularly the mathematical description developed by Erwin Schrödinger, led to the conclusion that instead of focusing on the position of electrons in terms of shells, it is more accurate to assign them energy levels. These energy levels are numbered 1, 2, 3, etc. and correspond broadly to the shells. All but level 1 are subdivided and given letters that characterise them in more detail. The letters used are s, p, d and f.

The lowest electron energy level in an atom is called '1s'. This is the only option for shell 1. Stepping upwards in energy, the next energy level is '2s'. However, shell 2 has a further energy subdivision called a sub-shell and labelled '2p'. Level 3 has 3s, 3p and also 3d sub-shells. Level 4 has 4s, 4p, 4d and 4f sub-shells.

An energy level labelled 'p' is of higher energy than the equivalent 's' level. Therefore, 2p has higher energy than 2s; 3p has higher energy than 3s and so on.

It is important to know the order of increasing energy of the electron levels because this indicates the electron configuration of the element in its ground state (i.e. the lowest energy state that it can achieve). These energies do not always follow the sequence you might expect, but fortunately there is an easy way of remembering the order (Figure 2.11).

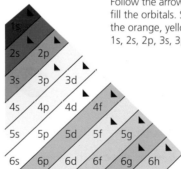

Follow the arrows to obtain the order that the electrons fill the orbitals. Start with the red and follow with the orange, yellow, green etc. to give the sequence, 1s, 2s, 2p, 3s, 3p, 4s, 3d, 4p, 5s, 4d, 5p etc.

**Figure 2.11** Key to remembering the order in which energy levels are filled.

Note that the 4s fills up before the 3d; likewise, the 5s fills up before the 4d.

## Electron orbitals

Chemists have used complex mathematics to develop a description of how electrons look in the various energy levels. It will seem strange when you first meet it, but the idea of an electron as a particle is replaced by a description based on the electron as a cloud of negative charge. This charge is spread like a mist around the nucleus of the element. The mist varies in its density at different distances from the nucleus. These regions of 'electron mist' are called orbitals and correspond to the energy levels used by an atom. Each orbital can hold one or two electrons and has a density of charge that depends on the energy level being described. If the orbital has two electrons the two electrons will differ in the direction in which they spin.

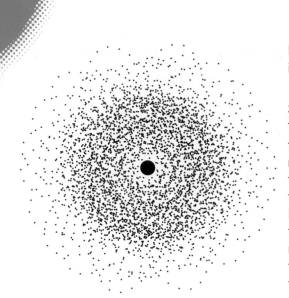

**Figure 2.12** A 1s orbital.

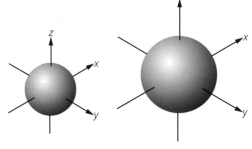

1s-orbital        2s-orbital

**Figure 2.13** The shape of 1s- and 2s-orbitals.

In Figure 2.12, the plot is darkest where the electron charge density is greatest and lightest where the electron charge density is less.

## s-orbitals

The simplest type of orbital to describe is the orbital corresponding to an 's' energy level. It consists of a spherical volume of negative charge with the nucleus at its centre. All s-orbitals have this shape. The difference between a 1s-orbital and a 2s-orbital is the distance from the nucleus to where the major density of charge is concentrated. In the 1s-orbital, the electron density is closer to the nucleus. As the levels (principal quantum numbers) increase, the radius at which the charge is most dense becomes further away from the nucleus. For example, a 4s-orbital is larger than a 3s because the point at which the 4s-orbital has its greatest density of negative charge is further from the nucleus than it is for the 3s-orbital.

## p-orbitals

The p-orbitals are more complex and harder to visualise (Figure 2.14). They have an elongated dumbbell shape and a variable charge density, with the area of greatest concentration increasing with the distance from the nucleus as the principal quantum number increases. For each principal quantum number there are three p-orbitals. They are identical and have the same energy, differing only in their orientation in space. They are labelled '$x$', '$y$' and '$z$' to correspond to the three principal axes.

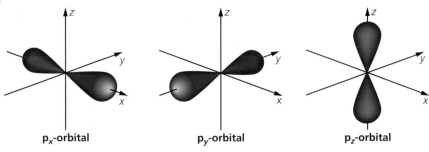

p$_x$-orbital        p$_y$-orbital        p$_z$-orbital

**Figure 2.14** The shape of 2p-orbitals.

> **Tip**
>
> Orbitals that have the same energy are said to be degenerate.

> **Tip**
>
> For each principal quantum number of 3 and above there are five d-orbitals. It is not necessary to know their shapes at this stage.

> **Tip**
>
> The reasons why the orbitals are filled in this order are complex and you will not be asked to provide any explanation.

## Representing the electron configuration of the elements

Electron energy levels can be represented in 'box' diagrams, as shown in Figure 2.15.

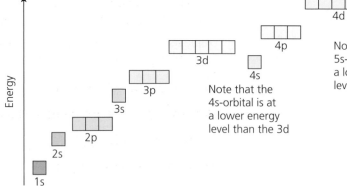

Note that the 4s-orbital is at a lower energy level than the 3d

Note that the 5s-orbital is at a lower energy level than the 4d

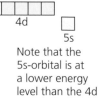

**Figure 2.15** Energy-level diagram.

Each box represents an orbital and each orbital can contain a maximum of two electrons.

Some examples of electron configurations using 'electrons in box' diagrams are given in Figure 2.16. It has been mentioned that where an orbital contains two electrons they differ in their direction of spin. To distinguish between the two electrons they are shown with an upward arrow (↑) and a downward arrow (↓).

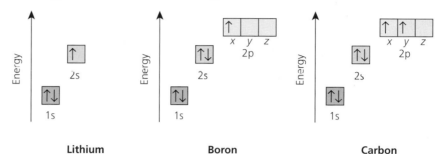

**Lithium**  **Boron**  **Carbon**

**Figure 2.16** 'Electrons in box' diagrams for the electron configurations of lithium, boron and carbon.

Box diagrams are a useful visual way of showing how electrons are distributed but it is often more helpful to identify the orbitals using the orbital names. The system for doing this is as follows.

- An atom of hydrogen has one electron, which in its ground state is in the 1s-orbital. It is represented as $1s^1$. (You would say this as 'one s one')

- An atom of helium has two electrons that occupy the 1s-orbital. Its electron configuration is represented as $1s^2$. (You would say this as 'one s two')

- An atom of lithium has three electrons and its electron structure is represented as $1s^2 2s^1$.

- The electron structure of beryllium is $1s^2 2s^2$.

- An atom of boron has five electrons and the next highest energy orbital, 2p, is used. The electron configuration is represented as $1s^2 2s^2 2p_x^1$, although the subscript $x$ is not really necessary because any of the three 2p-orbitals could be used as they all have the same energy. The label $x$ has no direct meaning until the $y$- and $z$-orbitals are also occupied. Labelling is only necessary to distinguish between orbitals of equivalent energy.

Each orbital of equivalent energy is occupied by one electron before the second electron is added. The reason for this is that two electrons within the same orbital experience a degree of repulsion that makes the pairing of electrons slightly less favourable.

- Carbon, atomic number 6, therefore has the ground state $1s^2 2s^2 2p_x^1 2p_y^1$.

- Nitrogen, atomic number 7, has the ground state $1s^2 2s^2 2p_x^1 2p_y^1 2p_z^1$.

After nitrogen, the electrons in the p-orbitals pair up:

- oxygen is $1s^2 2s^2 2p_x^2 2p_y^1 2p_z^1$

- neon is $1s^2 2s^2 2p_x^2 2p_y^2 2p_z^2$.

**Tip**

In diagrams of this kind you can also represent the electrons using half arrows.

Once the 2p-orbitals have been filled, the 3s-orbital and the 3p-orbitals are occupied in a similar way so that, for example, the ground state of silicon (14 electrons) is $1s^2 2s^2 2p^6 3s^2 3p_x^1 3p_y^1$ and that of chlorine with 17 electrons is $1s^2 2s^2 2p^6 3s^2 3p_x^2 3p_y^2 3p_z^1$.

Note that once the 3p-orbitals have been filled, the next orbital to be occupied is the 4s (not the 3d). Therefore, the ground state of potassium (19 electrons) is $1s^2 2s^2 2p^6 3s^2 3p^6 4s^1$.

The 3d-orbitals are filled in the atoms of scandium (21 electrons) to zinc (30 electrons). These elements have a number of properties in common and are called **d-block elements**. For example, scandium would be written as $1s^2 2s^2 2p^6 3s^2 3p^6 4s^2 3d^1$ or $1s^2, 2s^2, 2p^6, 3s^2, 3p^6, 3d^1, 4s^2$.

'Electrons in box' representations of the electron ground states of some elements are given in Figure 2.17.

| Element | Number of electrons | 1s | 2s | 2p | 3s | 3p | 3d | 4s | 4p | 5s |
|---|---|---|---|---|---|---|---|---|---|---|
| Hydrogen | 1 | ↑ | | | | | | | | |
| Helium | 2 | ↑↓ | | | | | | | | |
| Lithium | 3 | ↑↓ | ↑ | | | | | | | |
| Carbon | 6 | ↑↓ | ↑↓ | ↑ ↑ | | | | | | |
| Neon | 10 | ↑↓ | ↑↓ | ↑↓ ↑↓ ↑↓ | | | | | | |
| Sodium | 11 | ↑↓ | ↑↓ | ↑↓ ↑↓ ↑↓ | ↑ | | | | | |
| Sulfur | 16 | ↑↓ | ↑↓ | ↑↓ ↑↓ ↑↓ | ↑↓ | ↑↓ ↑ ↑ | | | | |
| Argon | 18 | ↑↓ | ↑↓ | ↑↓ ↑↓ ↑↓ | ↑↓ | ↑↓ ↑↓ ↑↓ | | | | |
| Potassium | 19 | ↑↓ | ↑↓ | ↑↓ ↑↓ ↑↓ | ↑↓ | ↑↓ ↑↓ ↑↓ | | ↑ | | |
| Scandium | 21 | ↑↓ | ↑↓ | ↑↓ ↑↓ ↑↓ | ↑↓ | ↑↓ ↑↓ ↑↓ | ↑ | ↑↓ | | |
| Iron | 26 | ↑↓ | ↑↓ | ↑↓ ↑↓ ↑↓ | ↑↓ | ↑↓ ↑↓ ↑↓ | ↑↓ ↑ ↑ ↑ ↑ | ↑↓ | | |
| Zinc | 30 | ↑↓ | ↑↓ | ↑↓ ↑↓ ↑↓ | ↑↓ | ↑↓ ↑↓ ↑↓ | ↑↓ ↑↓ ↑↓ ↑↓ ↑↓ | ↑↓ | | |
| Bromine | 35 | ↑↓ | ↑↓ | ↑↓ ↑↓ ↑↓ | ↑↓ | ↑↓ ↑↓ ↑↓ | ↑↓ ↑↓ ↑↓ ↑↓ ↑↓ | ↑↓ | ↑↓ ↑↓ ↑ | |
| Krypton | 36 | ↑↓ | ↑↓ | ↑↓ ↑↓ ↑↓ | ↑↓ | ↑↓ ↑↓ ↑↓ | ↑↓ ↑↓ ↑↓ ↑↓ ↑↓ | ↑↓ | ↑↓ ↑↓ ↑↓ | |
| Strontium | 38 | ↑↓ | ↑↓ | ↑↓ ↑↓ ↑↓ | ↑↓ | ↑↓ ↑↓ ↑↓ | ↑↓ ↑↓ ↑↓ ↑↓ ↑↓ | ↑↓ | ↑↓ ↑↓ ↑↓ | ↑↓ |

**Figure 2.17** The electronic ground states of some elements.

**Test yourself**

8 Copy and complete the following information for the quantum shell with principal quantum number 3.
   a) Total number of sub-shells = _____
   b) Total number of orbitals = _____
   c) Number of different types of orbital = _____
   d) Maximum number of electrons in the shell = _____
9 Give the electron orbital configuration for the ground state of the following atoms or ions:
   a) Na
   b) K
   c) N
   d) $O^{2-}$
   e) $Ca^{2+}$
   f) $Al^{3+}$
   g) $Cl^-$
   h) $P^{3-}$

# Practice questions

## Multiple choice questions 1–10

**1** Which one of the following has more neutrons than electrons and more electrons than protons?

A $^{19}_{9}F^-$

B $^{37}_{17}Cl^-$

C $^{9}_{4}Be$

D $^{9}_{4}Be^{2+}$

**2** Successive ionisation enthalpies of an element X in $kJ\,mol^{-1}$ are as follows:

578, 1817, 2745, 11 578, 14 831, 18 378

Which one of the following is X?

A boron

B carbon

C aluminium

D silicon

**3** Which one of the rows giving information about the fourth period of the periodic table is correct?

|   | Total number of orbitals | The number of different types of orbital | Maximum number of electrons in the shell |
|---|---|---|---|
| A | 4 | 2 | 8 |
| B | 9 | 2 | 18 |
| C | 9 | 3 | 8 |
| D | 9 | 3 | 18 |

**4** Chlorine exists as two isotopes, $^{35}_{17}Cl$ with an abundance of 75.5% and $^{37}_{17}Cl$ with an abundance of 24.5%.

Phosphorus has only one isotope, $^{31}_{15}P$.

The mass spectrum of $PCl_3$ has four lines at $m/z$ = 136, 138, 140 and 142.

Which one these lines will have the smallest height?

A 136      B 138

C 140      D 142

**5** Antimony has two isotopes $^{121}Sb$ and $^{123}Sb$.

The relative atomic mass of a naturally occurring sample of antimony is measured as 121.75.

Which one of the following is the best approximate estimate of the percentage of $^{121}Sb$ present in the naturally occurring sample?

A 20%

C 60%

B 40%

D 80%

**6** When sulfur, $^{32}_{16}S$ is bombarded with neutrons $^{0}_{1}n$, two particles are formed.

One of them is a hydrogen atom, $^{1}_{1}H$ and the other is an element, X.

$$^{32}_{16}S + ^{1}_{0}n \rightarrow ^{1}_{1}H + X$$

Which one of the following correctly represents X?

A $^{32}_{15}S$

B $^{32}_{15}P$

C $^{33}_{16}S$

D $^{33}_{15}P$

**7** Chlorine has two isotopes, $^{35}Cl$ and $^{37}Cl$. Bromine has two isotopes, $^{79}Br$ and $^{81}Br$.

How many lines would you expect to observe on the mass spectrum of the molecule ClBr?

A 5            B 6

C 7            D 8

Use the key below to answer Questions 8, 9 and 10.

| A | B | C | D |
|---|---|---|---|
| 1, 2 & 3 correct | 1, 2 correct | 2, 3 correct | 1 only correct |

**8** Which of the following elements have atoms that contain only one unpaired p-orbital electron?

1 phosphorus, $_{15}P$

2 bromine, $_{35}Br$

3 aluminium, $_{13}Al$

**9** Which of the following statements is true for elements in the third period of the periodic table?

1 They all have electrons in at least six different orbitals.

2 They all have some p-orbitals in their electron structures.

3 Only two of the elements have electron structures with all of their orbitals containing pairs of electrons.

**10** Which of the following has the sum of the number of neutrons and the number of electrons equal to 12?

1 three helium atoms ($^{4}_{2}He$)

2 two lithium ions ($^{7}_{3}Li^+$)

3 six hydrogen molecules ($^{1}_{1}H_2$)

**11** Give the numbers of protons, neutrons and electrons present in each of the following atoms.

a) $^{40}_{18}Ar$       b) $^{127}_{53}I$       c) $^{127}_{53}I^-$

d) $^{197}_{79}Au$      e) $^{197}_{79}Au^+$      f) $^{52}_{24}Cr^{3+}$

(6)

**12** This question concerns the following five species:

$^{16}_{8}O^{2-}$ $\quad$ $^{19}_{9}F$ $\quad$ $^{20}_{10}Ne$ $\quad$ $^{23}_{11}Na$ $\quad$ $^{25}_{12}Mg^{2+}$

a) Which two species have the same number of neutrons?

b) Which two species have the same ratio of neutrons to protons?

c) Which two species do not have 10 electrons?

d) What do the numbers 16, 8 and 2– represent in the symbol $^{16}_{8}O^{2-}$? $\quad$ (6)

**13** The isotopes in naturally occurring silicon – $^{28}_{14}Si$, $^{29}_{14}Si$ and $^{30}_{14}Si$ – can be separated by mass spectrometry.

a) Explain what you understand by the term **isotope**.

b) Copy and complete the table below to show the composition of isotopes $^{28}_{14}Si$ and $^{30}_{14}Si$.

| Isotope | Protons | Neutrons | Electrons |
|---------|---------|----------|-----------|
| $^{28}_{14}Si$ | | | |
| $^{30}_{14}Si$ | | | |

c) Why do samples of silicon extracted from different samples of clay have slightly different relative atomic masses? $\quad$ (5)

**14** The element rhenium (Re) has two main isotopes, $^{185}Re$ with an abundance of 37.1% and $^{187}Re$ with an abundance of 62.9%.

a) Calculate the weighted mean atomic mass of rhenium.

b) Why is the atomic mass described as 'relative'? $\quad$ (3)

**15** Antimony has two main isotopes, $^{121}Sb$ and $^{123}Sb$. A forensic scientist was asked to help a crime investigation by analysing the antimony in a bullet. This was found to contain 57.3% of $^{121}Sb$ and 42.7% of $^{123}Sb$.

a) Calculate the relative atomic mass of the sample of antimony from the bullet. Write your answer to three significant figures.

b) State one similarity and one difference between the isotopes in terms of subatomic particles. $\quad$ (5)

**16** Draw the electrons in their orbits for each of the following elements:

a) $^{12}_{6}C$

b) $^{19}_{9}F$

c) $^{27}_{13}Al^{3+}$

d) $^{12}_{16}S^{2-}$ $\quad$ (4)

**17** The first five successive ionisation energies in $kJ\,mol^{-1}$ of an element X are 578, 1817, 2745, 11 578 and 14 831.

a) How many electrons are there in the outer shell of X?

b) Explain how you obtained your answer. $\quad$ (3)

**18** Give the electron orbital configuration for the ground state of the following atoms or ions:

a) $_5B$

b) $_{13}Al$

c) $_4Be^{2+}$

d) $_{16}S^{2-}$

e) $_{21}Sc^{3+}$ $\quad$ (5)

**19** The element, thallium, Tl, has two isotopes, $^{203}Tl$ and $^{205}Tl$, and the relative atomic mass is 204.37. Calculate the percentage of each isotope present in a naturally occurring sample of thallium. $\quad$ (6)

**20** The first five ionisation energies of an element are 738, 1451, 7733, 10 541 and 13 629 $kJ\,mol^{-1}$, respectively. Explain why the element cannot have an atomic number less than 12. $\quad$ (4)

**21** a) Explain what you understand by the word 'orbital' as applied to electrons.

b) Draw an s- and a p-orbital.

c) How does a 1s-orbital differ from a 2s-orbital?

d) Place the following orbitals in order of increasing energy: 2p, 3d, 3s, 2p, 4p and 3p.

e) How many electrons are there in:
  i) a 3d-orbital
  ii) orbitals with a principal quantum number of 3? $\quad$ (7)

**22** Bromine exists as a molecule with two bromine atoms combined together. Bromine has two isotopes: bromine-79 and bromine-81.

a) A molecule of bromine containing two atoms of bromine both with a mass number of 79 can be written as $^{79}_{35}Br_2$. Write the formulae for the two other possible molecules of bromine.

b) The mass spectrum of molecules of bromine is shown on the next page.

  i) Explain why these peaks are observed.
  ii) The peaks at 79 and 81 are the same height. What does this tell you about the relative abundances of the two isotopes?
  iii) Explain why the peak at 160 is twice the height of the peaks at 158 and 162. $\quad$ (7)

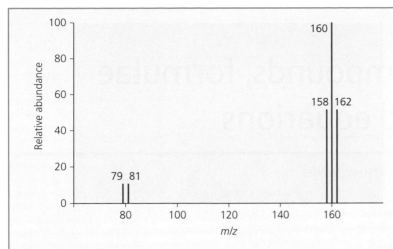

## Challenge

**23** Chlorine exists as a molecule with two chlorine atoms combined together. Chlorine has two isotopes: chlorine-35 and chlorine-37.
The mass spectrum of chlorine is shown below.

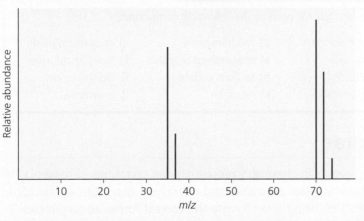

a) The peak at 35 is three times as high as the peak at 37. Calculate the relative atomic mass of chlorine.

b) Explain why peaks are observed at 70, 72 and 74.

c) The heights of the peaks at 70, 72 and 74 are in the ratio 9:6:1. Explain why the heights are in this ratio. **(9)**

**24** You have already learned that s-, p- and d-orbitals can hold two, six and ten electrons respectively. You might have noticed that there is a pattern to the total number of electrons that can be held in the orbitals of each shell. For example, when $n = 1$, only the 1s energy level is possible and it holds two electrons. When $n = 2$, there are the 2s and 2p energy levels, so level 2 can hold a total of eight electrons. Level 3 has the 3d-orbitals, which means that level 3 can hold a total of 18 electrons. The total number of electrons that can be held at the various shells is given by the formula $2n^2$.

a) Deduce the total number of electrons that can be held in the fourth and fifth shells.

b) The extra numbers of electrons are possible because level 4 has f-orbitals and level 5 has f- and g-orbitals. How many different f- and g-orbitals are there?

In November 1952, the first large thermonuclear explosion was carried out in the Pacific Ocean. It was codenamed 'Mike' and was the prototype of the 'hydrogen bomb'.

In the fallout of that event, a new transuranic element was identified. The element is called einsteinium and has the symbol Es. It took until 1961 before enough of the element – 0.01 mg – had been made to study its properties. It was found that its atomic number was 99 and it had at least two isotopic forms, with mass numbers 253 and 254. Since then, it has been established that einsteinium has 14 isotopes.

c) Give the numbers of protons, neutrons and electrons found in an atom of the einsteinium-253 isotope.

d) What is the difference in the number of particles present in the atoms of einsteinium-253 and einsteinium-254?

e) Use your answer to part (a) to decide how many 5f electrons there are in an atom of einsteinium.

All elements with such high atomic numbers have nuclei that break down spontaneously. This is the cause of radioactivity. Usually this breakdown (known as decay) occurs in several steps, ultimately producing the element lead. In the first step of its decay einsteinium releases alpha radiation, which consists of particles containing two protons and two neutrons.

f) Suggest what happens to an atom of einsteinium when it loses an alpha particle. **(16)**

# Chapter 3

# Compounds, formulae and equations

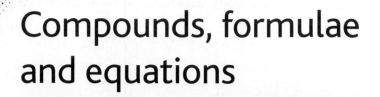

> ## Prior knowledge
>
> *In this chapter it is assumed you will be able to:*
> - recall the symbols and the charges carried by the ions of elements in Groups 1, 2, 16 and 17
> - recall the symbols and the charges carried by the ions of aluminium (3+), zinc (2+), iron (both 2+ and 3+), copper (2+) and silver (1+)
> - write the formulae of simple compounds based on the elements above
> - recall the formulae $H_2O$, $CO_2$, $H_2$, $O_2$, $N_2$, $Cl_2$, $Br_2$, $I_2$, $NH_3$.

> ## Test yourself on prior knowledge
>
> Write the formula for each of the following compounds:
>
> | | | |
> |---|---|---|
> | **a)** sodium chloride | **b)** calcium oxide | **c)** potassium oxide |
> | **d)** zinc bromide | **e)** magnesium bromide | **f)** iron(III) chloride |
> | **g)** silver iodide | **h)** sodium sulfide | **i)** iron(II) oxide |
> | **j)** aluminium oxide | **k)** nitrogen | **l)** ammonia. |

## Formulae

Many experiments in chemistry require precise quantities of reagents to be used. It is important to understand how these quantities are decided. This involves first being able to write the correct formulae for simple compounds, and then using these formulae to produce balanced equations for chemical reactions. An equation does more than provide information about what reacts together and what is produced; it gives exact details of the numbers of particles – atoms, molecules or ions – that are involved.

This chapter should be studied carefully because much of what follows in this book depends on your ability to apply the ideas met here. The final section, covering ionic equations, can be studied now or left until the work on structure and bonding (Chapter 7) has been completed.

The approach taken in this book uses the word 'valency' but you do not need to know a definition or use the word in an exam. It is introduced here as a helpful way to make sure the formulae you write are correct. If you have learnt to obtain the formulae using another method this does not matter: it is not the method you use but the answers that you obtain that is important. You may be already familiar with the work covered in this section of the book but it is of such importance that you should make absolutely certain that this really is the case.

Although you may be able to write the formulae of the simple compounds given in the 'Test yourself' section above you must also be able to construct the formulae of a range of other compounds.

The extensive range of compounds containing carbon is the basis of organic compounds that are studied in later chapters of this book. Other compounds based on metals are considered here.

Metal atoms usually lose electrons; non-metals usually gain or share electrons. The number of electrons involved is a characteristic property of the element known as its **valency**. For example, an atom of sodium can lose only one electron, so sodium has valency 1. An atom of oxygen can gain two electrons, so oxygen has valency 2.

A valency can also be given to groups of atoms. For example, a nitrate group ($NO_3$) consists of one nitrogen atom and three oxygen atoms combined together and this whole unit has valency 1.

Valency does not specify whether electrons are gained or lost, but it is a helpful concept that can be used to generate correct formulae.

Table 3.1 is a list of the valencies that you should learn. The groups referred to in the table are those of the periodic table (page 116). Notice that some metals can have more than one valency.

> **Tip**
>
> A copy of the periodic table is provided in the examination.

**Table 3.1** Valencies of elements and groups of elements.

| Valency | Element/group losing electrons | Element/group gaining electrons |
|---|---|---|
| 1 | All Group 1 elements, hydrogen (H), silver (Ag), ammonium ($NH_4$), copper (Cu) | All Group 17 elements, hydroxide (OH), nitrate ($NO_3$), hydrogencarbonate ($HCO_3$) |
| 2 | All Group 2 elements, iron (Fe), copper (Cu), zinc (Zn), lead (Pb) | Oxygen (O), sulfur (S), sulfate ($SO_4$), carbonate ($CO_3$) |
| 3 | All Group 13 elements, iron (Fe) | |
| 4 | Lead (Pb) | |

> **Tip**
>
> You can use the groups of the periodic table (see Chapter 8) to help you remember some of these valencies correctly.

> **Tip**
>
> Elements with valency 1 are called **univalent**. Those with valency 2 are **divalent** and those with valency 3 are called **trivalent**.

## Writing formulae

The second column in Table 3.1 contains electron givers (but also including ammonium). Electron givers make compounds with the electron receivers (non-metals or groups of non-metals) listed in the third column. If each has the same valency, they combine directly. For example, the formula of potassium bromide is KBr (both potassium and bromine have valency 1) and the formula of barium oxide is BaO (both barium and oxygen have valency 2).

If the valencies differ, the numbers of the components have to be balanced until each of their valencies is satisfied. For example, sodium has valency 1 and oxygen has valency 2. In sodium oxide, the valency of oxygen must be satisfied by the use of two sodium. The formula is $Na_2O$. The subscript 2 indicates that two sodiums are required for each oxygen. This balancing of valencies is essential. Sodium cannot form a compound of formula NaO. In terms of the electrons that are involved, each oxygen can react only by receiving two electrons and each sodium can provide only one electron. Therefore, it is necessary to have two sodiums for each oxygen.

Some further examples of using the valency table are given below.

- The formula of aluminium oxide is $Al_2O_3$. Aluminium has valency 3 and oxygen has valency 2 and to balance the valencies requires two aluminium and three oxygen. Each then has a total valency of 6. In terms of electrons, each of two aluminium atoms provides three electrons (total six) and each of three oxygen atoms receives two electrons.

- The formula of magnesium nitrate is $Mg(NO_3)_2$.

Notice the use of brackets in the formula for magnesium nitrate. Magnesium with valency 2 requires two nitrate units of valency 1. The whole nitrate unit must be kept intact and brackets are used to make this clear. They are essential to avoid the ambiguity of writing $MgNO_{32}$.

Some elements can have more than one valency – for example, iron can have a valency of 2 or 3 (Table 3.1). The name given to the compounds reflects the valency of the element: for example, iron(II) chloride indicates a compound containing iron with valency 2; iron(III) sulfate contains iron with valency 3. The formulae are, therefore, $FeCl_2$ and $Fe_2(SO_4)_3$ respectively.

Once you have obtained the correct formulae you must understand that metallic compounds are composed of ions. The charge on metal ions is always positive and the size of the positive charge can often be predicted using the Periodic Table on page 116.

Group 1 metals have a charge of 1+, whilst group 2 and 3 metals have charges of 2+ and 3+ respectively. You will also have to recall the names and formulae of the ammonium ion, $NH_4^+$ and the transition metal ions $Cu^{2+}$, $Zn^{2+}$ and $Ag^+$.

The charge on non-metallic ions is always negative. The size of the negative charge can also often be predicted using the periodic table. Group 17 ions have a charge of 1- whilst the charge of group 16 ions is 2-. You will also have to recall the names and formulae of nitrate, $NO_3^-$, carbonate, $CO_3^{2-}$, sulfate, $SO_4^{2-}$ and hydroxide, $OH^-$.

The formulae you generate are such that they always have the positive and negative charges of the ions in balance making the overall compound neutral. For example magnesium nitrate is $Mg(NO_3)_2$ because the magnesium has a charge of 2+ ($Mg^{2+}$) whilst the nitrate ion has a charge of 1- ($NO_3^-$), so two nitrate ions are required to balance the charge on the magnesium ion. Ammonium sulfate is $(NH_4)_2SO_4$ because the ammonium has a charge of 1+ ($NH_4^+$) whilst the sulfate ion has a charge of 2- ($SO_4^{2-}$), so two ammonium ions are required to balance the charge on the sulfate ion.

## Test yourself

1 Select the correct formula for the following compounds and explain your choice.
   a) calcium hydroxide: $CaOH$, $CaOH_2$, $Ca(OH)_2$
   b) sodium sulfate: $NaSO_4$, $(Na)_2SO_4$, $Na_2SO_4$, $Na_2(SO)_4$
   c) iron(III) chloride: $Fe_3Cl$, $FeCl_3$

2 Use the data in Table 3.1 to answer the following questions.
   a) The formula of strontium bromide is $SrBr_2$. Deduce the valency of strontium.
   b) The formula of sodium phosphide is $Na_3P$. Deduce the valency of phosphorus.

3 Write the correct chemical formula for each of the following compounds and state the ions that each contains.
   a) silver sulfate
   b) sodium carbonate
   c) zinc nitrate
   d) iron(II) bromide
   e) iron(III) oxide
   f) aluminium sulfate

4 Write the formula for each of the following:
   a) lead(IV) oxide
   b) tin(II) chloride
   c) iron(III) sulfate

5 Name the following compounds:
   a) $Fe_2O_3$
   b) $PbO$
   c) $PbO_2$
   d) $Sn(NO_3)_2$
   e) $PbI_4$
   f) $SnI_4$

# Equations

Being able to state the correct formulae for substances is essential when writing equations for chemical reactions. An equation includes the reactants used and the products obtained and also indicates the number of particles of each substance required. The following examples illustrate the procedure involved.

## Example 1

Write a balanced equation for the reaction of magnesium and oxygen to make magnesium oxide.

**Answer**

First, write the correct formulae for the reactants and products:

$$Mg + O_2 \rightarrow MgO$$

Now examine the formulae to decide what numbers of each particle are required and how much product is formed.

Oxygen is $O_2$ (not 'O') and the formula of magnesium oxide is MgO, so two particles of MgO are made. Therefore, the balanced equation is:

$$2Mg + O_2 \rightarrow 2MgO$$

When balancing an equation, you must not alter the formulae of the reactants or products. These are fixed by the valency rules and must not be changed.

The equations can also be written using fractions to balance the equation.

## Example 2

Write a balanced equation for the reaction between sodium hydroxide and sulfuric acid to form sodium sulfate and water.

**Answer**

The unbalanced equation is:

$$NaOH + H_2SO_4 \rightarrow Na_2SO_4 + H_2O$$

Balancing is required, as follows:

$$2NaOH + H_2SO_4 \rightarrow Na_2SO_4 + 2H_2O$$

The formula of sodium sulfate indicates that 2NaOH are required.

## Example 3

Write a balanced equation for the reaction between aluminium oxide and sulfuric acid to form aluminium sulfate and hydrogen.

**Answer**

The unbalanced equation is:

$$Al_2O_3 + H_2SO_4 \rightarrow Al_2(SO_4)_3 + H_2O$$

Balancing this equation is trickier. Note that to form the aluminium sulfate, 2Al are required and $3H_2SO_4$ are needed to provide the sulfate. The balanced equation is:

$$Al_2O_3 + 3H_2SO_4 \rightarrow Al_2(SO_4)_3 + 6H_2O$$

In summary, the procedure to be adopted in balancing equations is:

- Use valencies to determine the correct formula of each reactant and product.
- Write down the unbalanced equation with the reactants on the left-hand side of the arrow and the products on the right-hand side.
- Make sure that the number of each atom on the left-hand side equals the number on the right-hand side.

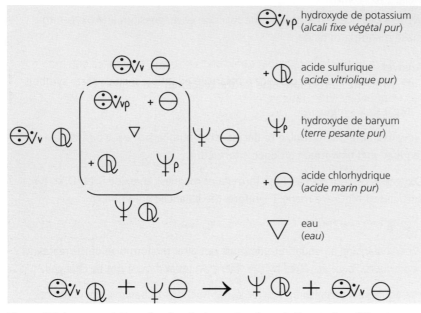

**Figure 3.1** A representation of a chemical equation from *A dissertation of Elective Attractions* by Torben Bergman published in 1775.

## State symbols

Equations can be made more informative by including state symbols. The equation for the reaction between silver nitrate and potassium bromide is as follows:

$$AgNO_3 + KBr \rightarrow AgBr + KNO_3$$

This is correctly balanced, but it does not indicate that the reaction occurs only if the silver nitrate and potassium bromide are both in aqueous solution. Nor does it give the useful information that mixing the solutions produces solid silver bromide. The physical state of the substances in the equation can be shown by including state symbols:

$$AgNO_3(aq) + KBr(aq) \rightarrow AgBr(s) + KNO_3(aq)$$

The symbol (aq) indicates aqueous solution; (s) indicates that the silver bromide is a solid.

There are four state symbols:

- (s) for a solid
- (l) for a liquid
- (g) for a gas
- (aq) for an aqueous solution.

The equation for the reaction of calcium carbonate with dilute hydrochloric acid uses all four state symbols:

$$CaCO_3(s) + 2HCl(aq) \rightarrow CaCl_2(aq) + CO_2(g) + H_2O(l)$$

**Test yourself**

6 Balance the following equations:
   a) $Zn + O_2 \rightarrow ZnO$
   b) $Na + Cl_2 \rightarrow NaCl$
   c) $CaO + HCl \rightarrow CaCl_2 + H_2O$
   d) $Zn + AgNO_3 \rightarrow Zn(NO_3)_2 + Ag$
   e) $Fe + Cl_2 \rightarrow FeCl_3$

7 For each of the following reactions, first write an unbalanced equation and then amend it to produce a balanced equation. Include state symbols in your balanced equations.
   a) sodium + oxygen → sodium oxide
   b) zinc + hydrochloric acid → zinc chloride(solution) + hydrogen
   c) magnesium carbonate(solid) + hydrochloric acid → magnesium chloride(solution) + carbon dioxide + water
   d) silver nitrate(solution) + copper chloride(solution) → silver chloride(solid) + copper nitrate(solution)
   e) copper(II) oxide(solid) + nitric acid → copper(II) nitrate(solution) + water
   f) iron(III) chloride(solution) + sodium hydroxide(solution) → iron(III) hydroxide(solid) + sodium chloride(solution)

# Ionic equations

**Tip**

This section may be more fully understood once Chapter 7 on bonding and structure has been studied.

In reactions involving ionic compounds (many acids and most compounds containing a metal) it is not the whole compound that reacts but rather it is one or more of the ions that they contain. It therefore makes sense when writing an equation to show only those ions that are actually reacting. This is what an ionic equation does; it indicates those particles that react and excludes those that, although present, take no part in the reaction. Remember the ionic bonds in ionic compounds are formed by the electrical attraction between the positive and negative ions, but the ions are still separate from each other. When the compound is melted or dissolved in water the ions become detached from each other and move around freely and separately. This gives them the capacity to react independently.

By contrast, covalent substances have bonds that involve the sharing of electrons. The shared electrons bind the atoms together and the molecules stay as complete entities even when the substance is melted or dissolved.

In an aqueous solution of sodium chloride, the sodium ions and chloride ions are moving freely and independently. In a solution of glucose – which is covalent – the glucose molecules would be intact.

## Writing an ionic equation

Aqueous silver nitrate reacts with aqueous sodium chloride to produce insoluble silver chloride. However, it is more correct to say that silver ions from the aqueous silver nitrate react with the chloride ions from the aqueous sodium chloride to produce a precipitate of solid silver chloride. It does not matter that it is a nitrate of silver or the chloride of sodium – the reaction takes place as a consequence of silver ions and chloride ions

> **Tip**
>
> This reaction is the basis of a method for identifying a chloride in aqueous solution by adding aqueous silver nitrate (page 139).

> **Tip**
>
> Note that in the reaction between silver nitrate and sodium chloride, the product, silver chloride, is included in the ionic equation but, in this example, sodium chloride is absent. The difference is that when silver chloride is formed as a solid, the silver ions and chloride ions held together in a crystal lattice, whereas when sodium chloride is in solution, its ions are free and independent.

being made mobile by the dissolving process. The reaction still occurs if the metal ion associated with the chloride is, for example, potassium, magnesium or copper and if the silver ion is dissolved as a sulfate.

A standard equation does not make this point clear. The reaction between silver nitrate and sodium chloride could be written as:

$$AgNO_3(aq) + NaCl(aq) \rightarrow AgCl(s) + NaNO_3(aq)$$

But this equation does not show that it is only the silver ion and the chloride ion that react. The ionic equation shown below is more precise.

$$Ag^+(aq) + Cl^-(aq) \rightarrow AgCl(s)$$

It indicates that only aqueous silver ions and aqueous chloride ions are involved in the reaction. Nitrate and sodium ions are present but do not change. They move freely and independently at the start of the reaction and also when the reaction has finished. They are sometimes referred to as **spectator ions**.

Ionic equations may seem strange because it is not possible to have silver or chloride ions in isolation; they have to be associated with other ions. Nevertheless, an ionic equation is a truer representation of what is occurring and they are widely used in chemistry.

## Examples of ionic equations

### Reaction between aqueous sodium hydroxide and hydrochloric acid

Aqueous sodium hydroxide reacts with hydrochloric acid to produce aqueous sodium chloride and water. This can be represented by the equation:

$$NaOH(aq) + HCl(aq) \rightarrow NaCl(aq) + H_2O(l)$$

Water is covalent and is, therefore, present as intact covalent molecules. Sodium hydroxide and sodium chloride are ionic and so is the hydrochloric acid. Their ions can behave independently.

The sodium ions and chloride ions are not changed by the reaction – they remain in solution as free, independent ions. It is the hydroxide ions and the hydrogen ions that react. They combine to form covalent water molecules.

This is shown by the ionic equation:

$$H^+(aq) + OH^-(aq) \rightarrow H_2O(l)$$

This ionic equation demonstrates the important fact that all hydroxides in aqueous solution react with a source of hydrogen ions (i.e. an acid) to produce water.

### Reaction between a carbonate and an acid

Aqueous sodium carbonate reacts with hydrochloric acid to produce sodium chloride, carbon dioxide and water:

$$Na_2CO_3(aq) + 2HCl(aq) \rightarrow 2NaCl(aq) + CO_2(g) + H_2O(l)$$

The sodium ions and chloride ions are not involved in the reaction – they remain unchanged in solution. The carbonate ions and hydrogen ions react to form the two covalent products, carbon dioxide and water. This is represented by the ionic equation:

$$CO_3{}^{2-}(aq) + 2H^+(aq) \rightarrow CO_2(g) + H_2O(l)$$

Ionic equations must be balanced. In the example above, a '2' has to be placed before the 'H⁺'. The equation indicates that all carbonates react with acids to form carbon dioxide and water. For sulfuric acid, the equation is:

$$Na_2CO_3(aq) + H_2SO_4(aq) \rightarrow Na_2SO_4(aq) + CO_2(g) + H_2O(l)$$

However the ionic equation is the same as for the reaction between sodium carbonate and hydrochloric acid:

$$CO_3^{2-}(aq) + 2H^+(aq) \rightarrow CO_2(g) + H_2O(l)$$

Do not be tempted to write '$H_2^+(aq)$' or '$H_2^{2+}(aq)$' to represent the hydrogen ions arising from $H_2SO_4$. Once in solution, the ions are detached and are simply $H^+$. The equation shows that two hydrogen ions are needed for each carbonate ion. '$H_2^+(aq)$' would imply that the two hydrogen atoms are bonded covalently and have acquired a positive charge; '$H_2^{2+}(aq)$' indicates the same, but with a charge of '2⁺'. Neither representation is correct.

## Reaction between zinc and aqueous copper sulfate

Zinc metal reacts with aqueous copper sulfate to form copper metal and aqueous zinc sulfate, according to the equation:

$$Zn(s) + CuSO_4(aq) \rightarrow Cu(s) + ZnSO_4(aq)$$

Copper sulfate and zinc sulfate are both ionic, so the ionic equation for this reaction is:

$$Zn(s) + Cu^{2+}(aq) \rightarrow Cu(s) + Zn^{2+}(aq)$$

## Reaction between solid magnesium oxide and hydrochloric acid

Remember that ionic substances must be in solution to exist as free ions. The equation for the reaction of solid magnesium oxide with hydrochloric acid to form magnesium chloride and water is as follows:

$$MgO(s) + 2HCl(aq) \rightarrow MgCl_2(aq) + H_2O(l)$$

In this case, the ions in the solid magnesium oxide are held in an ionic lattice and are not free to move. The only ions that are 'free' throughout this reaction are the chloride ions. Therefore, the ionic equation is:

$$MgO(s) + 2H^+(aq) \rightarrow Mg^{2+}(aq) + H_2O(l)$$

During this reaction the magnesium oxide lattice breaks down, releasing magnesium ions into the solution. The oxide ions combine with the hydrogen ions from the acid to form water.

### Test yourself

8 Write ionic equations for each of the reactions below. Include state symbols.
   a) reacting aqueous silver nitrate and aqueous sodium bromide to produce a precipitate of silver bromide and a solution of sodium nitrate
   b) reacting aqueous barium chloride and aqueous sodium sulfate to produce a precipitate of barium sulfate and a solution of sodium chloride
   c) reacting sulfuric acid and aqueous sodium hydroxide
   d) reacting hydrochloric acid and aqueous potassium carbonate
   e) reacting magnesium and aqueous zinc chloride
   f) reacting solid zinc oxide and sulfuric acid

# Practice questions

## Multiple choice questions 1–10

**1** Which one of the following is the correct formula of chromium(III) hydroxide ?

A $Cr_3OH$

B $Cr_3(OH)$

C $CrOH_3$

D $Cr(OH)_3$

**2** Which one of the following is the correct formula of cerium(IV) sulfate?

(The symbol for cerium is Ce.)

A $Ce(SO)_4$

B $Ce(SO_4)_2$

C $Ce(SO_4)_4$

D $Ce_4SO_4$

**3** When magnesium hydroxide reacts with hydrochloric acid magnesium chloride and water are formed.

Which one of the following gives the correct balancing numbers in the equation?

|   | Magnesium hydroxide | Hydrochloric acid | Magnesium chloride | Water |
|---|---|---|---|---|
| A | 1 | 1 | 1 | 1 |
| B | 1 | 2 | 1 | 1 |
| C | 1 | 2 | 1 | 2 |
| D | 1 | 2 | 2 | 2 |

**4** Aqueous potassium hydroxide reacts with aqueous sulfurous acid $(H_2SO_3)$ to form aqueous potassium sulfite and water.

Which one of the following gives the correct ionic equation for this reaction?

A $H^+(aq) + OH^-(aq) \rightarrow H_2O(l)$

B $2H^+(aq) + OH^-(aq) \rightarrow H_2O(l)$

C $2H^+(aq) + O^{2-}(aq) \rightarrow H_2O(l)$

D $2K^+(aq) + SO_3^{2-}(aq) \rightarrow K_2SO_3(aq)$

**5** Aqueous barium nitrate reacts with aqueous magnesium sulfate to form aqueous magnesium nitrate and a precipitate of barium sulfate.

Which one of the following gives the correct ionic equation for this reaction?

A $Ba^+(aq) + SO_4^{2-}(aq) \rightarrow Ba_2SO_4(s)$

B $Ba^{2+}(aq) + SO_4^{2-}(aq) \rightarrow BaSO_4(s)$

C $Ba^+(aq) + SO_4^-(aq) \rightarrow BaSO_4(s)$

D $Ba^{2+}(aq) + SO_4^-(aq) \rightarrow BaSO_4(s)$

**6** If the formula of nitrous acid is $HNO_2$, which one of the following is the formula of calcium nitrite?

A $CaNO_2$

B $Ca_2NO_2$

C $Ca(NO_2)_2$

D $Ca(NO)_2$

**7** If the formula of iron(III) arsenate is $FeAsO_4$, the formula of iron(II) arsenate will be

A $Fe(AsO_4)_2$

B $Fe(AsO_4)_3$

C $Fe_2(AsO_4)_3$

D $Fe_3(AsO_4)_2$

Use the key below to answer Questions 8, 9 and 10.

| A | B | C | D |
|---|---|---|---|
| 1, 2 & 3 correct | 1, 2 correct | 2, 3 correct | 1 only correct |

**8** Gallium is in Group 13 of the periodic table. Which of the following give correct information about gallium compounds?

1 Gallium hydroxide has three times as many hydroxide ions as gallium ions.

2 Gallium oxide has 1½ times as many oxide ions as gallium ions.

3 Gallium bromide has three times as many bromide ions as gallium ions.

**9** Which of the following ions are present in a mixture of aqueous sodium sulfate and aqueous sodium nitrate?

1 $Na_2^+(aq)$

2 $SO_4^{2-}(aq)$

3 $NO_3^-(aq)$

**10** The correct ionic equation for the reaction of solid zinc carbonate with hydrochloric acid is:

$ZnCO_3(s) + 2H^+(aq) \rightarrow Zn^{2+}(aq) + CO_2(g) + H_2O(l)$

Which of the following give a correct explanation of this equation?

1 The zinc carbonate is written as its full formula, $ZnCO_3(s)$, as its ions are not free to move.

2 The carbon dioxide is written as its full formula as it does not contain ions.

3 The chloride ion from the hydrochloric acid is excluded from the equation because it is not present when the reaction has finished.

11 Identify which of the following substances are ionic and give the formulae of the ions formed when the substances are dissolved in an aqueous solution.
   a) $NaNO_3$
   b) $NO$
   c) $CCl_4$
   d) $Pb(NO_3)_2$
   e) $K_2SO_4$
   f) $CH_4$
   g) $H_2SO_4$
   h) $MgCl_2$ (8)

12 Write equations for each of the following reactions:
   a) hydrogen + copper oxide → copper + water
   b) potassium hydroxide + nitric acid → potassium nitrate + water
   c) calcium hydroxide + hydrochloric acid → calcium chloride + water
   d) sodium + water → sodium hydroxide + hydrogen
   e) chlorine + potassium iodide → potassium chloride + iodine
   f) magnesium + aluminium sulfate → magnesium sulfate + aluminium (6)

13 A surprising use of copper is to remove hydrogen sulfide from beer. During fermentation, hydrogen sulfide is produced as a natural by-product. However, because of its unpleasant odour, even tiny amounts can spoil the flavour of the beer. Low concentrations of copper ions are sufficient to give a precipitate of copper sulfide, which is then removed, leaving the beer free from this impurity. Write an ionic equation for this reaction. (2)

14 a) The formula of rubidium chloride is RbCl. What is the formula of rubidium sulfate?
   b) The formula of manganese sulfate is $MnSO_4$. What is the formula of manganese bromide?
   c) The formula of chromium chloride is $CrCl_3$. What is the formula of chromium sulfate?
   d) The formula of silver phosphate is $Ag_3PO_4$. What is the formula of phosphoric acid? (4)

15 When iron reacts with hydrochloric acid, iron(II) chloride and hydrogen are produced. If chlorine gas is passed into the iron(II) chloride solution it reacts to form iron(III) chloride. The iron(III) chloride reacts with potassium iodide to form iron(II) chloride, potassium chloride and a precipitate of iodine is obtained.

a) Write equations, including state symbols, for the sequence of reactions that are taking place.
b) Write ionic equations for each reaction in the sequence. (6)

16 Write ionic equations for the following reactions. Include appropriate state symbols.
   a) silver nitrate solution + potassium iodide solution → silver iodide precipitate + potassium nitrate solution
   b) potassium hydroxide solution + nitric acid → potassium nitrate solution + water
   c) zinc metal + hydrochloric acid → zinc chloride solution + hydrogen gas
   d) magnesium metal + copper sulfate solution → magnesium sulfate solution + copper metal
   e) lithium carbonate solution + hydrochloric acid → lithium chloride solution + carbon dioxide + water
   f) sodium hydroxide solution + copper nitrate solution → copper hydroxide precipitate + sodium nitrate solution
   g) sodium hydroxide solution + aluminium sulfate solution → aluminium hydroxide precipitate + sodium sulfate solution
   h) zinc oxide solid + hydrochloric acid → zinc chloride solution + hydrogen (8)

17 The chemical industry processes large amounts of limestone (calcium carbonate). It is heated to produce calcium oxide for use in cement production and in neutralising excess acidity in soils (liming). Carbon dioxide is given off when limestone is heated.
Calcium oxide is also used in a reaction with carbon to form an unusual compound, calcium carbide, $CaC_2$, with carbon dioxide as a by-product. Calcium carbide, $CaC_2$, reacts with water to produce the flammable gas ethyne (acetylene), $C_2H_2$.
Calcium oxide reacts with water to form calcium hydroxide. Calcium hydroxide reacts with chlorine to produce an aqueous paste of bleaching powder. This is a mixture of water, calcium chloride and calcium hypochlorite, $Ca(OCl)_2$. Calcium hydroxide is also used for liming and, in one industrial process, in a reaction with sodium carbonate to produce sodium hydroxide and a precipitate of calcium carbonate.
Write equations for the following reactions:
   a) the action of heat on calcium carbonate
   b) the reaction between water and calcium oxide

**c)** the reaction between calcium hydroxide and sodium carbonate

**d)** the reaction between calcium oxide and carbon

**e)** the reaction between calcium carbide and water

**f)** the production of bleaching powder. (6)

## Challenge

**18** One of the biggest challenges for firework manufacturers is creating satisfactory blue colours for stars and cascades. Copper(II) chloride can give this colour when mixed with a reagent such as potassium chlorate, but it tends to be unstable at high temperature and the colour is then lost. Copper chloride is manufactured either by the reaction of chlorine directly with copper metal or by dissolving copper oxide in dilute hydrochloric acid. The potassium chlorate ($KClO_3$) releases oxygen to the firework mixture by decomposing to form potassium chloride.

**a)** Write an equation to represent:

**i)** the reaction between chlorine and copper metal

**ii)** the reaction between copper oxide and dilute hydrochloric acid

**iii)** the decomposition of potassium chlorate.

A mixture that has been proposed to intensify the blue colour includes a pigment called Paris Green. This pigment is a salt containing one part of copper ethanoate and three parts of copper arsenite. The ethanoate ion has the formula $CH_3CO_2^-$. The formula of the arsenite ion is $AsO_2^-$.

**b)** Write the formula for:

**i)** copper ethanoate

**ii)** copper arsenite. (5)

**19** Scheele's Green (copper hydrogenarsenate, $CuHAsO_3$) was developed in the nineteenth century to improve the variety and quality of dyes. It was used by wallpaper manufacturers; William Morris included it as a dye. They were either unaware of or indifferent to the fact that the formulation is poisonous. Miners at the Devon Great Consuls mine near Tavistock who extracted the arsenic ore from which Scheele's Green was made suffered badly from the effects of their work. A recipe from the nineteenth century describes a process for making Scheele's Green:

- A mixture of arsenic(III) oxide is boiled with potassium carbonate and water until no more carbon dioxide is produced. In this reaction potassium hydrogenarsenate is also formed.
- A solution of copper sulfate is then added to this mixture and an impure precipitate of Scheele's Green is formed.

Some purification is then required before a usable powder is obtained.

**a)** Deduce the formula of potassium hydrogenarsenate.

**b)** Suggest an equation for the reaction described in the first bullet point above. (Note: water is required as a reactant.)

**c)** Suggest an equation for the reaction in the second bullet point. (5)

# Chapter 4

# Amount of substance – moles in solids and gases

## Test yourself on prior knowledge

1 Write equations for the following reactions, including state symbols:
  a) heating calcium carbonate to produce calcium oxide
  b) copper carbonate and nitric acid to form copper nitrate, carbon dioxide and water
  c) magnesium and aluminium oxide
  d) zinc and aqueous silver nitrate
  e) the decomposition of iron(III) oxide to iron(II) oxide and oxygen.
2 If $n = m/M$,
  a) write $m$ in terms of $n$ and $M$
  b) write $M$ in terms of $n$ and $m$.
3 Write each of the following numbers to three significant figures.
  a) 20.78        e) 0.6577
  b) 18.85        f) 0.05232
  c) 15.448       g) 0.003987
  d) 207.341
4 Write each of the numbers in Question 3(d)–(f) in standard (index) form.
5 Write each of the following numbers in non-standard form.
  a) $3.02 \times 10^2$      c) $2.25 \times 10^{-2}$
  b) $4.57 \times 10^3$      d) $4.11 \times 10^{-4}$

## Relative atomic mass

Nineteenth-century chemists established the formulae of common substances and also calculated the relative masses of the atoms from which they are made. It was by no means an easy task and it took several decades before agreement was reached. What they discovered was *not* the individual masses of the atoms – they knew that atoms were small and would have assumed that it would never be possible to know these values – but how the masses of atoms compared with each other. For example, it was established that a sulfur atom was twice as heavy as an oxygen atom and that an oxygen atom was 16 times heavier than a hydrogen

**Relative atomic mass** is the weighted mean mass of an atom of an element compared with $\frac{1}{12}$ the mass of an atom of carbon-12, which is taken as exactly 12.

**Tip**

As explained in Chapter 2 the expression 'weighted mean mass' takes into account isotopes that are present in different amounts.

**Tip**

It is a common mistake in exams to give relative atomic mass a unit such as grams. Remember that it is just a number with no units. When using the Periodic Table, make sure that you do not use the atomic number by mistake. For example, the atomic number of oxygen is 8, but its relative atomic mass is 16.

**Key terms**

**Relative isotopic mass** is the mass of an atom of an isotope of the element compared with $\frac{1}{12}$ the mass of an atom of carbon-12, which is taken as exactly 12.

**Relative molecular mass** is the weighted mean mass of a molecule of the compound compared with $\frac{1}{12}$ the mass of an atom of carbon-12, which is taken as exactly 12.

**Relative formula mass** is the weighted mean mass of a formula unit of the compound compared with $\frac{1}{12}$ the mass of an atom of carbon-12, which is taken as exactly 12.

atom. A nominal value of 1 was assigned to the mass of the lightest atom, hydrogen, and the masses of other elements were reported relative to this standard. Thus, the relative mass of an oxygen atom is approximately 16 and that of a sulfur atom, 32. Every element can be given a number that indicates its atomic mass relative to that of a hydrogen atom. These numbers are now known to relate to the sum of the numbers of protons and neutrons present in the nucleus of an atom (Chapter 2).

For many years this was the basis on which the masses of atoms were compared. The choice of the lightest element as the standard may seem the most obvious thing to do, but the situation becomes more complex when isotopes are taken into account. This is one reason why the standard element used for comparison today is the isotope of carbon that has mass number 12.

The mass of an element compared to carbon-12 is called the relative atomic mass.

Relative atomic masses have no units because they are a comparison between the element and carbon-12.

Relative atomic masses are listed in most versions of the Periodic Table (page 116). They will be available to you in an examination, so you do not have to learn them.

## Definitions linked to relative atomic mass

Other definitions follow from the definition for relative atomic mass. The relative isotopic mass for a particular isotope of an element is closely related.

A similar statement can be made for compounds. If the compound is covalent (Chapter 7), the smallest unit is a molecule, so the term relative molecular mass is used.

For ionic substances (such as those containing a metal) the smallest units are the ions indicated by their formulae and the expression relative formula mass should be used. Relative formula mass can also be used when referring to covalent compounds.

**Tip**

It is incorrect to call the smallest unit of an ionic compound a molecule. If you are in doubt as to whether a compound is covalent or ionic, always use the expression relative formula mass or where appropriate use the formula of the substance.

## Obtaining relative molecular mass and relative formula mass

The relative molecular or formula mass of a compound is obtained by adding together the relative atomic masses of its component atoms. In each of the following examples, the values of the relative atomic masses have been obtained from the Periodic Table (page 116).

## Example 1

Glucose has the formula $C_6H_{12}O_6$. What is its relative molecular mass?

**Answer**

The relative molecular mass of glucose is (6 × 12.0) + (12 × 1.0) + (6 × 16.0) = 180.0.

## Example 2

Magnesium carbonate has the formula $MgCO_3$. What is its relative formula mass?

**Answer**

The relative formula mass of magnesium carbonate is 24.3 + 12.0 + (3 × 16.0) = 84.3.

## Example 3

Copper sulfate crystals have the formula $CuSO_4{\cdot}5H_2O$. What is the relative formula mass of copper sulfate?

**Answer**

The formula $CuSO_4{\cdot}5H_2O$ means that there are five molecules of water for each $CuSO_4$. Therefore its relative formula mass is 63.5 + 32.1 + (4 × 16.0) + (5 × 18.0) = 249.6.

### Tip

Many students misread the formula of crystals such as $CuSO_4{\cdot}5H_2O$ and wrongly interpret the formula to contain one Cu, one S, four oxygen, ten hydrogen and one more oxygen. The '5' in front of the $H_2O$ shows that the formula unit contains five $H_2O$ molecules.

Note that 'relative formula mass' is used in Examples 2 and 3 instead of relative molecular mass, because magnesium carbonate and copper sulfate are ionic.

With more complicated formulae, be careful to use the correct number of atoms of each type. For example, calcium nitrate has the formula $Ca(NO_3)_2$, which means that it contains one calcium, two nitrogens and six oxygens – the brackets mean that there are two $NO_3^-$ ions.

## Test yourself

**1** For each of the following, write the formula and calculate the relative formula mass:

**a)** lithium chloride
**b)** potassium bromide
**c)** magnesium hydroxide
**d)** sulfuric acid
**e)** sodium sulfate
**f)** barium nitrate
**g)** iron(II) chloride
**h)** iron(III) sulfate
**i)** sodium carbonate crystals ($Na_2CO_3{\cdot}10H_2O$)
**j)** iron(II) sulfate crystals ($FeSO_4{\cdot}7H_2O$).

**2** Calculate the relative molecular mass of each of the following covalent compounds:

**a)** ethane, $C_2H_6$
**b)** ethanol, $C_2H_5OH$
**c)** ethanoic acid, $CH_3COOH$
**d)** chloromethane, $CH_3Cl$
**e)** aminoethane, $C_2H_5NH_2$
**f)** nitrogen gas.

**Figure 4.1** One mole amounts of copper (Cu), carbon (C), iron (Fe), aluminium (Al), mercury (Hg) and sulfur (S).

## Tip

Do not make the mistake of reading the amount in moles of $O_2$ as being '2 mol' just because the formula of oxygen has two atoms combined together. This would be incorrect. It is the balancing coefficients in the equation that tell you the amount in moles. So in this case it is 1 mol of $O_2$ molecules.

# The mole

The relative formula masses that you have calculated are related by their definition to the mass of carbon-12 and each of these masses will contain a fixed number of particles which is known as the **mole**.

You can have a mole of atoms, molecules or any other unit. When writing 1 mole of something you should be precise about what it is the mole refers to. So you can refer to 1 mole of oxygen atoms (meaning O) or 1 mole of oxygen molecules (meaning $O_2$). When considering an ionic compound such as sodium chloride, you should write 1 mole of NaCl.

## Example 4

The balanced equation for the reaction between magnesium and oxygen is:

$$2Mg(s) + O_2(g) \rightarrow 2MgO(s)$$

State what this indicates about the amounts in mol required for the reaction and the amount in mol of MgO obtained.

**Answer**

$$2Mg(s) + O_2(g) \rightarrow 2MgO(s)$$

This means that 2 mol of magnesium atoms reacts with 1 mol of oxygen molecules to form 2 mol of magnesium oxide.

## Example 5

The balanced equation for the reaction between hydrogen and oxygen is:

$$2H_2(g) + O_2(g) \rightarrow 2H_2O(l)$$

State what this indicates about the amounts in mol required for the reaction and the amount in mol of water molecules obtained.

**Answer**

The equation tells us that 2 mol of hydrogen molecules react with 1 mol of oxygen molecules to form 2 mol of water molecules.

## Test yourself

3 For the following balanced equations, write down the amounts in mol of each reactant used and the amount in mol of the product obtained.
   a) $Mg + S \rightarrow MgS$
   b) $S + O_2 \rightarrow SO_2$
   c) $Zn + I_2 \rightarrow ZnI_2$
   d) $N_2 + O_2 \rightarrow 2NO$
   e) $N_2 + 3H_2 \rightarrow 2NH_3$
   f) $2Al + 3O_2 \rightarrow Al_2O_3$

# The Avogadro Constant, $N_A$

The scientists who painstakingly compiled the relative masses of the atoms in the 19th century might not have thought it possible to count how many atoms there were in a mole of substance. However, advances in understanding and the availability of sophisticated measuring devices have since made this possible.

It turns out that $6.02 \times 10^{23}$ atoms ($602\,000\,000\,000\,000\,000\,000\,000$ atoms) of carbon are needed to obtain a mass of $12.0\,g$ and this number is referred to as the **Avogadro constant** and given the symbol $N_A$. Since the mole of any substance contains the same number of particles as carbon-12 it follows that the Avogadro constant is also the number of particles that are present in 1 mole of any substance. For example 1 mol of sodium atoms is the mass that contains $N_A$ atoms of sodium and this has a mass of $23.0\,g$.

Knowing the Avogadro constant allows you to determine the number of particles that are present in a given mass.

### Example 6

What is the number of atoms in $0.391\,g$ of potassium?

**Answer**

1 mole of potassium atoms has a mass of $39.1\,g$ and this mass contains Avogadro's number ($N_A$) of atoms i.e $6.02 \times 10^{23}$.

$0.391\,g$ is $\dfrac{1}{100}$th of the mass of 1 mole of potassium.

So the number of atoms of potassium is $6.02 \times 10^{23} \times \dfrac{1}{100} = 6.02 \times 10^{21}$ atoms.

### Key term

The **molar mass** of a substance is the mass of one mole of a substance. It has units of $g\,mol^{-1}$.

## Molar mass ($M$)

The molar mass of a substance is the same as the formula mass expressed in grams per mole. So the molar mass of sodium is $23\,g\,mol^{-1}$ and the molar mass of methane, $CH_4$, is $16.0\,g\,mol^{-1}$.

## The relationship between amounts in moles and masses in grams

You cannot measure the amount of substance in moles in the laboratory so in order to decide what mass you would need for a balanced chemical reaction you need to know how the amount of substance measured in moles relates to the mass of that substance. The mass of substance is related to the amount of substance using the equation:

Mass in grams ($m$) = the amount of substance in mol ($n$) × the molar mass of the substance ($M$)

From this, it follows that

$$M = \frac{m}{n}$$

and

$$n = \frac{m}{M}$$

These relationships enable you to calculate the masses to use in a reaction and to determine the masses of the products that would be obtained.

### Tip

To emphasise how the terms are related and to show how careful you must be to supply the correct units, note the following:
The relative molecular mass of methane, $CH_4$, is 16.0.
The mass of 1 mol of methane molecules is $16.0\,g$.
The molar mass of methane is $16.0\,g\,mol^{-1}$.
The relative formula mass of zinc sulfide is 97.5.
The mass of 1 mol of ZnS is $97.5\,g$.
The molar mass of ZnS is $97.5\,g\,mol^{-1}$.

**Example 7**

Calculate the amount in moles of 4.52 g of carbon atoms.

**Answer**

The molar mass of carbon is 12.0 g mol$^{-1}$.

$$\text{amount in moles } (n) = \frac{\text{mass in grams } (m)}{\text{molar mass } (M)} = \frac{4.52\,g}{12.0\,g\,mol^{-1}} = 0.377\ mol$$

**Example 8**

Calculate the amount in moles of 5.00 g of chlorine gas molecules.

**Answer**

The molar mass of $Cl_2$ = 71.0 g mol$^{-1}$.

$$\text{amount in moles } (n) = \frac{\text{mass in grams } (m)}{\text{molar mass } (M)} = \frac{5.00\,g}{71.0\,g\,mol^{-1}} = 0.0704\ mol$$

**Example 9**

Calculate the amount in moles of 8.48 g of $CaSO_4$.

**Answer**

The molar mass of calcium sulfate, $CaSO_4$ = 40.1 + 32.1 + (4 × 16.0)
$$= 136.2\ g\,mol^{-1}$$

$$n = \frac{m}{M} = \frac{8.48\,g}{136.2\,g\,mol^{-1}} = 0.0623\ mol$$

**Example 10**

Calculate the mass (in grams) of 0.25 mol of $CaCO_3$.

**Answer**

The molar mass of calcium carbonate, $CaCO_3$ = 40.1 + 12.0 + (3 × 16.0)
$$= 100.1\ g\,mol^{-1}$$

mass in grams (m) = amount in moles (n) × molar mass (M)

mass of 0.25 mol = 0.25 g mol × 100.1 g mol$^{-1}$ = 25.025 g

**Example 11**

Calculate the molar mass of an element if 0.10 mol has a mass of 6.54 g.

**Answer**

$$\text{molar mass } (M) = \frac{\text{mass in grams } (m)}{\text{amount in moles } (n)} = \frac{6.54\,g}{0.10\,mol} = 65.4\ g\,mol^{-1}$$

The balanced equation for a reaction gives information about the amounts in moles required. It is possible to determine the masses required for the reaction and the masses of the products obtained using the relationship between $m$, $n$ and $M$.

**Test yourself**

4 How many atoms are there in 2.00 g of magnesium?

5 Determine the amount in moles in each of the following:
   a) 2.4 g of magnesium
   b) 4.8 g of oxygen molecules
   c) 1.68 g of calcium oxide.

6 Calculate the mass (in grams) of each of the following:
   a) 0.5 mol of argon atoms
   b) 0.2 mol of KCl.

7 Calculate the molar mass of each of the following and use the Periodic Table to identify the element:
   a) 0.1 mol of an element that has a mass of 10.79 g
   b) 0.05 mol of an element that has a mass of 2.60 g.

8 Calculate the molar mass of each of the following compounds in which:
   a) 0.03 mol has a mass of 1.92 g
   b) 0.75 mol has a mass of 43.88 g.

# Relative atomic mass

## Example 12

What mass of sulfur will react with 1.86 g of iron and what mass of iron(III) sulfide will be formed? The equation for this reaction is:

$$2Fe + 3S \rightarrow Fe_2S_3$$

**Answer**

The molar mass ($M$) of iron is 55.8 g mol$^{-1}$ so the amount in mol used is $m/M = 1.86/55.8 = 0.033$ mol.

The equation indicates that 2 mol of iron atoms react with 3 mol of sulfur atoms to make 1 mol of $Fe_2S_3$.

Therefore 0.033 mol of iron atoms will react with $0.033 \times 3/2 = 0.05$ mol of sulfur atoms.

The molar mass of sulfur is 32.1 g mol$^{-1}$.

$m = n \times M$ so 0.05 mol of sulfur has a mass of $0.05 \times 32.1 = 1.605$ g.

The molar mass of iron(III) sulfide is $(2 \times 55.8 + 3 \times 32.1) = 207.9$ g mol$^{-1}$.

So the mass of iron(III) sulfide formed would be $(0.033/2) \times 207.9 = 3.465$ g.

Since no mass is lost during the reaction, the mass of iron(III) sulfide could also have been obtained by adding together the masses of iron and sulfur used.

## Example 13

Calculate the mass of silver formed when 3.27 g of zinc reacts completely with excess silver nitrate solution.

**Answer**

The equation for this reaction is:

$$Zn(s) + 2AgNO_3(aq) \rightarrow Zn(NO_3)_2(aq) + 2Ag(s)$$

The molar mass of zinc is 65.4 g mol$^{-1}$ so the amount in mol used is $m/M = 3.27/65.4 = 0.05$ mol.

The equation indicates that 1 mol of zinc atoms make 2 mol of silver atoms.

Therefore 0.05 mol of zinc will make $2 \times 0.05 = 0.10$ mol of silver.

The molar mass of silver is 107.9 g mol$^{-1}$.

$m = n \times M$ so 0.10 mol of silver has a mass of $0.10 \times 107.9 = 10.79$ g.

## Example 14

Calculate how many tonnes of iron would be produced in a blast furnace from 39.9 tonnes of iron ore, $Fe_2O_3$.

$$2Fe_2O_3(s) + 3C(s) \rightarrow 4Fe(s) + 3CO_2(g)$$

**Answer**

The molar mass of $Fe_2O_3$ is $(2 \times 55.8) + (3 \times 16.0) = 159.6$ g mol$^{-1}$.

If 39.9 g had reacted, the amount in mol used, $n = m/M = 39.9/159.6 = 0.25$ mol.

The equation indicates that 2 mol of $Fe_2O_3$ react to produce 4 mol of iron.

Therefore, 1 mol of $Fe_2O_3$ produces 2 mol of Fe.

Therefore, 0.25 mol of $Fe_2O_3$ will produce 0.50 mol of Fe.

The molar mass of iron is 55.8 g mol$^{-1}$.

$m = n \times M$ so 0.50 mol of iron has a mass of $0.5 \times 55.8 = 27.9$ g.

It follows that 39.9 tonnes of iron(III) oxide produces 27.9 tonnes of iron.

**Tip**

If questions involve quantities in tonnes (or any units of mass such as kg, mg or μg), it is easier to do the equivalent calculation in grams and then change this into tonnes (or kg, mg or μg). Using tonnes instead of grams is common in exam questions.

## Test yourself

9 What mass of aluminium oxide can be obtained by reacting 8.00 g of aluminium with excess oxygen?
$$4Al(s) + 3O_2(g) \rightarrow 2Al_2O_3(s)$$
10 What mass of zinc chloride is obtained by reacting 3.25 mg of zinc with excess chlorine?
$$Zn(s) + Cl_2(g) \rightarrow ZnCl_2(s)$$
11 What mass of calcium oxide would be obtained by heating 2002 tonnes of limestone ($CaCO_3$)?
$$CaCO_3(s) \rightarrow CaO(s) + CO_2(g)$$

# Empirical formulae

An experiment to find an empirical formula involves measuring the masses of elements that combine in the compound. From these masses, it is possible to calculate the amounts in moles of the atoms that react and this gives their ratio. This then can be converted into the simplest whole number ratio for the atoms of the different elements in the compound.

This can be summarised as:

mass in grams $\xrightarrow{\text{convert into}}$ amount in moles $\xrightarrow{\text{convert into}}$ simplest whole number ratio

**Tip**

Determining empirical formulae is often important in establishing the formulae of organic compounds. You will find further examples of the procedure in Chapter 12.

## Example 15

Analysis of 20.1 g of iron bromide showed that it contained 3.8 g of iron and 16.3 g of bromine. What is its empirical formula?

**Answer**
The calculation is summarised in the table below.

| | Fe | Br |
|---|---|---|
| Mass in grams | 3.8 | 16.3 |
| Amount in moles | 3.8/55.8 = 0.068 | 16.3/79.9 = 0.204 |
| Simplest ratio | 0.068/0.068 = 1 | 0.204/0.068 = 3 |
| Empirical formula | $FeBr_3$ | |

## Water of crystallisation

### Anhydrous and hydrated salts

All salts exist in the solid state either as pure substances or in the form of crystals that contain water molecules as part of their structure. The pure substance is anhydrous, which means that it contains no water. The crystals containing water are said to possess water of crystallisation and are, therefore, hydrated.

**Key terms**

Water of crystallisation is the water present in hydrated salts.
A hydrated compound contains water as part of its structure.
Anhydrous is the term used to describe a hydrated compound after it has lost its water of crystallisation.

The water in hydrated salts is indicated by writing the formula of the substance followed by a full stop and the number of molecules of water. For example, iron(II) sulfate crystals have the formula $FeSO_4 \cdot 7H_2O$.

Given the relationship between mass in grams and the amount of substance in moles (page 43), you should be able to use experimental data to determine the formula of a hydrated salt, as illustrated in the worked example below.

(page 43)

## Example 16

A sample of copper sulfate crystals has a mass of 6.80 g. The sample is heated to drive off all the water of crystallisation. The mass is reduced to 4.35 g.

Calculate the formula of the copper sulfate crystals.

**Answer**

mass of water in the crystals driven off by the heating = 6.80 − 4.35 = 2.45 g

amount (in moles) of water of crystallisation $= \dfrac{2.45}{18} = 0.136$ mol

molar mass of anhydrous copper sulfate = 63.5 + 32.1 + (4 × 16.0) = 159.6 g

amount (in moles) of anhydrous copper sulfate $= \dfrac{4.35}{159.6} = 0.02726$ mol

ratio of moles of anhydrous copper sulfate to moles of water = 0.02726:0.136 or 1:5

Therefore, the formula of hydrated copper sulfate is $CuSO_4 \cdot 5H_2O$.

## Example 17

A sample of magnesium sulfate crystals contains 9.78% $Mg^{2+}$, 38.69% $SO_4^{2-}$ and 51.53% $H_2O$ by mass.

Determine the formula of the crystals.

**Answer**

amount (in moles) of magnesium $= \dfrac{9.78}{24.3} = 0.4025$ mol

amount (in moles) of sulfate $= \dfrac{38.69}{96.1} = 0.4026$ mol

amount (in moles) of water $= \dfrac{51.53}{18} = 2.863$ mol

ratio = 0.4025:0.4026:2.863 or 1:1:7

Therefore, the formula of magnesium sulfate crystals is $MgSO_4 \cdot 7H_2O$.

**Experiment to determine the number of molecules of water of crystallisation in $FeSO_4 \cdot xH_2O$**

Crystalline salts often contain water bound up in the crystal. There is a fixed ratio between the number of molecules of water and the number of formula units of the salt. The apparatus is set up as shown below.

*Method*

a) Weigh the crucible and record the mass.
b) Half fill the crucible with hydrated iron (II) sulphate and reweigh.
c) Support the crucible on a pipe clay triangle on a tripod and heat using a clean blue flame which should be low at first and gradually increased.
d) Heat strongly for about five minutes until no further change is observed.
e) Allow to cool and then reweigh the crucible together with its contents.
f) Reheat the crucible, then cool and reweigh.
g) Repeat (f) until successive weighings are identical.

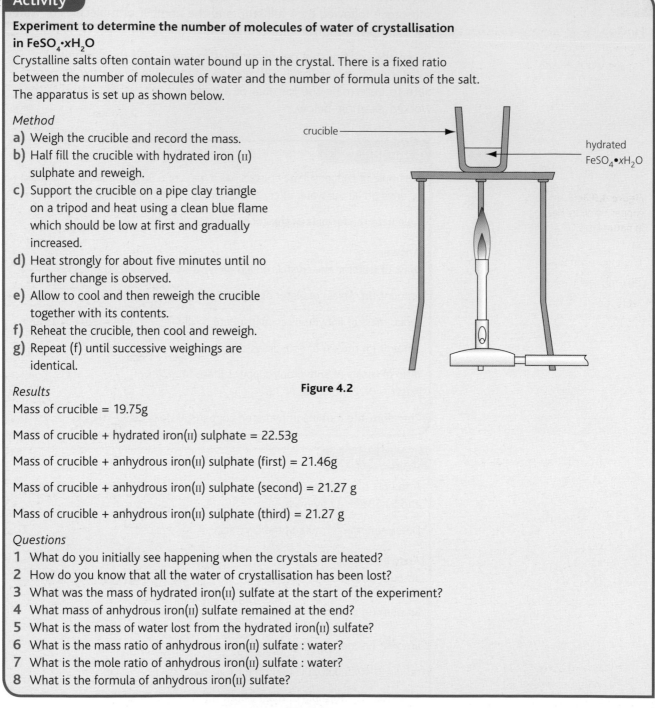

crucible

hydrated $FeSO_4 \bullet xH_2O$

**Figure 4.2**

*Results*

Mass of crucible = 19.75g

Mass of crucible + hydrated iron(II) sulphate = 22.53g

Mass of crucible + anhydrous iron(II) sulphate (first) = 21.46g

Mass of crucible + anhydrous iron(II) sulphate (second) = 21.27 g

Mass of crucible + anhydrous iron(II) sulphate (third) = 21.27 g

*Questions*

1 What do you initially see happening when the crystals are heated?
2 How do you know that all the water of crystallisation has been lost?
3 What was the mass of hydrated iron(II) sulfate at the start of the experiment?
4 What mass of anhydrous iron(II) sulfate remained at the end?
5 What is the mass of water lost from the hydrated iron(II) sulfate?
6 What is the mass ratio of anhydrous iron(II) sulfate : water?
7 What is the mole ratio of anhydrous iron(II) sulfate : water?
8 What is the formula of anhydrous iron(II) sulfate?

**Test yourself**

12 When 8.94 g of hydrated sodium carbonate are heated, 3.32 g of anhydrous sodium carbonate is formed. Calculate the formula of the hydrated sodium carbonate.

## Activity

### Finding the formula of red copper oxide

A group of students investigated the formula of red copper oxide by reducing it to copper using natural gas as shown in Figure 4.3.

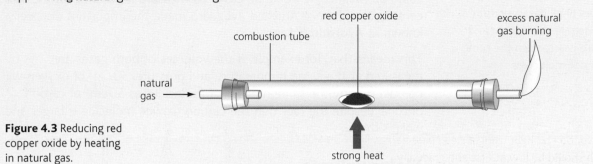

**Figure 4.3** Reducing red copper oxide by heating in natural gas.

The experiment was carried out five times, starting with different amounts of red copper oxide. The results are shown in Table 4.1.

**Table 4.1**

| Experiment number | Mass of copper in the oxide/g | Mass of red copper oxide/g |
|---|---|---|
| 1 | 1.43 | 1.27 |
| 2 | 2.14 | 1.90 |
| 3 | 2.86 | 2.54 |
| 4 | 3.55 | 3.27 |
| 5 | 4.29 | 3.81 |

1 Look at Figure 4.3. What safety precautions should the students take during the experiments?

2 What steps should the students take to ensure that all the copper oxide is reduced to copper?

3 Start a spreadsheet program on a computer and open up a new spreadsheet for your results. Enter the experiment numbers and the masses of copper oxide and copper in the first three columns of your spreadsheet, as in Table 4.1.

4 a) Enter a formula in column 4 to work out the mass of oxygen in the red copper oxide used.
   b) Enter a formula in column 5 to find the amount of copper in moles in the oxide.
   c) Enter a formula in column 6 to find the amount of oxygen in moles in the oxide.

5 From the spreadsheet, plot a line graph of amount of copper ($y$-axis) against amount of oxygen ($x$-axis). Print out your graph. If you cannot plot graphs directly from the spreadsheet, draw the graph by hand.

6 Which of the points should be disregarded in drawing the line of best fit?

7 a) What, from your graph, is the average value of the ratio: $\dfrac{\text{amount of copper/mol}}{\text{amount of oxygen/mol}}$?
   b) How much copper, in moles, combine with one mole of oxygen in red copper oxide?
   c) What is the formula of red copper oxide?

8 Give reasons why the students could claim that their answer for the formula of the oxide was valid.

9 Write a word equation, and then a balanced equation, for the reduction of red copper oxide to copper using methane ($CH_4$) in natural gas. (Hint: The only solid product is copper.)

# Volumes of gases

## Key term

Avogadro's Law states that equal volumes of all gases measured at the same temperature and pressure will contain the same number of molecules.

## Tip

The requirement that the measurements are made under the same conditions is important because the volume of gas varies with changes in temperature and pressure.

## Tip

Remember there are $1000\,cm^3$ in $1\,dm^3$.

If a substance is a solid, mass is used to measure the amount needed for a reaction. However, if a reactant or product is a gas, it is unlikely that it could be weighed; it would be easier to measure its volume. In the 19th century, the chemist Amedeo Avogadro made the important discovery known as Avogadro's Law.

This means that, for example, if the volumes of both gases are measured at the same temperature and pressure, $50\,cm^3$ of oxygen will contain exactly the same number of molecules as $50\,cm^3$ of carbon dioxide, despite the fact that the carbon dioxide molecule is larger and weighs more.

## Molar volume

If a volume of gas contains a fixed number of molecules, then it is useful to know what volume contains the Avogadro number of particles. This volume is known as the **molar volume**. There are two useful benchmarks:

- Under standard conditions of 0 °C and 100 kPa pressure, the molar volume is $22.7\,dm^3\,mol^{-1}$. (Standard conditions are abbreviated to STP.)

- Under normal laboratory conditions, the molar volume is taken as $24000\,cm^3\,mol^{-1}$. This value is used in calculations for reactions at room temperature and pressure, although it should be appreciated that it does vary according to the exact conditions. (Room temperature and pressure is abbreviated to RTP.)

Molar volumes can be expressed in $dm^3$:

- At STP 1 mol of a gas occupies $22.7\,dm^3$.

- At RTP 1 mol of a gas occupies $24.0\,dm^3$.

The key relationship between the amount in moles of gas molecules and the volume of the gas is:

$$\text{amount of gas in moles } (n) = \frac{\text{volume of gas } (V)}{\text{molar volume at that temperature and pressure}}$$

$$\text{At RTP, } n = \frac{V}{24000} \text{ if } V \text{ is measured in } cm^3.$$

or

$$V \text{ (in } cm^3) = 24000 \times n$$

### Example 18

Calculate the volume of 8.0 g of methane at RTP.

**Answer**

Molar mass of methane, $CH_4 = 12.0 + (4 \times 1.0) = 16.0\,g\,mol^{-1}$

Amount in mol of methane molecules $(m/M) = \dfrac{8.0}{16.0} = 0.50\,mol$

Volume of 1 mol of methane $= 24.0\,dm^3$ at RTP.

Volume of 0.50 mol of methane $= 0.50 \times 24.0 = 12.0\,dm^3$

> **Example 19**
>
> Calculate the mass of 500 cm³ of oxygen at RTP. Give your answer to three significant figures.
>
> **Answer**
>
> Volume of 1 mol of oxygen molecules at RTP = 24 000 cm³
>
> Amount in mol in $500\ \text{cm}^3 = \dfrac{500}{24\,000} = 0.0208\ \text{mol}$
>
> Molar mass of oxygen, $O_2 = 32.0\ \text{g mol}^{-1}$
>
> So mass of 0.0208 mol = 0.0208 × 32.0 = 0.667 g.

## Ideal gas equation

The volume of a gas at temperatures or pressures other than STP or RTP can be calculated using the ideal gas equation.

The difference between gases and solids or liquids is that the particles of a gas are very much further apart than the particles in a solid or a liquid.

Distance between the particles

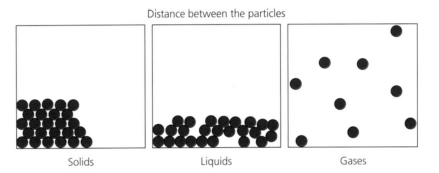

Solids          Liquids          Gases

Because of this the volume of a gas depends on the temperature, the pressure as well as the number of moles of the gas.

- If the amount in moles, $n$, of a gas is increased, the volume will increase. This is summarised by Avogadro's Law which can be written as $V \propto n$

- If the temperature, $T$, of a gas is increased, the volume will increase. This is summarised by Charles' Law which can be written as: $V \propto T$ (where $T$ is in kelvin, K)

- If the pressure, P, is increased, the volume will decrease. This is summarised by Boyle's Law which can be written as: $V \propto \dfrac{1}{P}$

By combining the three gas laws we obtain:

$V \propto n$

$V \propto T$

$V \propto \dfrac{1}{P}$ which when combined gives $V \propto n\dfrac{T}{P}$ or $PV \propto nT$

Adding a constant of proportionality results in the relationship $PV = nRT$ which is known as the **ideal gas equation**. The constant of proportionality, $R$, is known as **the gas constant**.

The value of $R$ can be calculated using standard conditions with the more accurate standard temperature of 273.16 K, and a pressure of 101.32 kPa and, for 1 mol of gas, a volume of 22.414 dm³.

$PV = nRT$ hence $R = \dfrac{PV}{nT} = \dfrac{101.31 \times 22.414}{1 \times 273.16}$

The units of the gas constant, $R$, are $JK^{-1}mol^{-1}$ (joules per kelvin per mole)

Note that in SI units pressure should be given in Pa and volume in $m^3$. In which case the pressure is 101320 Pa and the volume is 0.022414. PV would therefore be 101320 x 0.022414 (= 2270.99) but this is the same as measuring the pressure in kPa and the volume in $dm^3$ 101.32 x 22.414 (= 2270.99). As this is the case it is usual in calculations to use volume in $dm^3$ and pressure in kPa. All the examples in this book will do this.

**Tip**

Explanation of units of the gas constant R
The SI unit of pressure is Pa ($1\,Pa = 1\,N\,m^{-2}$ (1 newton per square metre) therefore $1\,kPa = 1 \times 10^3\,N\,m^{-2}$.
The SI unit of volume is $m^3$ (metres cubed) therefore $1\,dm^3 = 1 \times 10^{-3}\,m$.
The SI unit of temperature is K (Kelvin) To convert from °C to K add 273.
The units of $R$ can be worked out using
$$R = \frac{PV}{nT} = (N\,m^{-2})(m^3)/(mol)(K) = N\,m/(mol)(K) = N\,m\,mol^{-1}\,K^{-1}$$
In SI units 1 newton metre (N m) = 1 joule (J)
So the units of $R$ are $J\,mol^{-1}\,K^{-1}$.

**Tip**

The value and units of the gas constant will be on the data sheet which will be supplied in all examinations.

Another way of expressing the relationship between the temperature, pressure and volume of a gas is as follows:

By combining Charles' Law and Boyle's Law we have $\dfrac{PV}{T}$ = constant (for a given amount in mol of a gas)

It follows that

$$\frac{P_1V_1}{T_1} = \frac{P_2V_2}{T_2}$$ ($P_1$, $V_1$ and $T_1$ are the initial pressure, volume and temperature and $P_2$ $V_2$ and $T_2$ are the final pressure, volume and temperature.

Ideal gases follow these gas laws exactly but require very specific factors to apply. An ideal gas is one that is composed of independent particles (molecules or atoms) which are widely separated from each other. An ideal gas can be described by the following:

1   The molecules (or atoms) are in continuous random motion

2   There are no intermolecular forces between the individual molecules (or atoms)

3   All collisions are perfectly elastic with no exchange of kinetic energy

4   The molecules (or atoms) have no size (i.e. they occupy zero volume)

In reality gases deviate from these ideal conditions but gases do approach ideal behaviour at very high temperatures and at very low pressures. Calculations assume that all gases follow the ideal gas equation exactly.

## Example 20

If the volume of a gas collected at 50 °C and 110 kPa is 75 cm³, what would the volume be at STP?

**Answer**

For calculations of this type it is best to use the equation

$$\frac{P_1V_1}{T_1} = \frac{P_2V_2}{T_2}$$

and the only unknown is $V_2$ so the equation can be rearranged

to give $\dfrac{P_1V_1T_2}{P_2T_1} = V_2$

Pressure is in kPa. Standard pressure is 100 kPa.

Temperatures must be in K, $T_1 = 50\,°C + 273 = 323\,K$. Standard temperature is 0 °C = 273 K.

Whatever the units of volume used for $V_1$ will also be the units of the calculated volume $V_2$.

$$\frac{110 \times 75 \times 273}{100 \times 323} = \frac{2252250}{32300} = 70\,cm^3$$

## Example 21

5.6 g of liquid nitrogen was added to an evacuated 5.0 dm³ volumetric flask and the temperature was allowed to rise to 25 °C. Calculate the pressure inside the volumetric flask at 25 °C.

**Answer**

Use the ideal gas equation, $PV = nRT$ and rearrange to get $P = \dfrac{nRT}{V}$

$n = 5.6/28 = 0.20\,mol$

$R = 8.314\,J\,mol^{-1}\,K^{-1}$

$T = 25 + 273 = 298\,K$

$V = 5.0\,dm^3$

$P = (0.20 \times 8.314 \times 298)/5 = 99.1\,kPa$

## Example 22

1.68 g of a noble gas occupies a volume of 500 cm³ at RTP. Identify the noble gas.

**Answer**

Use the ideal gas equation, $PV = nRT$. Since we are using RTP, the only unknown is the amount in moles, $n$.

$$n = \frac{(100 \times 0.5)}{(8.314 \times 298)} = \frac{50}{2477.6} = 0.02\,mol$$

The ideal gas equation can be rearranged to give $n = \dfrac{PV}{RT}$

We now know that 0.02 mol of the noble gas atoms have a mass = 1.68 g, therefore the atomic mass of the noble gas is

$A_R = \dfrac{m}{n} = 1.68/0.02 = 84$, therefore the noble gas is krypton.

(Krypton has an atomic mass = 83.8 g mol⁻¹)

**Test yourself**

13 Under standard conditions the volume of 1 mole of a gas is $22.0\,dm^3$. Calculate the volume of 1 mole of gas at RTP

14 Calculate the following gas volumes when converted to STP.
a $200\,cm^3$ at $27\,°C$ and $100\,kPa$
b $5.60\,dm^3$ at $425\,K$ and $200\,kPa$

15 $1.73\,g$ of a halogen gas occupies a volume of $600\,cm^3$ at RTP. Identify the halogen gas.

# Volumes of gas produced in a reaction

To calculate the volume of gas produced in a reaction, use the balanced equation, as shown in the following examples. The molar volume appropriate to the conditions of the measurement is always given.

## Example 23

Calculate the volume of hydrogen produced when excess dilute sulfuric acid is added to $5.00\,g$ of zinc. Assume that under the conditions of the reaction the molar volume is $24000\,cm^3$. Give your answer to three significant figures.

**Answer**

The equation is:

$$Zn(s) + H_2SO_4(aq) \rightarrow ZnSO_4(aq) + H_2(g)$$

The equation shows that 1 mol of zinc atoms produces 1 mol of hydrogen molecules. (It also shows that 1 mol of $H_2SO_4$ is used and 1 mol of $ZnSO_4$ is formed, but this is not relevant here.)

Molar mass of zinc = $65.4\,g\,cm^{-3}$

Amount in moles of zinc atoms used, $n/M = \dfrac{5.00}{65.4} = 0.076452599\,mol$

It follows from the equation that $0.0765\,mol$ of hydrogen molecules is produced.

Molar volume = $24000\,cm^3$

Volume of hydrogen produced = $0.076452599 \times 24000 = 1834.862385\,cm^3$ which, to three significant figures, is $1830\,cm^3$.

The last example shows the importance of keeping all the digits in your calculator throughout the calculation. If $0.076452599$ had been rounded to $0.0765\,mol$ after the first step and then multiplied by $24000$ in the second step, the answer obtained would be $1836\,cm^3$ which, when rounded to three significant figures, would have given an answer of $1840\,cm^3$.

## Example 24

Calculate the volume of hydrogen produced when excess dilute sulfuric acid is added to 5.00 g of aluminium. Under the conditions of the reaction, the molar volume is 24 000 cm³.

**Answer**

The equation for the reaction is:

$$2Al(s) + 3H_2SO_4(aq) \rightarrow Al_2(SO_4)_3(aq) + 3H_2(g)$$

The equation indicates that 2 mol of aluminium atoms produces 3 mol of hydrogen molecules or 1 mol of aluminium atoms produces 1.5 mol of hydrogen molecules.

Amount in moles of Al atoms used $= \dfrac{5.00}{27.0} = 0.185$ mol

Amount in moles of $H_2$ molecules produced $= 1.5 \times 0.185 = 0.278$ mol

Volume of $H_2$ produced $= 0.278 \times 24\,000 = 6670$ cm³.

## Example 25

A gas syringe has a maximum capacity of 100 cm³. Calculate the mass of copper(II) carbonate that would have to be heated to produce enough carbon dioxide to just fill the syringe at RTP.

**Answer**

The equation for the reaction is:

$$CuCO_3(s) \rightarrow CuO(s) + CO_2(g)$$

Syringe volume of 100 cm³ is equivalent to $\dfrac{100}{24\,000} = 0.00417$ mol of gas molecules.

0.00417 mol of carbon dioxide is obtained by heating 0.00417 mol of copper carbonate.

Mass of 1 mol of copper carbonate $= 63.5 + 12.0 + (3 \times 16) = 123.5$ g

Mass of 0.00417 mol of copper carbonate $= 0.00417 \times 123.5 = 0.514995$ g which, to three significant figures, is 0.515 g.

The answer to the last example is the theoretical mass that could be used. In practice, it would be unwise to heat a mass as large as this because, although the gas when cooled would fit inside the syringe, the hot gas produced would have a volume greater than 100 cm³.

## Test yourself

16 What is the amount, in moles, of gas at room temperature and pressure in:
   a) 240000 cm³ chlorine
   b) 48 cm³ hydrogen
   c) 3 dm³ ammonia?
17 What are the volumes, in cm³, of the following amounts of gas at RTP:
   a) 2 mol nitrogen molecules
   b) 0.0002 mol neon atoms
   c) 0.125 mol carbon dioxide molecules?
18 What is the volume in cm³ of the following gases at RTP:
   a) 8.50 g of ammonia ($NH_3$)
   b) 1.10 g of carbon dioxide?
19 What is the mass of 250 cm³ of sulfur dioxide ($SO_2$) measured at RTP? Give your answer to three significant figures.
20 Calculate the volume of hydrogen produced at RTP when 0.802 g of calcium reacts with water.

$$Ca(s) + 2H_2O(l) \rightarrow Ca(OH)_2(aq) + H_2(g)$$

21 When reacted with excess dilute hydrochloric acid, what mass of calcium carbonate produces 1 dm³ of carbon dioxide measured at RTP?

## Activity

**Finding an equation for the reaction of magnesium with hydrochloric acid**

A small piece of magnesium was weighed and then added to excess dilute hydrochloric acid (HCl(aq)) in the apparatus shown in Figure 4.4. A vigorous reaction occurred and hydrogen gas ($H_2$) was produced. Eventually all the magnesium reacted and the reaction stopped.

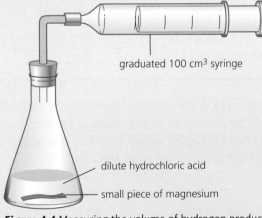

graduated 100 cm³ syringe

dilute hydrochloric acid

small piece of magnesium

**Figure 4.4** Measuring the volume of hydrogen produced when magnesium reacts with dilute hydrochloric acid.

Here are the results.

Mass of magnesium added = 0.061 g

Volume of hydrogen produced = 60 cm³

1 Why is it important to use excess hydrochloric acid?
2 What modifications could you make to prevent hydrogen escaping from the flask before the apparatus is reconnected after adding the magnesium?
3 How many moles of magnesium reacted? ($A_r$(Mg) = 24.3)
4 How many moles of hydrogen were produced? (Assume that the molar volume of hydrogen is 24.0 dm³ mol⁻¹.)
5 Copy and complete the following statements:

   _____ mol magnesium, Mg, produced
   _____ mol hydrogen, $H_2$.

   So 1 mol magnesium, Mg, produces _____ mol hydrogen, $H_2$.
6 Use your result from Question 5 to write a balanced equation, with state symbols, for the reaction of magnesium with hydrochloric acid to form hydrogen. (Assume that magnesium chloride ($MgCl_2$) is also produced.)

## Reactions between gases

Avogadro's Law is extremely useful when relating the volume of gas to an equation, as shown in the following examples.

### Example 26

Assuming the measurements are made at the same temperature and pressure, what volume of oxygen reacts exactly with 100 cm³ of hydrogen?

**Answer**

The equation for the reaction is: $2H_2(g) + O_2(g) \rightarrow 2H_2O(l)$

The equation shows that 2 mol of hydrogen molecules reacts with 1 mol of oxygen molecules.

The number of molecules in 100 cm³ of hydrogen is not known. However, applying Avogadro's Law it is known that 100 cm³ of oxygen contains the same number of molecules as there are in the 100 cm³ of hydrogen.

Since the equation indicates that half the number of oxygen molecules as hydrogen molecules are needed, it must be that 50 cm³ of oxygen reacts with 100 cm³ of hydrogen.

In this last example, if the temperature were such that the $H_2O$ was gaseous, then the volume produced would be 100 cm³. Be careful not to assume that you can simply add the volumes of the reactants to obtain the volume of the product. The total *mass* on the left-hand side of the equation must equal the total mass on the right-hand side, but this does not apply to volumes because the densities affect the mass. In this case, 100 cm³ of hydrogen reacts with 50 cm³ of oxygen but only 100 cm³ of steam is produced.

**Tip**

Remember
volume = mass/density

### Example 27

Assuming the measurements are made at the same temperature and pressure, what volume of oxygen reacts exactly with 30 cm³ of methane and what volume of carbon dioxide is obtained?

**Answer**

The equation for the reaction is: $CH_4(g) + 2O_2(g) \rightarrow CO_2(g) + 2H_2O(l)$

The equation shows that when 1 mol of methane molecules react with 2 mol of oxygen molecules, 1 mol of carbon dioxide is formed. Using Avogadro's Law, it follows that 30 cm³ of methane requires 60 cm³ of oxygen to produce 30 cm³ of carbon dioxide.

**Tip**

As water is a liquid, you cannot deduce its volume using Avogadro's Law.

### Test yourself

22 Assuming that all gas volumes are measured at the same temperature and pressure, what volume of oxygen is needed to react with 50 cm³ ethane, $C_2H_6$, when it burns and what volume of carbon dioxide forms?

23 What volume of oxygen reacts with 200 cm³ of propane, $C_3H_8$, and what volume of carbon dioxide is produced? (Assume that all volumes are measured at room temperature and pressure.)

# Practice questions

## Multiple choice questions 1–10

1 Which one of the following contains the greatest number of molecules?
   - A 30g of methane ($CH_4$)
   - B 1.5 mol of methane
   - C 30g of ethane ($C_2H_6$)
   - D 1.5 mol of ethane

2 Which one of the following contains a number of particles equal to the Avogadro constant?
   - A molecules in 1.0g of hydrogen gas
   - B atoms in 1.0g of carbon
   - C chloride ions in 1 mol of calcium chloride
   - D carbonate ions in 1 mol of calcium carbonate

3 Copper nitrate decomposes when heated as shown by the following equation:

   $$2Cu(NO_3)_2(s) \rightarrow 2CuO(s) + 4NO_2(g) + O_2(g)$$

   Which one of the following correctly describes the full decomposition of 1 mol of copper nitrate?
   - A A total of 3 mol of gas are formed
   - B The volume of $NO_2(g)$ produced is twice that of oxygen
   - C The total amount in mol of gas produced is $2\frac{1}{2}$ times the amount in mol of the copper nitrate used.
   - D A total of 7 mol of products is obtained.

4 Which one of the following is the molecular formula of a hydrocarbon containing 80% by mass of carbon?
   - A $CH_4$
   - B $C_2H_4$
   - C $C_2H_6$
   - D $C_3H_6$

5 When heated with sodium hydroxide, ammonium sulfate completely reacts to form ammonia gas.
   What is the maximum amount of ammonia that can be obtained at RTP if 0.2 mol of ammonium sulfate is completely reacted?
   (I mol of gas has a volume of 24.0 dm³ at RTP)
   - A 0.8 mol
   - B 0.4 mol
   - C 0.2 mol
   - D 0.1 mol

6 What volume of oxygen measured at the same temperature and pressure is needed to burn completely 100 cm³ of propane ($C_3H_8$)?
   - A 300 cm³
   - B 500 cm³
   - C 700 cm³
   - D 1100 cm³

7 At 50°C and 200kPa, the volume of a sample of a gas is $V$ cm³. Which one of the following changes to the temperature and pressure would cause the maximum increase in the volume of the gas?

|   | Temperature | Pressure |
|---|---|---|
| A | 50°C | 50kPa |
| B | 100°C | 100kPa |
| C | 200°C | 150kPa |
| D | 400°C | 200kPa |

**Key**

| A | B | C | D |
|---|---|---|---|
| 1, 2 and 3 correct | 1 and 2 correct | 2 and 3 correct | 1 only correct |

Use the key above to answer questions 8, 9 and 10.

8 Which of the following statements about ethane is correct?
   1 30g of ethane contains 1 mol of ethane molecules
   2 Ethane has a molar mass of 30 g mol⁻¹
   3 Ethane has a relative molecular mass of 30 g

9 The equation for the combustion of but-1-ene is:

   $$C_4H_8(g) + 6O_2(g) \rightarrow 4CO_2(g) + 4H_2O(g)$$

   Which of the following would be present in a closed container if 10 cm³ of but-1-ene was completely burned in 100 cm³ of oxygen? (Assume all volumes were measured at a temperature above 100°C)
   1 40 cm³ of carbon dioxide
   2 40 cm³ of water vapour
   3 40 cm³ of oxygen

10 When heated copper oxide reacts with gaseous ammonia to form copper;

   $$2CuO(s) + 2NH_3(g) \rightarrow 2Cu(s) + N_2(g) + 3H_2O(g)$$

   If the reaction takes place to completion using ammonia and an excess of copper oxide, which of the following will be found?
   1 At 400 °C, the volume of the products will be twice the volume of the ammonia that was used.
   2 If the reaction is cooled to room temperature, the volume of the products will be half the volume of ammonia that was used.

**3** The amount in mol of copper obtained will be half the amount in mol of copper oxide that was used.

**11 a)** Calculate the relative formula mass of sodium carbonate, $Na_2CO_3$. (1)

**b)** Calculate the relative formula mass of barium chloride crystals, $BaCl_2 \cdot 2H_2O$. (1)

**12** Determine the amount in moles present in each of the following:

**a)** 8.0 g of sulfur (1)

**b)** 1.68 g of calcium oxide. (1)

**13** Calculate the mass in grams for each of the following. Give your answers to 2 significant figures.

**a)** 0.040 mol of aluminium chloride (2)

**b)** 0.45 mol of aluminium hydroxide (2)

**14** Calculate the mass of each of the following volumes measured at RTP:

**a)** 4 dm³ of carbon dioxide (2)

**b)** 500 cm³ of ethane. (2)

**15** Calculate the molar mass of an element if 3.745 g is equivalent to 0.0500 mol of that element. Identify the element. (2)

**16** 0.02 mol of the hydroxide of X has a mass of 1.48 g. X is a divalent metal.

**a)** Calculate the molar mass of the hydroxide of X. (1)

**b)** What is the formula of the hydroxide of X? (1)

**c)** Identify the metal, X. (1)

**17** How many atoms are there in 5.0 g of silicon? (2)

**18** How many molecules are there in 5.0 g of oxygen gas? (2)

**19 a)** Write the equation, including state symbols, for the reaction of magnesium with aqueous zinc sulfate to form zinc metal and aqueous magnesium sulfate. (1)

**b)** What mass of zinc is obtained by reacting 3.0 g of magnesium with excess zinc sulfate in solution? Give your answer to 2 significant figures. (2)

**c)** Why is it necessary to specify that the aqueous zinc sulfate is in excess? (1)

**20** Zinc nitrate decomposes when heated to form zinc oxide:

$$2Zn(NO_3)_2(s) \rightarrow 2ZnO(s) + 4NO_2(g) + O_2(g)$$

**a)** What are the mole ratios in the equation? (2)

**b)** Calculate the mass of zinc nitrate that would have to be heated to produce 3.6 g of zinc oxide. (2)

**21** Calculate how many tonnes of lead could be produced in a furnace by reacting 50.0 tonnes of lead oxide, PbO, with carbon. Give your answer to 3 significant figures. (5)

**22** When 10.80 g of hydrated iron(II) sulfate were heated, the water of crystallisation was driven off. This left 5.91 g of anhydrous iron(II) sulfate.

**a)** How many moles of $FeSO_4$ and $H_2O$ are present in 10.80 g of hydrated iron(II) sulfate? (2)

**b)** What is the formula of hydrated iron(II) sulfate? (2)

**23** Calculate the volume in dm³ at RTP of each of the following gases:

**a)** 1 g of hydrogen (1)

**b)** 2 g of methane, $CH_4$ (1)

**c)** 8.8 g of carbon dioxide (1)

**d)** 4.0 g of nitrogen monoxide, NO. (1)

**24** Calculate the mass of each of the following volumes. Assume that the conditions are RTP. i.e. 1 mol of gas has a volume of 24 dm³.

**a)** 12 dm³ of oxygen (1)

**b)** 4 dm³ of carbon dioxide (1)

**c)** 500 cm³ of ethane, $C_2H_6$ (1)

**d)** 100 cm³ of nitrogen (1)

**25** What is the volume of each of the following gases at 25°C and 100kPa?

**a)** 5.60 dm³ of gas at 425°C and 80kPa (1)

**b)** 100 cm³ of gas at 103kPa and 120°C (1)

**26** 100 cm³ of oxygen is collected in a syringe at 150°C. If the pressure remains unchanged, by how much will the volume contract if it is allowed to cool to a room temperature of 20°C? (2)

**27** 0.363g of a solid element, Y, is vapourised at 400K and 100kPa. Its volume is found to be 47 cm³. Calculate the relative atomic mass of the element and identify Y. (3)

**28 a)** Compound X is a volatile liquid and is known to contain 52.17% C; 13.0% H and 34.78% O by mass. Calculate the empirical formula of compound X. Show your working. (2)

**b)** 0.25g of X is vapourised at 90°C and 101kPa pressure and is found to have a volume of 162 cm³. Calculate the molecular formula of X. (3)

**29** In parts (a)–(c), correct any errors that you find.

**a)** Consider the reaction:

$$Zn(s)+2HCl(aq) \rightarrow ZnCl_2(aq)+H_2(g)$$

**i)** The equation indicates that 1 mol of zinc reacts with 2 mol of hydrochloric acid to produce 1 mol of zinc chloride and 2 mol of hydrogen atoms.

**ii)** The equation indicates that 1 mol of zinc produces 48 dm³ of hydrogen at room temperature and pressure. (2)

**b)** Consider the reaction:

$$CH_4+2O_2 \rightarrow CO_2+2H_2O$$

The equation indicates that, at room temperature and pressure, 24 dm³ of methane reacts with 48 dm³ of oxygen to produce 24 dm³ of carbon dioxide and 48 dm³ of water. (1)

**c)** Gaseous nitrogen monoxide (NO) reacts with oxygen to form gaseous nitrogen dioxide (NO₂), according to the equation:

$$2NO(g)+O_2(g) \rightarrow 2NO_2(g)$$

The equation indicates that 2 mol of nitrogen monoxide react with 1 mol of oxygen. So 48 dm³ of nitrogen monoxide reacts with 24 dm³ of oxygen and 72 dm³ of nitrogen dioxide is obtained. (2)

**30** Calculate the volume of hydrogen in cm³ at RTP that is produced by reacting 1.00 g of magnesium with excess hydrochloric acid. (3)

**31** Calculate the volume of carbon dioxide in cm³ at RTP that could be obtained when 3.00 g of carbon is burned in excess oxygen. (2)

**32** Calculate the volume of hydrogen in cm³ at RTP that is needed to convert 7.95 g of copper oxide into copper metal. The equation for the reaction is:

$$CuO(s)+H_2(g) \rightarrow Cu(s)+H_2O(l)$$
(2)

**33** Calculate the mass of magnesium that could be reacted with excess sulfuric acid at RTP such that the hydrogen produced would fill a 250 cm³ measuring cylinder with hydrogen. (3)

**34** Dinitrogen monoxide can decompose to nitrogen and oxygen when heated.

$$2N_2O(g) \rightarrow 2N_2(g) + O_2(g)$$

What volumes of oxygen and nitrogen would be obtained when 50 cm³ of dinitrogen monoxide was decomposed? (2)

**35** Carbon monoxide reacts with oxygen to form carbon dioxide:

$$2CO(g)+O_2(g) \rightarrow 2CO_2(g)$$

What volume of oxygen is required to react with 40 cm³ of carbon monoxide and what volume of carbon dioxide would be formed? Assume that all volumes are measured at the same temperature and pressure. (2)

**36** When a group (II) carbonate is heated, 0.74 g of an oxide is obtained and 172 cm³ of carbon dioxide is collected at room temperature and pressure. Identify the group (II) carbonate and explain how you obtained your answer. (4)

**37** Ammonia can be manufactured by combining nitrogen and hydrogen under high pressure and with a controlled temperature. Even so, the process is not very efficient. Some of the ammonia produced is converted in a reaction with sulfuric acid into ammonium sulfate for use as a fertiliser. A process using 15 000 dm³ of nitrogen operates such that the reaction is only 22% efficient. The ammonia produced is then all converted into ammonium sulfate.

**a)** What volume of hydrogen would be required to react completely with the 15 000 dm³ of nitrogen and what volume of ammonia would be produced? (3)

**b)** What mass in tonnes (1000 kg) of ammonium sulfate would be formed if the reaction was only 22% efficient? Give your answer to 3 significant figures. (3)

## Challenge

The work of Avogadro was important because it provided the means of establishing the formulae of some gases. This question shows you how this might have been done. You have to establish the formula of a gas, $C_xH_y$, from experimental data. Compound $C_xH_y$ burns in oxygen to form carbon dioxide and water, provided that enough oxygen is supplied.

In an experiment, $40\,cm^3$ of $C_xH_y$ and $500\,cm^3$ of oxygen are mixed together at room temperature and pressure. An electrical spark is then applied which causes the $C_xH_y$ to react completely. When cooled to room temperature and pressure, droplets of water are observed and the volume of gas is $420\,cm^3$. This gas consists of the carbon dioxide formed and unreacted oxygen.

The $420\,cm^3$ of gas is shaken with sodium hydroxide solution. This absorbs the carbon dioxide by forming sodium carbonate and water. The volume of gas remaining is $260\,cm^3$.

a) Write a balanced equation for the reaction between carbon dioxide and sodium hydroxide solution. Include state symbols.
b) What volume of carbon dioxide is formed in the reaction?
c) What volume of oxygen is used in reacting with $C_xH_y$?
d) The equation for the reaction can be written in terms of $x$ and $y$.

$$C_xH_y + \ldots\ldots O_2 \rightarrow x\,CO_2 + \frac{y}{2}H_2O$$

Copy and complete the equation by providing the balancing number for the oxygen molecules.
e) Use the volumes you have established in parts (b) and (c) and the equation in part (d) to determine the value of $x$ and then the value of $y$.
f) Write the formula of $C_xH_y$.
g) Use the balanced equation to determine the mass of water produced when $40\,cm^3$ (measured at RTP) of $C_xH_y$ is burned.
h) The density of water is $1\,g\,cm^{-3}$. What volume of water is obtained in the experiment? (12)

# Chapter 5

# Amount of substance – moles in solution

## Prior knowledge

*This chapter extends the ideas met in Chapter 4 and you should make sure the work covered in that chapter is fully understood before proceeding. You will therefore need to be familiar with:*

- writing equations
- interpreting the information in an equation in terms of the amount in moles required for a complete reaction
- the relationship between the amount in moles, the mass in grams and the molar mass of a substance ($n = m/M$)
- the meanings of the terms **solute**, **solvent** and **solution**.

## Test yourself on prior knowledge

1 Solid sodium carbonate reacts with dilute hydrochloric acid to form sodium chloride, carbon dioxide and water.
   a) Write an equation for the reaction including state symbols.
   b) What does the equation tell you about the amount in moles of the reactants that would be required for a complete reaction to take place?
   c) What does the equation tell you about the amount in moles of the products that would be formed when a complete reaction had taken place?
   d) What amount in moles of hydrochloric acid would be required to react completely with 0.32 mol of sodium carbonate?
   e) What amount in moles of hydrochloric acid would be required to react completely with 0.32 g of sodium carbonate? Give your answer to 2 significant figures.
   f) What mass of sodium chloride would be formed from the reaction in (e)? Give your answer to 2 significant figures.
2 If sodium chloride is shaken with water until it dissolves, which is the solvent and which is the solute?

## Concentrations of solutions

Many chemical reactions only occur if the reactants involved are in aqueous solution. The common acids, for example, usually only react when dissolved in water. In order to decide on the correct amounts to use for a chemical reaction it is necessary to know how much of the reacting substance is present in a given volume of the solution. This is the concentration of the solution and solutions for quantitative work are specified by how many grams or moles of the substance there are in a certain volume of solution. This volume is usually $1\,dm^3$ ($1000\,cm^3$).

For example, consider a solution of sodium carbonate containing $2\,mol$ of $Na_2CO_3$ in every $1\,dm^3$ of the solution. The concentration is $2\,mol$ per $dm^3$, which is written as $2\,mol\,dm^{-3}$. It follows that $500\,cm^3$ ($0.5\,dm^3$) of this solution contains $1\,mol$. Different volumes of solution contain different numbers of moles.

In general:

amount in moles = concentration of solution × volume of solution

units:      (mol)            $(mol\,dm^{-3})$           $(dm^3)$

equation:             $n = cV$

## Example 1

What is the amount in moles of NaOH in $50\,cm^3$ of a solution whose concentration is $2\,mol\,dm^{-3}$?

**Answer**

The volume given is in $cm^3$ and must be converted to $dm^3$ by dividing by 1000.

Then using $n = cV$:

$$\text{amount in moles } (n) = 2 \times \frac{50}{1000} = 0.1\,mol$$

## Example 2

What volume of a solution of hydrochloric acid of concentration $0.5\,mol\,dm^{-3}$ contains $0.15\,mol$?

Give the answer in $cm^3$.

**Answer**

Rearrange the equation $n = cV$:

$$V = \frac{n}{c}$$

$$V = \frac{0.15}{0.5} = 0.3\,dm^3$$

To convert $0.3\,dm^3$ to $cm^3$, multiply by 1000.

$$\text{volume of solution} = 300\,cm^3$$

## Example 3

If $0.4\,mol$ of $Na_2CO_3$ is dissolved to make $200\,cm^3$ of solution, what is its concentration in $mol\,dm^{-3}$?

**Answer**

Rearrange the equation $n = cV$:

$$c = \frac{n}{V}$$

$$c = \frac{0.4}{0.2} = 2\,mol\,dm^{-3}$$

## Example 4

8.00 g of sodium hydroxide is dissolved to make 100 cm³ of solution.

What is its concentration of the solution in mol dm⁻³?

**Answer**

The molar mass of sodium hydroxide is 40.0 g mol⁻¹.

So the amount used in moles is:

$$\frac{m}{M} = \frac{8.00}{40.0} = 0.200 \, \text{mol}$$

This is dissolved in 100 cm³ (0.1 dm³) so the concentration in mol dm⁻³ is:

$$\frac{n}{V} = \frac{0.200}{0.1} = 2.00 \, \text{mol dm}^{-3}$$

# Diluting solutions

**Tip**

A common mistake is to think that 200 cm³ of concentrated acid would require 1000 cm³ of water in order to dilute it five times. That would create 1200 cm³ of solution, which is a six-fold dilution.

**Tip**

Diluting concentrated acids is an exothermic process. If a small volume of water is added to concentrated acid, the heat produced may cause the water to boil, which can be dangerous. Concentrated acid should always be added slowly to water; water should *never* be added to concentrated acid.

Hydrochloric acid is often sold as a solution of concentration 10 mol dm⁻³. If a solution of concentration 2 mol dm⁻³ is required, then it has to be diluted five times. For example, 200 cm³ of the concentrated acid requires 800 cm³ water to make 1000 cm³ of dilute acid of concentration 2 mol dm⁻³.

The dilution that is required is calculated using the following using equation A shown below:

Equation A

$$\text{dilution} = \frac{\text{concentration of the original solution}}{\text{concentration of the diluted solution}}$$

So to make 200 cm³ of a solution of concentration 0.5 mol dm⁻³ from a stock solution of concentration 2 mol dm⁻³, the dilution required is $\frac{2}{0.5} = 4$, which is a four-fold dilution.

Therefore, the stock solution has to be diluted four times. To make 200 cm³ of concentration 0.5 mol dm⁻³ requires 50 cm³ of the stock solution and 150 cm³ of water.

The volume of the original stock solution can be calculated by using equation B shown below.

Equation B

$$\text{volume of original stock solution needed} = \frac{\text{total volume of diluted solution required}}{\text{dilution (calculated in equation A)}}$$

## Example 5

How can 500 cm³ of a 0.4 mol dm⁻³ solution of sulfuric acid be prepared from a solution of concentration 2 mol dm⁻³?

**Answer**

Equation A dilution required $= \frac{2}{0.4} = 5$

Equation B volume of original stock solution $= \frac{500}{5} = 100 \, \text{cm}^3$

Therefore, 100 cm³ of the dilute sulfuric acid of concentration 2 mol dm⁻³ should be further diluted by the addition of 400 cm³ of water.

## Example 6

What is the concentration in mol dm⁻³ of 375 cm³ aqueous sodium hydroxide that has been prepared by diluting 75 cm³ of sodium hydroxide of concentration 0.6 mol dm⁻³ with 300 cm³ of water?

**Answer**

In exams, a solution of a substance such as sodium hydroxide will be referred to as 'aqueous sodium hydroxide', which indicates that it has been dissolved in water.

volume of dilute solution = 375 cm³

$$\text{concentration} = \frac{75}{375} \times 0.6 = 0.12\,\text{mol dm}^{-3}$$

## Test yourself

1 State the amount in moles in each of the following:
   a) 2 dm³ of sodium hydroxide solution of concentration 0.1 mol dm⁻³
   b) 200 cm³ of sulfuric acid of concentration 0.5 mol dm⁻³.
2 Calculate the volume of 0.5 mol dm⁻³ aqueous potassium chloride that contains:
   a) 1 mol
   b) 0.2 mol.
3 Give the concentration of each of the following solutions in mol dm⁻³:
   a) 100 cm³ of potassium chloride containing 0.1 mol
   b) 50 cm³ of sodium nitrate containing 0.01 mol.
4 Calculate the mass of solid that on dissolving would produce:
   a) 250 cm³ of aqueous sodium hydroxide of concentration 0.2 mol dm⁻³
   b) 100 cm³ of aqueous sodium carbonate of concentration 0.4 mol dm⁻³.
5 Describe how each of the following solutions of sulfuric acid could be prepared from a solution of acid of concentration 1 mol dm⁻³:
   a) 1 dm³ of solution of concentration 0.1 mol dm⁻³
   b) 500 cm³ of solution of concentration 0.2 mol dm⁻³.
6 What is the concentration obtained by diluting 50 cm³ of 2 mol dm⁻³ aqueous sodium carbonate with water to make:
   a) 500 cm³ of solution
   b) 650 cm³ of solution?

# Titrations

In general, a titration is a procedure used to determine the concentration of a solution by reacting it with another solution whose concentration is known. An example is the titration of an alkali with an acid, which requires the use of an acid–base indicator, such as methyl orange, to show when the reaction is complete.

When an acid neutralises an alkali, the pH of the solution changes. We can follow the pH change using an indicator. The indicator shows how much acid is needed to react with all of the alkali.

In an experiment, a known volume of one solution is pipetted into a conical flask. The second solution is put into a burette and added bit by bit to the first solution until the reaction is complete.

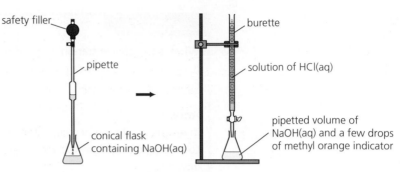

safety filler

pipette

burette

solution of HCl(aq)

conical flask containing NaOH(aq)

pipetted volume of NaOH(aq) and a few drops of methyl orange indicator

**Figure 5.1** The apparatus used for a titration.

# Performing an acid–base titration

The following are the steps to take when 25.0 cm³ of an aqueous solution of sodium hydroxide is titrated against hydrochloric acid using methyl orange as the indicator.

1   Using a pipette, take exactly 25.0 cm³ of the aqueous sodium hydroxide.

2   Transfer the solution to a conical flask.

3   Add a few drops of methyl orange indicator. As aqueous sodium hydroxide is alkaline, the methyl orange appears yellow.

4   Place the hydrochloric acid in a burette and note the initial reading of the burette.

5   Add the hydrochloric acid carefully from the burette until the indicator turns orange. This shows that the neutral point has been reached.

6   Note the final reading of the burette and subtract the initial reading to determine the volume of hydrochloric acid that has been added.

7   Repeat the titration until concordant (consistent) results for the volume of hydrochloric acid added are obtained. With care, it is possible to obtain reproducible results that are accurate to within 0.1 cm³.

The point at which the two solutions have reacted completely is called the **end point** of the titration; the volume of solution added from the burette is known as the **titre**.

# Tabulating the results of a titration

The results of a titration should be recorded in an orderly way. It is best to tabulate the data. For each titration, record the initial burette reading and the final reading, and then calculate the volume used.

Table 5.1 shows how the results of a titration should be recorded.

**Table 5.1** Results of a titration.

| Titration | Rough | 1 | 2 | 3 |
|---|---|---|---|---|
| Final burette reading/cm³ | 19.10 | 18.55 | 18.85 | 37.50 |
| Initial burette reading/cm³ | 0.00 | 0.00 | 0.00 | 18.85 |
| Volume (titre) used/cm³ | 19.10 | 18.55 | 18.85 | 18.65 |
| Titre volumes used in calculations | ✗ | ✓ | ✗ | ✓ |

A burette reading is usually accurate to 0.05 cm³ and the results recorded should indicate this. It is usual, but not essential, to do a 'rough' titration first as a guide to the end point of the titration. This is then followed by obtaining more accurate readings taking particular care to add the solution from the burette drop by drop as the end point is approached. The 'rough' value is not used in working out the titre.

The results in Table 5.1 show that the titres from the first and third accurate titrations are in close agreement. The titre from the second accurate titration is out of line and is not used in calculating the mean titre. Therefore, the mean value for the results of this titration is 18.60 cm³.

(Chapter 1 gives more details on the precision that you should use when recording results.)

## Activity

**Comparing the acidity of different vinegars**

Two students were asked to compare the acidity of different white wine vinegars. The acid in vinegar is ethanoic acid, $CH_3COOH$. The students decided to use 25.0 cm³ of each vinegar and then measure how much 1.0 mol dm⁻³ sodium hydroxide just reacts with each vinegar using phenolphthalein indicator. Phenolphthalein is colourless in acid and pink/red in alkali.

25.0 cm³ of the first vinegar was measured into a conical flask using a pipette (Figure 5.2).

After adding 5 drops of phenolphthalein, the sodium hydroxide solution was slowly added from the burette (Figure 5.3) while the contents of the conical flask were kept well mixed. When the first permanent tinge of pink/red appeared in the mixture, the students recorded the volume of sodium hydroxide added.

**Figure 5.2** Measuring 25.0 cm³ of vinegar into a conical flask using a pipette.

**Figure 5.3** Adding sodium hydroxide solution from a burette to 25.0 cm³ of white wine vinegar.

This method of adding one solution from a burette to a measured volume of another solution is called a titration. Titrations enable chemists to determine how much of the two solutions will just react.

The first titration is only a rough titration and the students then carried out three accurate titrations. To do this, they added 1 cm³ less sodium hydroxide solution than the total amount required in their rough titration. They then added the sodium hydroxide solution one drop at a time until the first permanent tinge of pink/red appeared in the mixture.

The students' results are shown in Table 5.2, together with the results for the second white wine vinegar.

**Table 5.2** Results of the titration.

| Experiment | Rough titration | Accurate titrations | | |
|---|---|---|---|---|
| **First wine vinegar** | | | | |
| Final burette reading/cm³ | 20.50 | 40.15 | 19.60 | 39.30 |
| Initial burette reading/cm³ | 0.00 | 20.50 | 0.00 | 19.60 |
| **Second wine vinegar** | | | | |
| Final burette reading/cm³ | 13.70 | 26.80 | 40.65 | 13.10 |
| Initial burette reading/cm³ | 0.00 | 13.70 | 26.80 | 0.00 |

1 Why does the mixture in the conical flask suddenly change from colourless to pink/red during the addition of sodium hydroxide solution?
2 Why must the contents of the conical flask be well mixed during the titration?
3 Calculate the volumes of sodium hydroxide solution added in the rough and accurate titrations for both wine vinegars in Table 5.2.
4 Calculate an average value for the accurate titration with the first wine vinegar.
5 What did the students do to increase the reliability of their results?
6 a) What value will you take for the accurate titration with the second wine vinegar?
  b) Explain why you have taken this value.
7 Compare the acidity of the two vinegars.
8 State three factors the students must control in their experiments in order to compare the two vinegars fairly.

## Calculation of an unknown concentration from a titration

Suppose the concentration of the pipetted sodium hydroxide used in the titration is $0.100 \, \text{mol dm}^{-3}$ and the mean titre obtained is $18.60 \, \text{cm}^3$, the concentration of the hydrochloric acid can be calculated as follows.

Step 1 Calculate the amount (in moles) of NaOH used.

This is possible because the concentration, $c$, and the volume, $V$, are known.

concentration of sodium hydroxide = $0.100 \, \text{mol dm}^{-3}$ (i.e. $0.100 \, \text{mol}$ are present in $1000 \, \text{cm}^3$ of solution)

$25.0 \, \text{cm}^3$ was taken in the pipette and this sodium hydroxide contains

$$0.100 \times \frac{25.0}{1000} = 0.00250 \, \text{mol}$$
(or $n = cV = 0.100 \times 0.0250$)

Step 2 Refer to the balanced equation to find the amount in moles of HCl needed to react with $0.00250 \, \text{mol}$ of NaOH.

The equation for the reaction is:

$$NaOH + HCl \rightarrow NaCl + H_2O$$

which shows that $1 \, \text{mol}$ of NaOH reacts with $1 \, \text{mol}$ of HCl.

So $0.00250 \, \text{mol}$ of HCl must react with $0.00250 \, \text{mol}$ NaOH.

Since $18.60 \, \text{cm}^3$ of HCl was added from the burette, it follows that $18.60 \, \text{cm}^3$ contains the $0.00250 \, \text{mol}$ of acid.

Step 3 Calculate the concentration of hydrochloric acid in $\text{mol dm}^{-3}$.

volume of hydrochloric acid = $18.60 \, \text{cm}^3 = \frac{18.60}{1000} \, \text{dm}^3$

amount in moles of hydrochloric acid = $0.00250 \, \text{mol}$

$$= 0.0186 \, \text{dm}^3$$

concentration of hydrochloric acid = $\frac{0.00250}{0.0186} = 0.134 \, \text{mol dm}^{-3}$

## Example 7

25.0 cm³ of aqueous sodium carbonate is neutralised by 27.20 cm³ of hydrochloric acid of concentration 0.120 mol dm⁻³. What is the concentration in mol dm⁻³ of the aqueous sodium carbonate?

**Answer**

The concentration, $c$, and the volume, $V$, of the hydrochloric acid are known.

concentration of hydrochloric acid $= 0.120$ mol dm⁻³

27.20 cm³ of acid contains $\dfrac{27.20}{1000} \times 0.120$ mol $= 0.003\,264$ mol

The equation for the reaction is:

$$Na_2CO_3 + 2HCl \rightarrow 2NaCl + H_2O + CO_2$$

2 mol of hydrochloric acid reacts with 1 mol of sodium carbonate, so for every 1 mol of hydrochloric acid that reacts, 0.5 mol of sodium carbonate is required. Therefore 0.003 264 mol of hydrochloric acid reacts with $0.5 \times 0.003\,264 = 0.001\,632$ mol of sodium carbonate.

This amount must have been present in the 25.0 cm³ of sodium carbonate used.

volume of sodium carbonate solution $= 25.0$ cm³ $= \left(\dfrac{25.0}{1000}\right) = 0.025$ dm³

Therefore:

concentration of sodium carbonate $= \dfrac{0.001632}{0.025} = 0.0653$ mol dm⁻³

It is sometimes possible to deduce other information from a titration, as illustrated by the following example.

## Example 8

5.00 g of an impure sample of potassium carbonate is weighed out and dissolved to make 1.00 dm³ of solution. 25.0 cm³ of this solution is titrated against 0.0500 mol dm⁻³ hydrochloric acid.

28.80 cm³ of the hydrochloric acid is needed to neutralise the aqueous potassium carbonate.

a) Use the titration results to calculate the concentration in mol dm⁻³ of the aqueous potassium carbonate.
b) Deduce the percentage purity of the solid sample of potassium carbonate.

**Answer**

a) concentration of hydrochloric acid $= 0.0500$ mol dm⁻³
28.80 cm³ of hydrochloric acid contains

$\dfrac{28.80}{1000} \times 0.0500$ mol $= 0.001\,44$ mol

The equation for the reaction is:

$$K_2CO_3 + 2HCl \rightarrow 2KCl + H_2O + CO_2$$

2 mol of hydrochloric acid reacts with 1 mol of potassium carbonate, so for every 1 mol of hydrochloric acid that reacts, 0.5 mol of potassium carbonate is required.

▶▶▶

0.00144 mol of hydrochloric acid reacts with $0.5 \times 0.00144 = 0.000720$ mol of potassium carbonate

This amount must have been present in the 25.00 cm³ of potassium carbonate used.

volume of potassium carbonate $= 25.0\,\text{cm}^3 = \left(\dfrac{25.0}{1000}\right) = 0.025\,\text{dm}^3$

Therefore the concentration of potassium carbonate $= \dfrac{0.000720}{0.025}$

$= 0.0288\,\text{mol}\,\text{dm}^{-3}$

b) The relative formula mass of potassium carbonate,

$K_2CO_3 = (2 \times 39.1) + 12.0 + (3 \times 16.0) = 138.2$

mass of $K_2CO_3$ in the $0.0288\,\text{mol}\,\text{dm}^{-3}$ solution is therefore
$0.0288 \times 138.2 = 3.98\,\text{g}$

This is the mass of pure potassium carbonate in the original 5.00 g.

So the percentage purity $= \dfrac{3.98}{5.00} = 79.6\%$

## Test yourself

7  25.0 cm³ of nitric acid was neutralised by 18.0 cm³ of 0.15 mol dm⁻³ sodium hydroxide solution.
   a)  Write an equation for the reaction.
   b)  Calculate the amount in moles of sodium hydroxide present in the 18.0 cm³.
   c)  Calculate the concentration of the nitric acid in mol dm⁻³.

8  2.65 g of anhydrous sodium carbonate was dissolved in water and the solution was made up to 250 cm³. In a titration, 25.0 cm³ of this solution was pipetted into a flask and the end point was reached after adding 22.5 cm³ of hydrochloric acid.
   a)  Write an equation for the reaction.
   b)  Calculate the concentration of the sodium carbonate solution in mol dm⁻³.
   c)  Calculate the amount in moles present in 25.0 cm³ of the sodium carbonate solution
   d)  Calculate the concentration of the hydrochloric acid in mol dm⁻³.

9  45.0 cm³ of 0.200 mol dm⁻³ NaOH just react with 10.0 cm³ of 0.300 mol dm⁻³ $H_3PO_4$ (phosphoric acid).
   a)  How many moles of NaOH react?
   b)  How many moles of $H_3PO_4$ react?
   c)  How many moles of NaOH react with 1 mole of $H_3PO_4$?
   d)  Write an equation for the reaction.

# Practice questions

## Multiple choice questions 1–10

1 Which one of the following solutions would have the highest number of hydroxide ions?
   A 500 cm³ of 0.01 mol dm⁻³ sodium hydroxide
   B 400 cm³ of 0.01 mol dm⁻³ calcium hydroxide
   C 300 cm³ of 0.02 mol dm⁻³ potassium hydroxide
   D 200 cm³ of 0.02 mol dm⁻³ lithium hydroxide

2 What volume of 0.40 mol dm⁻³ nitric acid will react with 20.0 cm³ of 0.20 mol dm⁻³ potassium carbonate?
   The equation is:
   $K_2CO_3(aq) + 2HNO_3(aq) \rightarrow 2KNO_3(aq) + H_2O(l) + CO_2(g)$

   A 5.0 cm³
   B 10.0 cm³
   C 20.0 cm³
   D 40.0 cm³

3 A solution is made by dissolving 5.72 g of sodium carbonate crystals, $Na_2CO_3 \bullet 10H_2O$ to make 200 cm³ of solution. (The molar mass of $Na_2CO_3$ is 106.0 g mol⁻¹.)
   The concentration of this solution to 1 significant figure will be:
   A 0.004 mol dm⁻³
   B 0.008 mol dm⁻³
   C 0.05 mol dm⁻³
   D 0.1 mol dm⁻³

4 25.0 cm³ of 0.100 mol dm⁻³ sodium hydroxide is neutralised by 20.0 cm³ of nitric acid. The equation is:
   $NaOH + HNO_3 \rightarrow NaNO_3 + H_2O$
   The concentration of the nitric acid is therefore:
   A 0.080 mol dm⁻³
   B 0.120 mol dm⁻³
   C 0.125 mol dm⁻³
   D 0.150 mol dm⁻³

5 30.0 cm³ of a 0.100 mol dm⁻³ solution of hydrochloric is diluted by adding 15.0 cm³ of water. The whole of this diluted solution is then titrated against a 0.100 mol dm⁻³ solution of sodium hydroxide.
   Which one of the following will be the volume of the sodium hydroxide in cm³ required to reach the end point of the titration?
   A 15.0        B 22.5
   C 30.0        D 45.0

6 100 cm³ of 0.100 mol dm⁻³ potassium hydroxide is diluted by adding $x$ cm³ of water. The new concentration of the potassium hydroxide is 0.0800 mol dm⁻³.
   Which one of the following gives the value of $x$ in cm³?
   A 20
   B 25
   C 40
   D 80

7 9.0 g of an acid $H_2X$ is dissolved to make 1 dm³ of solution.
   When 25.0 cm³ of this solution is titrated against 0.200 mol dm⁻³ sodium hydroxide, 25.0 cm³ is required to reach the end-point of the titration.
   Which one of the following gives the relative molecular mass of $H_2X$?
   A 90
   B 45
   C 20.5
   D 11.25

   Use the key below to answer Questions 8, 9 and 10.

| A | B | C | D |
|---|---|---|---|
| 1, 2 & 3 correct | 1, 2 correct | 2, 3 correct | 1 only correct |

8 Sodium carbonate solution is sucked into a pipette but is incorrectly filled so that the top of the meniscus of the solution is level with the mark on the pipette. A titration against hydrochloric acid is then correctly carried out. Which of the following would be a consequence of the error in the use of the pipette?
   1 The concentration of the sodium carbonate used would be less than it should be.
   2 The volume of hydrochloric acid required in the titration would be less than it should be.
   3 The concentration of the hydrochloric acid calculated from the titration would be greater than it should be.

9 25 cm³ of 0.04 mol dm⁻³ of sodium sulfate is added to 25 cm³ of 0.04 mol dm⁻³ barium nitrate. A reaction occurs and some barium sulfate is precipitated. Which of the following describes the mixture that is obtained?
   1 The number of sodium ions in the solution will be the same as the number of nitrate ions in the solution.
   2 0.001 mol of barium sulfate is precipitated.
   3 No sulfate ions remain in the solution.

**10** When $10\,cm^3$ of $0.1\,mol\,dm^{-3}$ of magnesium chloride solution is added to $10\,cm^3$ of $0.1\,mol\,dm^{-3}$ silver nitrate, silver chloride is precipitated.

Which of the following apply to this reaction?

1 The solution obtained after the reaction will not contain any nitrate ions.

2 The solution obtained after the reaction will contain $0.001\,mol$ of chloride ions.

3 $0.001\,mol$ of silver chloride will be precipitated.

**11** State the amount in moles in each of the following:

a) $50\,cm^3$ of aqueous sodium carbonate of concentration $0.05\,mol\,dm^{-3}$

b) $25\,cm^3$ of hydrochloric acid of concentration $0.1\,mol\,dm^{-3}$. (2)

**12** Calculate the volume of $0.5\,mol\,dm^{-3}$ aqueous copper sulfate that contains:

a) $0.05\,mol$

b) $0.36\,mol$. (2)

**13** Give the concentration of each of the following in $mol\,dm^{-3}$:

a) $5\,dm^3$ of hydrochloric acid containing $2\,mol$

b) $25\,cm^3$ of aqueous magnesium sulfate containing $0.027\,mol$. (2)

**14** Calculate the mass of solid that on dissolving would produce:

a) $50\,cm^3$ of potassium sulfate solution of concentration $0.15\,mol\,dm^{-3}$

b) $75\,cm^3$ of sodium chloride solution of concentration $0.4\,mol\,dm^{-3}$. (4)

**15** A solution of potassium hydroxide has a concentration of $0.100\,mol\,dm^{-3}$. $25.0\,cm^3$ of this solution is neutralised by $22.50\,cm^3$ of hydrochloric acid.

a) Write a balanced equation for the reaction between potassium hydroxide and hydrochloric acid.

b) What amount in moles of potassium hydroxide is contained in $25.0\,cm^3$ of solution?

c) What amount in moles of hydrochloric acid will neutralise the potassium hydroxide present in $25.0\,cm^3$ of solution?

d) What is the concentration of the hydrochloric acid in $mol\,dm^{-3}$? (4)

**16** A solution of potassium carbonate has a concentration of $0.04\,mol\,dm^{-3}$. $25.0\,cm^3$ of this solution is neutralised by $28.10\,cm^3$ of hydrochloric acid.

a) Write a balanced equation for the reaction between potassium carbonate and hydrochloric acid.

b) What amount in moles of potassium carbonate is contained in $25.0\,cm^3$ of solution?

c) What amount in moles of hydrochloric acid is needed to neutralise the number of moles of potassium carbonate in $25.0\,cm^3$ of solution?

d) What is the concentration of the hydrochloric acid in $mol\,dm^{-3}$? (4)

**17** Describe how each of the following solutions of sulfuric acid could be prepared from a solution of acid of concentration $1\,mol\,dm^{-3}$:

a) $100\,cm^3$ of solution of concentration $0.15\,mol\,dm^{-3}$

b) $25\,cm^3$ of solution of concentration $0.15\,mol\,dm^{-3}$. (6)

**18** What is the concentration obtained by diluting $25\,cm^3$ of $0.50\,mol\,dm^{-3}$ aqueous sodium carbonate with water to make:

a) $1\,dm^3$ of solution

b) $320\,cm^3$ of solution? (4)

**19** A solution of potassium hydroxide contains $5.61\,g\,dm^{-3}$. This solution is used to neutralise $20.0\,cm^3$ of a solution of sulfuric acid of concentration $0.0500\,mol\,dm^{-3}$.

a) Write an equation for the reaction.

b) Determine the volume of potassium hydroxide solution that is required to neutralise the sulfuric acid. (5)

**20** The concentration of calcium hydroxide in lime water can be determined by titrating a measured volume of lime water with a standard solution of hydrochloric acid.

In an experiment, $25.0\,cm^3$ of limewater was titrated against $0.050\,mol\,dm^{-3}$ hydrochloric acid.

a) Copy and complete the results below.

In the flask: $25.0\,cm^3$ of lime water

In the burette: $0.050\,mol\,dm^{-3}$ _____

The indicator was _____

| Titration number | 1 (rough) | 2 (accurate) | 3 (accurate) |
|---|---|---|---|
| Final burette reading/cm³ | 22.00 | 42.70 | 22.15 |
| Initial burette reading/cm³ | 0.45 | 21.90 | 1.25 |
| Titre/cm³ | | | |

**b)** Calculate the average accurate titre.

**c)** Write an equation for the reaction of calcium hydroxide with hydrochloric acid.

**d)** Calculate the amount in moles of hydrochloric acid in the average accurate titre.

**e)** Calculate the concentration of calcium hydroxide in lime water in $mol\,dm^{-3}$.  (7)

**21** Magnesium oxide neutralises nitric acid. In an experiment, it is found that 1.20 g of magnesium oxide is neutralised by $24.20\,cm^3$ of nitric acid.

**a)** Write an equation for the reaction between magnesium oxide and nitric acid.

**b)** What is the amount in moles of 1.20 g of magnesium oxide?

**c)** What amount in moles of nitric acid is contained in $24.20\,cm^3$?

**d)** What is the concentration of the nitric acid in $mol\,dm^{-3}$?  (4)

**22 a)** Write an equation for the reaction between zinc oxide and hydrochloric acid.

**b)** What is the amount in moles of 4.07 g of zinc oxide?

**c)** Calculate the volume of $2.00\,mol\,dm^{-3}$ hydrochloric acid that would react exactly with 4.07 g of zinc oxide.

**d)** If $2.00\,mol\,dm^{-3}$ sulfuric acid is used instead of $2\,mol\,dm^{-3}$ hydrochloric acid, what volume of sulfuric acid would be required?  (6)

**23** $100\,cm^3$ of hydrochloric acid is diluted to form $1\,dm^3$ of solution. When the diluted acid is titrated against $0.500\,mol\,dm^{-3}$ sodium carbonate solution, it is found that $27.20\,cm^3$ of the diluted hydrochloric acid is needed to neutralise $25.0\,cm^3$ of the sodium carbonate solution.

**a)** Write the equation for the reaction.

**b)** Calculate the amount in moles of sodium carbonate in $25.0\,cm^3$ of solution.

**c)** Calculate the concentration of the diluted hydrochloric acid.

**d)** What is the concentration of the hydrochloric acid before dilution?  (5)

**24** $25\,cm^3$ of $0.050\,mol\,dm^{-3}$ sodium hydroxide solution, NaOH(aq), just reacts with $27.8\,cm^3$ of $0.045\,mol\,dm^{-3}$ ethanoic acid, $CH_3COOH$(aq).

**a)** Calculate how many moles there are in each solution and use this to state how many moles of sodium hydroxide react with one mole of ethanoic acid.

**b)** Write an equation for the reaction.  (4)

**25** A solution is made by dissolving 5.00 g of sodium hydroxide to make $1\,dm^3$ of solution. A solution of an acid, $H_2X$, is made by dissolving 7.66 g to make $1.0\,dm^3$ of solution. $25.0\,cm^3$ of the solution of sodium hydroxide is neutralised by $20.0\,cm^3$ of the solution of the acid $H_2X$.

**a)** Write an equation for the reaction.

**b)** Calculate the concentration of the sodium hydroxide in $mol\,dm^{-3}$.

**c)** Calculate the amount in moles of sodium hydroxide in $25.0\,cm^3$.

**d)** Calculate the amount in moles of $H_2X$ in $20.0\,cm^3$ of solution.

**e)** Calculate the concentration of $H_2X$ in $mol\,dm^{-3}$.

**f)** Calculate the mass of the anion, $X^{2-}$.

**g)** Suggest an identify for the anion $X^{2-}$.  (8)

**26** $10.00\,g$ of sodium carbonate crystals $Na_2CO_3 \cdot xH_2O$ are dissolved to make $1\,dm^3$ of solution. $25.0\,cm^3$ of this solution is neutralised by $17.50\,cm^3$ of hydrochloric acid of concentration $0.100\,mol\,dm^{-3}$.

**a)** Calculate the amount in moles of hydrochloric acid present in $17.50\,cm^3$.

**b)** What is the amount in moles of sodium carbonate present in $25.0\,cm^3$ of solution?

**c)** Calculate the concentration in $mol\,dm^{-3}$ of the sodium carbonate solution.

**d)** Use your answer to (c) to calculate the molar mass of the sodium carbonate crystals.

**e)** Deduce the value of $x$ in $Na_2CO_3 \cdot xH_2O$  (7)

**27** $41.0\,g$ of the acid $H_3PO_3$ was dissolved in water and the solution was made up to $1.0\,dm^3$. $20.0\,cm^3$ of this solution just reacted with $25.0\,cm^3$ of $0.080\,mol\,dm^{-3}$ sodium hydroxide. What is the equation for the reaction? (H = 1, O = 16, P = 31)  (6)

## Challenge

**28** Beachy Head, near Eastbourne in East Sussex, is a dominating chalk cliff characteristic of the scenery of that part of the south coast. The soft chalk is subject to erosion by the sea. Areas of the cliff edge have to be cordoned off when they become unstable; they will eventually fall into the sea. Constant monitoring of the state of the cliff is essential to ensure public safety.

A soil sample taken from the area consists of large amounts of chalk (calcium carbonate) with some clay and other silicate materials, as well as decomposing organic matter. Assuming that the chalk is the only material that reacts with dilute acid, there are several ways of determining the percentage of chalk in the soil sample.

**A** Collection of carbon dioxide

A 1.00 g sample of a soil is reacted with excess hydrochloric acid and the carbon dioxide evolved is collected over water. After cooling to room temperature, 89 cm$^3$ of carbon dioxide is obtained.

**a)** Write a balanced equation for the reaction between calcium carbonate and hydrochloric acid.

**b)** Calculate the amount in moles of carbon dioxide obtained.

**c)** Deduce the amount in moles of calcium carbonate that react.

**d)** Calculate the mass of calcium carbonate that reacts.

**e)** Calculate the percentage of chalk in the soil sample.

**B** Back titration

10.00 g of soil is added to 100.0 cm$^3$ of hydrochloric acid of concentration 1.00 mol dm$^{-3}$. Once all the chalk has reacted, the residue is allowed to settle. Excess acid remains in the solution. Using a pipette, 25.00 cm$^3$ of this solution is carefully removed and titrated against 0.0500 mol dm$^{-3}$ sodium carbonate solution. 23.10 cm$^3$ is required to neutralise the acid.

**a)** Write a balanced equation for the reaction between sodium carbonate and hydrochloric acid.

**b)** Calculate the amount in moles of sodium carbonate in the 23.10 cm$^3$ required by the titration.

**c)** Use your answer to (b) to deduce the amount in moles of hydrochloric acid contained in 25.00 cm$^3$ of the solution.

**d)** Deduce the amount in moles in 100.0 cm$^3$ of the hydrochloric acid after it reacts with the soil sample.

**e)** Calculate the amount in moles of hydrochloric acid in the 100.0 cm$^3$ before it reacts with the soil sample.

**f)** Use your answers to (d) and (e) to deduce the amount in moles of hydrochloric acid that react with the soil sample.

**g)** Write the equation for the reaction between calcium carbonate and hydrochloric acid.

**h)** Use the equation in (g) and your answer to (f) to state the amount in moles of calcium carbonate in 10.00 g of soil.

**i)** Deduce the mass of calcium carbonate in 10.00 g of soil.

**j)** Calculate the percentage of chalk in the soil sample.

**C** Consider the experimental methods in Questions A and B. Suggest reasons for the difference in the answers for the percentage of chalk in the soil sample obtained by the two methods.

**D** Two other possible methods of determining the percentage of chalk in a sample of the soil are as follows:

- A weighed sample of the soil is heated strongly and the volume of carbon dioxide evolved is measured.
- A weighed sample of the soil is reacted with dilute hydrochloric acid until all the chalk dissolves. The solution is filtered and the mass of the unreacted residue is found.

Consider each of these methods and suggest, with reasons, whether they would be likely to produce an accurate result. (20)

# Chapter 6

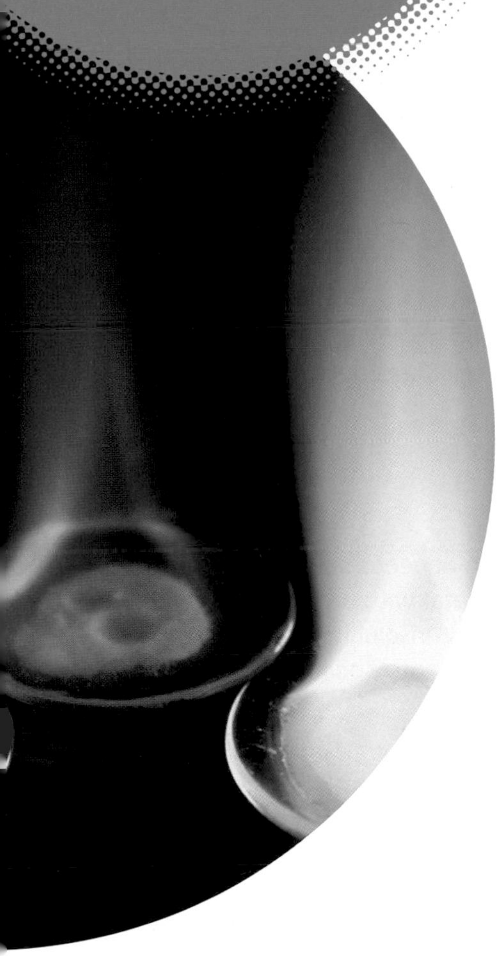

# Types of reaction – precipitation, acid–base and redox

## Prior knowledge

*In this chapter it is assumed that you are familiar with:*

- formulae
- balancing equations
- mole calculations.

For example, you should be aware that reactions occur because atoms can lose, gain or share electrons with other atoms. The number of electrons gained, lost or shared by an atom is known as its valency and the valency can be used to determine the formula of compounds. Valencies for common elements and groups are shown in Table 3.1, page 29.

## Test yourself on prior knowledge

1 Write the formula of:
   a) sodium carbonate
   b) calcium hydroxide
   c) ammonium sulfate
   d) magnesium hydrogencarbonate.

2 Calculate the molar mass of:
   a) $Ca(OH)_2$
   b) $C_6H_{12}O_6$
   c) $CuSO_4 \cdot 5H_2O$
   d) $K_2Cr_2O_7$.

3 a) Calculate the mass of 0.25 mol of NaOH.
   b) Calculate the amount in moles in 25.0 cm³ of 0.2 mol dm⁻³ solution of $H_2SO_4$.
   c) Calculate the volume, at room temperature and pressure, of 0.25 mol of $H_2(g)$.

4 Divide the following compounds into two groups: ionic and covalent.
   copper(II) hydroxide    silver nitrate    phosphorus trichloride
   chromium(III) oxide    ammonium chloride    sulfur dioxide
   silicon dioxide    carbon disulfide    barium chloride    magnesium oxide

## Tip

If you are unfamiliar with simple ideas of bonding, you should look at Chapter 7 pages 89–92 and 94–96.

# Classifying chemical reactions

Classification of chemical reactions helps chemists to understand the way in which elements or compounds react. There are a limited number of reaction types:

- precipitation reactions
- redox reactions
- acid–base reactions.

## Precipitation reactions

In a **precipitation reaction**, a cation and an anion in solution combine to form an insoluble substance that precipitates out as a solid.

Cations are positive ions such as $Na^+$, $Mg^{2+}$ and $Cu^{2+}$. They are usually metallic ions but do include $H^+$ and $NH_4^+$.

Anions are negative ions such as $Cl^-$, $Br^-$, $CO_3^{2-}$, $SO_4^{2-}$, $NO_3^-$ and $OH^-$.

If two solutions are mixed, one being an aqueous solution of a carbonate and the other containing a metal with an insoluble carbonate, the metal carbonate precipitates instantly. Insoluble copper(II) carbonate, for example, can be formed by mixing aqueous sodium carbonate with aqueous copper(II) sulfate:

$$Na_2CO_3(aq) + CuSO_4(aq) \rightarrow CuCO_3(s) + Na_2SO_4(aq)$$

The best way to summarise this is to use an ionic equation (Chapter 3, pages 33–35). The equation includes state symbols.

$$CO_3^{2-}(aq) + Cu^{2+}(aq) \rightarrow CuCO_3(s)$$

A second precipitation reaction involves the reaction between a solution of potassium chloride and solution of silver nitrate.

$$KCl(aq) + AgNO_3(aq) \rightarrow AgCl(s) + KNO_3(aq)$$

This can be simplified to:

$$Cl^-(aq) + Ag^+(aq) \rightarrow AgCl(s)$$

Precipitation reactions are used as a test for halide ions (Chapter 9, page 139).

(Chapter 3, pages 33–35)

> ### Tip
>
> Use the state symbols as a guide:
>
> - (aq) indicates an ionic substance and the ions are mobile and separate and therefore can be written as separate ions
> - (g) indicates a gas which is usually covalent
> - (l) indicates a liquid which is usually covalent
> - (s) indicates a solid in which the particles are not mobile; even if the substance is ionic the ions should not be written separately.

> ### Test yourself
>
> 1 Write an ionic equation for the following reactions:
>   a) $2NaOH(aq) + FeCl_2(aq) \rightarrow 2NaCl(aq) + Fe(OH)_2(s)$
>   b) $Na_2SO_4(aq) + BaCl_2(aq) \rightarrow BaSO_4(s) + 2NaCl(aq)$
>   c) $PbCl_2(aq) + K_2S(aq) \rightarrow PbS(s) + 2KCl(aq)$
>   d) formation of solid zinc hydroxide from solutions of zinc chloride and sodium hydroxide.

# Redox reactions

In a **redox reaction**, electrons are transferred from one substance to another. Some everyday examples of redox reactions include combustion of fuels and the rusting of iron.

Oxidation was originally defined as the gaining of oxygen. For example, when zinc reacts with oxygen to form zinc oxide, the zinc is oxidised:

$$Zn(s) + \frac{1}{2}O_2(g) \rightarrow ZnO(s)$$

Closer inspection shows that the zinc atom becomes a zinc ion and in doing so loses two electrons:

$$Zn \rightarrow Zn^{2+} + 2e^-$$

Therefore, a species is oxidised if it loses electrons; a species is reduced if it gains electrons. You may find the mnemonic OILRIG useful:

**O**xidation **I**s **L**oss **R**eduction **I**s **G**ain

## Oxidation number

Chemists use oxidation numbers to keep track of the electrons transferred or shared during a reaction. It is a convenient way of quickly identifying whether or not a substance has undergone oxidation or reduction. If oxidation numbers change during a reaction then oxidation and reduction has occurred and a redox reaction has taken place.

In order to work out the oxidation number, you must first learn a few simple rules (Table 6.1).

**Table 6.1** Rules for determining oxidation numbers.

| Rule | Example |
|------|---------|
| All elements in their natural state have the oxidation number zero | Hydrogen, $H_2$; oxidation number = 0 |
| Oxidation numbers of the atoms of any molecule add up to zero | Water, $H_2O$; sum of oxidation numbers = 0 |
| Oxidation numbers of the components of any ion add up to the charge on that ion | Sulfate, $SO_4^{2-}$; sum of oxidation numbers = −2 |

The oxidation numbers of some elements in a molecule or ion never change; other elements have more than one oxidation number.

When calculating the oxidation numbers of elements in either a molecule or an ion, you should apply the following order of priority:

1  The oxidation numbers of elements in Groups 1, 2 and 13 are always +1, +2 and +3 respectively.

2  The oxidation number of fluorine is always −1.

3  The oxidation number of hydrogen is usually +1.

4  The oxidation number of oxygen is usually −2.

5  The oxidation number of chlorine is usually −1.

Using these rules, it is possible to deduce any oxidation number. However, it is important that they are applied **in sequence**, as hydrogen, oxygen and chlorine can vary.

### Tip

If an element can form compounds in which its oxidation state varies, Roman numerals are used in the name to differentiate between them. For example iron chloride can be either $FeCl_2$ or $FeCl_3$. $FeCl_2$ is named iron (II) chloride whilst $FeCl_3$ is named iron (III) chloride.

In a metallic hydride such as NaH, H has oxidation number −1 and not the usual +1 (Na is in group 1 and has priority over the H).

In a peroxide such as $H_2O_2$, O has oxidation number −1 and not the usual −2 (H has priority over O).

The following worked examples show how the oxidation number of chlorine can also vary.

### Example 1

Deduce the oxidation number of Cl in NaCl.

**Answer**
The sum of the oxidation numbers in NaCl must add up to zero.

Using the order of priority given above, the oxidation number of Na must be +1.

The oxidation number of Cl must balance this.

Therefore, the oxidation number of Cl in NaCl is −1.

### Example 2

Deduce the oxidation number of Cl in NaClO.

**Answer**
The sum of the oxidation numbers in NaClO must add up to zero.

In order of priority, Na comes first and its oxidation number must be +1; O is second and its oxidation number is −2.

For the oxidation numbers to add up to zero, the oxidation number of Cl in NaClO must be +1.

### Example 3

Deduce the oxidation number of Cl in $ClO_3^-$.

**Answer**
The sum of the oxidation numbers in $ClO_3^-$ must add up to the charge on the ion, i.e. they must add up to −1

In order of priority, O comes first and its oxidation number is −2.

There are three oxygen atoms in $ClO_3^-$ so the total for oxygen is −6.

For the oxidation numbers to add up to −1 (the charge of the ion), the oxidation number of Cl in $ClO_3^-$ must be +5. This is reflected in the name of $ClO_3^-$ which is chlorate(v).

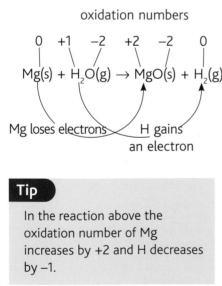

oxidation numbers

$$\overset{0}{Mg}(s) + \overset{+1}{H_2}\overset{-2}{O}(g) \rightarrow \overset{+2}{Mg}\overset{-2}{O}(s) + \overset{0}{H_2}(g)$$

Mg loses electrons    H gains an electron

Consider the following reaction:

$$Mg(s) + H_2O(g) \rightarrow MgO(s) + H_2(g)$$

This reaction occurs when magnesium is heated with steam. It is easy to see that magnesium is oxidised (it has gained oxygen) and that water is reduced (it has lost oxygen). The oxidation numbers for this reaction are shown on the left:

- An increase in oxidation number is due to oxidation. Magnesium changes from 0 to +2.

- A decrease in oxidation number is due to reduction. H changes from +1 to 0.

Consider the reaction between Mg(s) and $Cl_2(g)$:

oxidation numbers

$$\overset{0}{Mg}(s) + \overset{0}{Cl_2} \rightarrow \overset{+2\ -1}{MgCl_2}$$

### Tip

In the reaction above the oxidation number of Mg increases by +2 and H decreases by −1.

Electron transfer can be shown by half-equations. In this example, the half-equations are:

$Mg \rightarrow Mg^{2+} + 2e^-$ (loss of electrons = oxidation) (magnesium readily loses electrons)

$Cl_2 + 2e^- \rightarrow 2Cl^-$ (gain of electrons = reduction) (chlorine readily gains electrons)

- An **oxidising agent** encourages oxidation. It is a substance that receives electrons readily.

- A **reducing agent** encourages reduction. It is a substance that donates electrons readily.

Group 2 elements form 2+ ions, each atom releasing two electrons. Therefore, they are **reducing** agents.

Group 17 elements form 1– ions, each atom receiving one electron. Therefore, they are **oxidising** agents.

Electron loss occurs more readily on descending a group of the periodic table (page 120).

Electron loss occurs less readily across a period of the periodic table.

It follows from this that, in Groups 1 and 2, the elements at the bottom of the groups most easily lose electrons and are therefore:

- most easily oxidised

- the best reducing agents.

The opposite is true for the ability to gain an electron, so fluorine is the best oxidising agent.

The trends shown in Table 6.2 occur elsewhere in the periodic table.

**Table 6.2** Reducing and oxidising power in Groups 2 and 17.

| Group 2 | | Group 17 | |
| --- | --- | --- | --- |
| Be | A barium atom loses its outer electrons most easily, therefore it is:<br>• most easily oxidised<br>• the best reducing agent in Group 2 | $F_2$ | A fluorine atom is most likely to gain an electron, therefore it is:<br>• most easily reduced<br>• the best oxidising agent in Group 17 |
| Mg | | $Cl_2$ | |
| Ca | | $Br_2$ | |
| Sr | | $I_2$ | |
| Ba | | | |

**Test yourself**

2 Deduce the oxidation numbers of each element in the following:
   a) i) $H_2O$  ii) NaOH  iii) $KNO_3$  iv) $NH_3$  v) $N_2O$
   b) i) $SO_4^{2-}$  ii) $CO_3^{2-}$  iii) $NH_4^+$  iv) $MnO_4^-$  v) $Cr_2O_7^{2-}$
3 In the reaction $Ca(s) + 2HCl(aq) \rightarrow CaCl_2(aq) + H_2(g)$
   a) what has been oxidised
   b) what is the reducing agent?

# Acid–base reactions

## Testing for acids and bases

You will be aware of the pH scale and that:

- acids and acidic solution have pH below 7
- alkalis and alkaline solutions have pH above 7
- neutral solutions have a pH of 7.

Indicators are a convenient method for testing for acids and bases (Figure 6.1).

**Key term**

An indicator is a substance that changes colour with a change in pH.

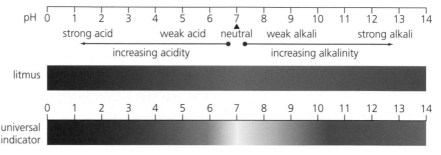

**Figure 6.1** Colours of litmus and universal indicator with solutions at different pH.

## Acids and bases

Reactions of acids involve the transfer of a hydrogen ion from one substance (the acid) to another (the base). Since the hydrogen ion, $H^+$, is in fact a proton, acids are described as **proton donors**.

Acids include:

- hydrochloric acid, HCl
- sulfuric acid, $H_2SO_4$
- nitric acid, $HNO_3$
- phosphoric acid, $H_3PO_4$
- ethanoic acid, $CH_3COOH$

**Key term**

An acid is a proton donor.

Acids such as HCl, $H_2SO_4$ and $HNO_3$ are strong acids and, when dissolved in water, completely dissociate into their ions. Equations for strong acids always contain the '$\rightarrow$' arrow symbol.

$$HCl(aq) \rightarrow H^+(aq) + Cl^-(aq)$$

Organic acids such as $CH_3COOH$ are weak acids and when dissolved in water only partially dissociate into their ions. Equations for weak acids always contain the '$\rightleftharpoons$' arrow symbol.

$$CH_3COOH(aq) \rightleftharpoons CH_3COO^-(aq) + H^+(aq)$$

**Key terms**

A strong acid is a proton donor that completely dissociates into its ions in water.
A weak acid is a proton donor that only partially dissociates into its ions in water.

Water plays an important role in acidity. In the absence of water, many of the substances called acids show none of the typical acid properties. Hydrochloric acid, for example, is an aqueous solution of the covalent gas hydrogen chloride. Non-metallic oxides such as $CO_2$ or $SO_2$ can react with water to form acidic solutions.

$$CO_2(g) + H_2O(l) \rightleftharpoons H_2CO_3(aq) \rightleftharpoons 2H^+(aq) + CO_3^{2-}(aq)$$

$$SO_2(g) + H_2O(l) \rightleftharpoons H_2SO_3(aq) \rightleftharpoons 2H^+(aq) + SO_3^{2-}(aq)$$

Both $H_2CO_3(aq)$ and $H_2SO_3(aq)$ are able to donate $H^+(aq)$ ions and so can behave as acids.

A base is a substance that can react with the hydrogen ion released by an acid. Bases can be regarded as **proton** (hydrogen ion) **acceptors**. Typical bases are oxides or hydroxides of metals.

If a base dissolves in water (most do not), it is referred to as an alkali. Examples include:

- sodium hydroxide, NaOH

- potassium hydroxide, KOH

- ammonia, $NH_3$.

An alkali releases hydroxide ions ($OH^-$) in aqueous solution. For example:

$$NaOH(aq) \rightarrow Na^+(aq) + OH^-(aq)$$

The inclusion of ammonia in the list of alkalis may seem odd because, as a gas, it does not contain hydroxide ions. However, when ammonia dissolves in water it reacts to produce the ionic compound ammonium hydroxide, $NH_4OH$ which is a weak base. The reaction is an example of an **equilibrium reaction** since, once dissolved, there is a mixture of all four components, $NH_3$, $H_2O$, $NH_4^+$ and $OH^-$:

$$NH_3(g) + H_2O(l) \rightleftharpoons NH_4OH(aq) \rightleftharpoons NH_4^+(aq) + OH^-(aq)$$

In this example it is easy to see why bases are regarded as proton acceptors. The ammonia molecule, $NH_3$, has received a hydrogen ion to become an ammonium ion, $NH_4^+$.

Some metallic oxides react with water to produce hydroxides:

$$K_2O(s) + H_2O(l) \rightarrow 2KOH(aq)$$

$$CaO(s) + H_2O(l) \rightarrow Ca(OH)_2(aq)$$

Most other metal oxides and hydroxides are insoluble in water and are therefore bases not alkalis.

## The role of water in acidity and alkalinity

The properties of acids and alkalis described above apply to their solutions in water. If water is not present dry hydrogen chloride and dry sulfuric acid have no effect on dry litmus paper, which means that substances that we call acids do not behave like acids in the absence of water. When water is added, they become acidic straight away. Blue litmus paper turns red immediately.

Similar experiments with dry calcium hydroxide show that water must be present before it will behave like an alkali.

As a general rule, water must be present for substances to act as acids and alkalis:

- when acids dissolve in water, $H^+$ ions are formed

- when the alkalis dissolve in water, $OH^-$ ions are formed.

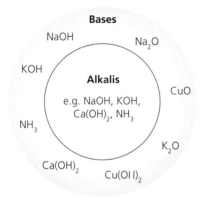

**Figure 6.2** Alkalis are a subsection of bases. They are soluble in water.

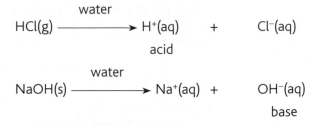

$$HCl(g) \xrightarrow{\text{water}} H^+(aq) \quad + \quad Cl^-(aq)$$
acid

$$NaOH(s) \xrightarrow{\text{water}} Na^+(aq) + \quad OH^-(aq)$$
base

In water, hydrogen ions, $H^+$, are attached to water molecules. Because of this, the $H^+$ ions are described as hydrated and are sometimes represented as $H_3O^+(aq)$, or more usually as $H^+(aq)$.

In water, the hydroxide ions, $OH^-(aq)$, are mobile and free to move. Water is a polar molecule and is attracted to the ions in the lattice. This attraction, hydration energy, is sufficient to break down the bonds within the lattice such that the hydrated ions are free to move (see Chapter 7, page 93).

(see Chapter 7, page 93)

## Salts

Acids react with metals, carbonates, bases or alkalis to form salts.

The formation of salts is illustrated by the following typical reactions of acids.

### Reaction of an acid and a base – neutralisation

Dilute hydrochloric acid reacts with magnesium oxide to produce magnesium chloride and water, according to the equation:

$$2HCl(aq) + MgO(s) \rightarrow MgCl_2(aq) + H_2O(l)$$

The ionic equation is:

$$2H^+(aq) + MgO(s) \rightarrow Mg^{2+}(aq) + H_2O(l)$$

### Reaction of an acid and an alkali – neutralisation

Dilute hydrochloric acid reacts with sodium hydroxide to produce sodium chloride and water, according to the equation:

$$HCl(aq) + NaOH(aq) \rightarrow NaCl(aq) + H_2O(l)$$

The ionic equation is:

$$H^+(aq) + OH^-(aq) \rightarrow H_2O(l)$$

Ammonia can react with an acid to produce a salt. Ammonium sulfate, $(NH_4)_2SO_4$, is used as a fertiliser. It is manufactured by reacting ammonia with sulfuric acid:

$$2NH_3(aq) + H_2SO_4(aq) \rightarrow (NH_4)_2SO_4(aq)$$

The ionic equation is:

$$2NH_3(aq) + 2H^+(aq) \rightarrow 2NH_4^+(aq)$$

---

**Key term**

A salt is formed when an acid has one or more of its hydrogen ions replaced by either a metal ion or an ammonium ion.

---

**Tip**

MgO is ionic but because it is in the (s) state we do not write the ions separately, as when solid the ions are not mobile.

---

Some organic molecules that contain the –COOH are also acidic. The H in the –COOH is the acidic hydrogen and such chemicals can donate a $H^+(aq)$.

$$CH_3COOH(aq) + NaOH(aq) \rightarrow CH_3COONa\ (aq) + H_2O(l)$$

The ionic equation is:

$$H^+(aq) + OH^-\ (aq) \rightarrow H_2O(l)$$

## Reaction of an acid and a carbonate – neutralisation

Dilute hydrochloric acid reacts with sodium carbonate to produce sodium chloride, water and carbon dioxide, according to the equation:

$$2HCl(aq) + Na_2CO_3(aq) \rightarrow 2NaCl(aq) + H_2O(l) + CO_2(g)$$

The ionic equation is:

$$2H^+(aq) + CO_3^{2-}(aq) \rightarrow H_2O(l) + CO_2(g)$$

## Reaction of an acid and a metal

Most metals react with an acid to produce a salt. Metals such as copper and silver only react with very concentrated acids. Some acids (e.g. nitric acid) produce complex products, but many acids (e.g. dilute hydrochloric and sulfuric acids) react with a metal to produce a salt and hydrogen.

Magnesium reacts with sulfuric acid to produce magnesium sulfate and hydrogen, according to the equation:

$$Mg(s) + H_2SO_4(aq) \rightarrow MgSO_4(aq) + H_2(g)$$

The ionic equation is:

$$Mg(s) + 2H^+(aq) \rightarrow Mg^{2+}(aq) + H_2(g)$$

It can be seen from the equations that this is a redox reaction. The magnesium atom loses two electrons to become an $Mg^{2+}$ ion; the two hydrogen ions from the acid each gain an electron to create a diatomic molecule of hydrogen gas. The magnesium is oxidised and the hydrogen ions are reduced.

In terms of oxidation number:

• the magnesium changes from oxidation number 0 to +2

• each hydrogen ion changes from oxidation number +1 to 0

All the reactions between acids and metals are redox reactions. A further illustration is the reaction of hydrochloric acid and zinc to produce zinc chloride and hydrogen:

$$Zn(s) + 2HCl(aq) \rightarrow ZnCl_2(aq) + H_2(g)$$

The ionic equation is:

$$Zn(s) + 2H^+(aq) \rightarrow Zn^{2+}(aq) + H_2(g)$$

The zinc is oxidised and the hydrogen is reduced.

# Salts formed from acids with more than one acidic hydrogen

There are a number of acids that contain more than one acidic hydrogen; the most common example is sulfuric acid, $H_2SO_4$. Sulfuric acid has two acidic protons and one mole of $H_2SO_4$ can react with two moles of a base such as NaOH(aq), hence the term 'dibasic acid' or 'diprotic acid'. It is possible to form two different salts by reacting sulfuric acid with sodium hydroxide. If one acid proton is replaced by the sodium ion we obtain sodium hydrogensulfate, $NaHSO_4$, but if both are replaced sodium sulfate is produced, $Na_2SO_4$.

Phosphoric acid, $H_3PO_4$, is a tribasic acid as it has three acidic hydrogens and it is possible to form three different salts:

1. sodium dihydrogenphosphate, $NaH_2PO_4$, in which one acidic hydrogen is replaced by the metal ion

2. sodium hydrogenphosphate, $Na_2HPO_4$, in which two acidic hydrogens are replaced by the metal ions

3. sodium phosphate, $Na_3PO_4$, in which three acidic hydrogens are replaced by the metal ions.

> **Tip**
>
> Sodium hydrogenphosphate could also be named as disodium hydrogenphosphate. Sodium phosphate could also be named as trisodium phosphate.

## Tests for ions

Ions in solution can be detected by a range of simple qualitative test-tube reactions. The sequence of testing should be test for carbonate first, then test for sulfate and finally test for halide as $BaCO_3$ and $Ag_2SO_4$ are both white insoluble compounds.

> **Tip**
>
> It is better to use $Ba(NO_3)_2$(aq) as the nitrate ion would not interfere with the subsequent halide test.

### Tests for carbonate ions

All carbonates react with dilute acids to produce a salt, water and carbon dioxide. The evolution of carbon dioxide can be observed by the effervescence (bubbles) (Table 6.4).

**Table 6.4** Tests for carbonate ions.

| Reaction | Observations |
|---|---|
| $CaCO_3(s) + 2HCl(aq) \rightarrow CaCl_2(aq) + H_2O(l) + CO_2(g)$ <br> $CaCO_3(s) + 2H^+(aq) \rightarrow Ca^{2+}(aq) + 2Cl^-(aq) + H_2O(l) + CO_2(g)$ | Solid reacts and goes into solution <br> Bubbles are given off / effervescence |
| $Na_2CO_3(aq) + 2HCl(aq) \rightarrow 2NaCl(aq) + H_2O(l) + CO_2(g)$ <br> $CO_3^{2-}(aq) + 2H^+(aq) \rightarrow H_2O(l) + CO_2(g)$ | Bubbles are given off / effervescence |

### Tests for sulfate ions

Sulfates can be detected by adding a solution of barium chloride, $BaCl_2$(aq) or $Ba(NO_3)_2$(aq). A white precipitate of $BaSO_4$(s) is produced (Table 6.5).

**Table 6.5** Test for sulfate ions.

| Reaction | Observations |
|---|---|
| $Na_2SO_4(aq) + Ba(NO_3)_2(aq) \rightarrow 2NaNO_3(aq) + BaSO_4(s)$ <br> $SO_4^{2-}(aq) + Ba^{2+}(aq) \rightarrow BaSO_4(s)$ | A white precipitate is formed |

## Tests for halide ions

Halides can be detected by adding a solution of silver nitrate, $AgNO_3(aq)$ (Table 6.6). Precipitates of the silver halides are formed except with the fluoride. Ammonia can be added to the precipitates; the chloride re-dissolves in dilute ammonia, the bromide re-dissolves in concentrated ammonia and the iodide remains insoluble.

**Table 6.6** Tests for halide ions.

| Reaction | Observations |
|---|---|
| $NaCl(aq) + AgNO_3(aq) \rightarrow NaNO_3(aq) + AgCl(s)$<br>$Cl^-(aq) + Ag^+(aq) \rightarrow AgCl(s)$ | A white precipitate is formed which re-dissolves in dilute ammonia |
| $NaBr(aq) + AgNO_3(aq) \rightarrow NaNO_3(aq) + AgBr(s)$<br>$Br^-(aq) + Ag^+(aq) \rightarrow AgBr(s)$ | A cream precipitate is formed which re-dissolves in concentrated ammonia |
| $NaI(aq) + AgNO_3(aq) \rightarrow NaNO_3(aq) + AgI(s)$<br>$I^-(aq) + Ag^+(aq) \rightarrow AgI(s)$ | A yellow precipitate is formed which remains insoluble |

## Tests for ammonium ions

Ammonium ions can be detected by warming $NH_4^+(aq)$ with dilute sodium hydroxide (Table 6.7). $NH_3(g)$ is produced which turns moist red litmus blue.

**Table 6.7** Tests for ammonium ions.

| Reaction | Observations |
|---|---|
| $NH_4Cl(aq) + NaOH(aq) \rightarrow NaCl(aq) + NH_3(g) + H_2O(l)$<br>$NH_4^+(aq) + OH^-(aq) \rightarrow NH_3(g) + H_2O(l)$ | 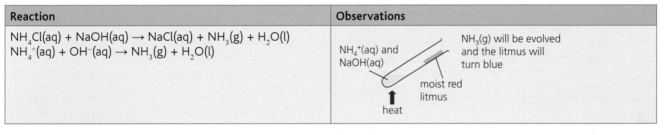 |

**Test yourself**

4 Write balanced equations for each of the following reactions:
 a) barium hydroxide(s) + nitric acid(aq) → barium nitrate(aq) + water(l)
 b) copper(II) carbonate(s) + sulfuric acid(aq) → copper(II) sulfate(aq) + water(l) + carbon dioxide(g)
 c) lithium(s) + hydrochloric acid → lithium chloride(aq) + hydrogen(g)
5 Write an ionic equation for each of the following reactions:
 a) barium hydroxide(s) + nitric acid(aq) → barium nitrate(aq) + water(l)
 b) copper(II) carbonate(s) + sulfuric acid(aq) → copper(II) sulfate(aq) + water(l) + carbon dioxide(g)
 c) lithium(s) + hydrochloric acid → lithium chloride(aq) + hydrogen(g)
6 Deduce the formula of all possible salts that could be formed from:
 a) reaction of KOH(aq) with $H_2S(aq)$
 b) reaction of $Ca(OH)_2$ with $H_2CO_3(aq)$

# Practice questions

## Multiple choice questions 1–10

Answers to Questions 1 to 6 relate to the following types of reaction:

    **A** precipitation only

    **B** redox only

    **C** acid–base only

    **D** a combination of more than one of the above.

Which type of reaction is illustrated by each of the following?

**1** $Br_2(aq) + 2NaI(aq) \rightarrow 2NaBr(aq) + I_2(aq)$

**2** $BaCl_2(aq) + H_2SO_4(aq) \rightarrow BaSO_4(s) + 2HCl(aq)$

**3** $CH_3COOH(aq) + NaOH(aq) \rightarrow$ $CH_3COO^-Na^+(aq) + H_2O(l)$

**4** $Ba(OH)_2(aq) + H_2SO_4(aq) \rightarrow BaSO_4(s) + 2H_2O(l)$

**5** $SrO(s) + 2HNO_3(aq) \rightarrow Sr(NO_3)_2(aq) + H_2O(l)$

**6** $Cu(s) + 4HNO_3(aq) \rightarrow Cu(NO_3)_2(aq) + 2NO_2(g)$ $+ 2H_2O(l)$

Use the key below to answer Questions 7 to 10.

| A | B | C | D |
|---|---|---|---|
| 1, 2 & 3 correct | 1, 2 correct | 2, 3 correct | 1 only correct |

**7** Which of the following underlined elements have oxidation state 6?

  **1** $Na\underline{Cl}O_3$   **2** $\underline{S}O_4^{2-}$   **3** $K_2\underline{Cr}_2O_7$

**8** In which of these conversions is the chlorine oxidised?

(The conversions shown below are not balanced and do not represent equations.)

  **1** $Cl_2 \rightarrow ClO^-$

  **2** $ClO^- \rightarrow ClO_3^-$

  **3** $Cl_2 + I_2 \rightarrow ICl$

**9** In which of the following reactions is the underlined the reducing agent?

  **1** $4\underline{Au}(s) + 8CN^-(aq) + 2H_2O(l) + O_2(g) \rightarrow$ $4[Au(CN)_2]^-(aq) + 4OH^-(aq)$

  **2** $2\underline{Cu}^{2+}(aq) + 4I^-(aq) \rightarrow 2CuI(s) + I_2(aq)$

  **3** $\underline{Ag}_2S(s) + 4CN^-(aq) \rightarrow 2[Ag(CN)_2]^-(aq) +$ $S^{2-}(aq)$

**10** Phosphoric acid, $H_3PO_4$, can react with magnesium hydroxide, $Mg(OH)_2$, to form a salt and water. The salt could have the formula:

  **1** $Mg(HPO_4)$   **2** $Mg_3(PO_4)_2$   **3** $Mg(H_2PO_4)_2$

**11** Deduce the oxidation numbers of the element underlined in each of the following:

  **a)** $\underline{Rb}_2O$

  **b)** $\underline{Sn}Cl_2$

  **c)** $\underline{Sn}O_2$

  **d)** $\underline{S}_2O_3^{2-}$            (4)

**12** What are the oxidation numbers of nitrogen in these molecules:

$N_2$, $NH_3$, $N_2H_4$, $HNO_3$, $NH_4Cl$, $HNO_2$, $NH_2OH$ and $NF_3$?     (8)

**13** Write equations, including state symbols, for each of the following reactions.

State what you would see, if anything, during each reaction.

  **a)** magnesium + sulfuric acid

  **b)** lithium carbonate solid + nitric acid

  **c)** aluminium oxide solid + hydrochloric acid (3)

**14** Chlorine forms a number of different chlorate ions with oxygen. In each of the ions below deduce the oxidation number of chlorine and name the ions.

$Cl^-$ oxidation number ...............................................

name .............................................................................

$ClO^-$ oxidation number ............................................

name .............................................................................

$ClO_2^-$ oxidation number ..........................................

name .............................................................................

$ClO_3^-$ oxidation number ..........................................

name .............................................................................

$ClO_4^-$ oxidation number ..........................................

name ......................................................................... (5)

**15** These are incomplete half-equations for changes involving reduction. Copy and complete and balance the half-equations. You need to add the electrons.

  **a)** $H^+(aq) \rightarrow H_2(g)$

  **b)** $Fe^{3+}(aq) \rightarrow Fe^{2+}(aq)$

  **c)** $Br_2(aq) \rightarrow 2Br^-(aq)$         (3)

**16** These are incomplete half-equations for changes involving oxidation. Copy and complete and balance the half-equations. You need to add the electrons.

  **a)** $Na(s) \rightarrow Na^+(aq)$

  **b)** $Zn(s) \rightarrow Zn^{2+}(aq)$

  **c)** $I^-(aq) \rightarrow I_2(aq)$         (3)

**17** Pick a reduction from Question 15 and an oxidation from Question 16 and combine them to give the full ionic equation for the reactions between:

a) bromine with iodide ions

b) iron(III) ions with iodide ions

c) hydrogen ions with zinc. (6)

**18** Ammonium sulfate was prepared by adding ammonia solution to $25\,cm^3$ of $2.0\,mol\,dm^{-3}$ sulfuric acid.

$$2NH_3(aq) + H_2SO_4(aq) \rightarrow (NH_4)_2SO_4(aq)$$

a) i) What volume of $2.0\,mol\,dm^{-3}$ ammonia solution will just neutralise the sulfuric acid?

ii) How could you test to check that enough ammonia had been added to neutralise all of the acid?

b) Explain why hydrogen chloride, HCl, acts as an acid and ammonia, $NH_3$, acts as a base in the equation below.

$$HCl(g) + NH_3(g) \rightarrow NH_4^+Cl^-(s)$$

c) Which of the reactants acts as an acid in the following equation?

$$HCO_3^-(aq) + H_2O(l) \rightarrow CO_3^{2-}(aq) + H_3O^+(aq)$$

d) Which of the reactants acts as a base in the following equation?

$$HCO_3^-(aq) + H_3O^+(aq) \rightarrow H_2CO_3(aq) + H_2O(l)$$
(5)

**19** Acids and bases are commonly used in our homes.

a) Baking powder contains sodium hydrogen-carbonate, $NaHCO_3$, mixed with an acid.

i) When water is added, the baking powder releases carbon dioxide. Why does the reaction not occur until water is added?

ii) The acid in baking powder can be written as $H_2X$. This produces two $H^+$ ions per molecule of $H_2X$. Write an equation to show the ions produced from $H_2X$ when water is added to it.

iii) Write an equation for the reaction between sodium hydrogencarbonate and $H_2X$ when water is added to baking powder.

b) Indigestion tablets contain bases that cure indigestion by neutralising excess hydrochloric acid in the stomach.

i) One type of indigestion tablet contains magnesium hydroxide. Write an equation, with state symbols, for the reaction of magnesium hydroxide with hydrochloric acid.

ii) How does the pH in the stomach change after a person takes the tablets?

iii) Explain why magnesium hydroxide acts as a base when it reacts with stomach acid. (10)

**20 a)** An aqueous solution of compound A reacts with an aqueous solution of hydrochloric acid and a vigorous effervescence in seen. In a second reaction compound B reacts with aqueous sodium hydroxide and the fumes turn moist red litmus blue. Write ionic equations for each reaction. Explain your reasoning.

b) An aqueous solution of compound B reacts separately with aqueous solutions of sulfuric acid and with silver nitrate to produce a white precipitate in each reaction. Write ionic equations for each reaction. Explain your reasoning. (10)

**21** Acids are proton donors and salts are formed when the $H^+$ in an acid is replaced by either a metal ion or an ammonium ion. The table below shows the formula of some acids:

| Acid | Carbonic acid | Phosphoric acid | Chloric(I) acid |
|---|---|---|---|
| Formula of acid | $H_2CO_3$ | $H_3PO_4$ | HOCl |

Deduce the formula of all possible salts that could be formed when:

a) potassium hydroxide reacts with carbonic acid

b) magnesium hydroxide reacts with carbonic acid

c) potassium hydroxide reacts with phosphoric acid

d) magnesium oxide reacts with phosphoric acid

e) potassium hydroxide reacts with chloric(I) acid

f) magnesium oxide reacts with chloric(I) acid (11)

**22 a)** Many naturally occurring organic compounds are acidic. Most contain the –COOH group in which the H can be replaced to form a salt. Ethanoic acid has the formula $CH_3COOH$ and it can react with NaOH, a base, to form the salt sodium ethanoate, $CH_3COO^-Na^+$, which can be written more simply as $CH_3COONa$.

What is the formula of magnesium ethanoate?

b) Ethanedioic acid (also known as oxalic acid), HOOC–COOH, can also form salts. What is the formula of:
   i) potassium ethanedioate (also known as potassium oxalate)
   ii) magnesium ethanedioate (also known as magnesium oxalate)? (3)

23 A sample of limestone (impure calcium carbonate) weighing 0.20 g was reacted with 50 cm³ of 0.10 mol dm⁻³ hydrochloric acid.

$$CaCO_3(s) + 2HCl(aq) \rightarrow CaCl_2(aq) + CO_2(g) + H_2O(l)$$

The hydrochloric acid was in excess, and the excess HCl(aq) reacted with 14 cm³ of 0.10 mol dm⁻³ sodium hydroxide.
Calculate the percentage by mass of calcium carbonate in the limestone. (7)

24 A patient suffering from a duodenal ulcer displays increased acidity in their gastric juices. The acid responsible for the acidity in the gastric juice is hydrochloric acid. One of the most common medications designed for the relief of excess stomach acidity is aluminium hydroxide.
a) Write an equation for the reaction between hydrochloric acid and aluminium hydroxide.

b) The gastric juice of the patient was found to contain the equivalent of a 0.05 mol dm⁻³ solution of hydrochloric acid. The patient produces 2 dm³ of gastric juice in a day. This volume of gastric juice is to be treated with tablets containing aluminium hydroxide. Each tablet contains 520 mg aluminium hydroxide per tablet.
   i) Calculate the mass of aluminium hydroxide required to neutralise the 2 dm³ of gastric juice.
   ii) Deduce the number of tablets that have to be taken. (6)

## Challenge

25 Lactic acid, $CH_3CH(OH)COOH$ is a naturally occurring acid that reacts with NaOH(aq) to produce the salt sodium lactate, $CH_3CH(OH)COO^-Na^+$.
Citric acid, $HOC(COOH)_3$, is also a naturally occurring acid and forms salts in the same way as lactic acid.
a) Deduce the formula of sodium citrate.
   Citric acid also reacts with calcium hydroxide.
b) Construct a balanced equation for the reaction between citric acid and calcium hydroxide. (5)

# Chapter 7

# Bonding and structure

## Prior knowledge

*In this chapter it is assumed that you are familiar with:*
- atomic number and mass number
- composition of atoms and ions in terms of protons, neutrons and electrons
- electronic configuration using the s, p and d notation.

## Test yourself on prior knowledge

1 Define atomic number and mass number.
2 Deduce the numbers of protons, neutrons and electrons in each of the following:
   a) $^{27}_{13}Al$          b) $^{204}_{81}Tl$          c) $^{127}_{52}Te^{2-}$          d) $^{50}_{23}V^{3+}$
3 Sodium and chlorine atoms both have three shells. Draw diagrams of each to show the distribution of electrons.
4 Write the full electronic configuration of:
   a) $^{75}_{33}As$          b) $^{32}_{16}S^{2-}$          c) $^{52}_{24}Cr$          d) $^{56}_{26}Fe^{3+}$

## Introducing bonding and structure

One of the central aims in chemistry is to explain the properties of different substances in terms of their bonding and structure. The bonding and structure is determined by the arrangement of electrons in the outer shells of the atoms involved in the bonding. Atoms bond or combine in two main ways:

- by losing or gaining electrons (**ionic bonding**)

- by sharing electrons (**covalent bonding**).

The noble gases are not reactive and in the early twentieth century it was suggested that this lack of reactivity was associated with the electron configuration of the noble gases. This led to the suggestion that when atoms of elements combine to form compounds, they do so by gaining, losing or sharing electrons to attain the same number of electrons as a noble gas.

The full electron configurations of the noble gases helium, neon, argon and krypton are:

- $_2He - 1s^2$

- $_{10}Ne - 1s^2 2s^2 2p^6$

- $_{18}Ar - 1s^2 2s^2 2p^6 3s^2 3p^6$

- $_{36}Kr - 1s^2 2s^2 2p^6 3s^2 3p^6 3d^{10} 4s^2 4p^6$ which can also be written as $1s^2 2s^2 2p^6 3s^2 3p^6 4s^2 3d^{10} 4p^6$.

Lighter elements, for example hydrogen and lithium, produce compounds with two electrons in their outer shells, so they have the same electron configuration as helium. Most other elements form compounds in which they have eight electrons in their outer shell and are **isoelectronic** with a noble gas.

The capacity of an atom to form chemical bonds is the valency of the element, see Chapter 3, page 29. This depends on the number of electrons in the outer electron (valence) shell. The periodic table can be used as a simple guide to predict the valency and, therefore, the number of chemical bonds formed by an element (Table 7.1).

**Table 7.1** The periodic table and valency.

|  | Group 1 | Group 2 | Group 13 | Group 14 | Group 15 | Group 16 | Group 17 |
|---|---|---|---|---|---|---|---|
| Period 3 | Na | Mg | Al | Si | P | S | Cl |
| Valency (number of chemical bonds) | 1 | 2 | 3 | 4 | 3 | 2 | 1 |

There are exceptions, so this table should be used only as a guide.

# Ionic bonding

**Key term**

An **ionic bond** is the electrostatic attraction between oppositely charged ions, the attraction between positive and negative ions.

**Ionic bonds** are usually formed between a metal and a non-metal.

A metal atom loses one or more electrons from its outer (valence) shell and the electrons are transferred to the non-metal atom or atoms. The resultant ions each have the electronic structure of a noble gas. (Apart from helium, this means eight electrons in their outer shells.)

Sodium chloride is an ionic compound formed by the transfer of an electron from sodium atoms to chlorine atoms.

Sodium is element 11 ($_{11}$Na). A sodium atom has the electron configuration $1s^2 2s^2 2p^6\ 3s^1$, with one electron in its outer third shell. It loses the single outer-shell electron and forms a positive ion with the electron configuration $1s^2 2s^2 2p^6$.

Chlorine is element 17 ($_{17}$Cl). A chlorine atom has the electron configuration $1s^2 2s^2 2p^6 3s^2 3p^5$, with seven electrons in its outer third shell. It follows that chlorine will gain one electron to achieve the electron configuration $1s^2 2s^2 2p^6 3s^2 3p^6$. As it has gained an electron, it is now an ion with 18 electrons and 17 protons. Therefore, it has a charge of 1−.

The formation of sodium chloride can be shown by a **dot-and-cross** diagram. Dot-and-cross diagrams are simplified structures showing the electrons in rings around the nucleus.

A sodium atom needs to lose one electron from the outer shell and a chlorine atom needs to gain one electron:

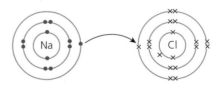

The resultant sodium and chloride ions each have eight electrons in their outer shells.

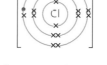

In examinations it is usual to show the outer electrons only:

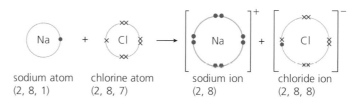

sodium atom (2, 8, 1)   chlorine atom (2, 8, 7)   sodium ion (2, 8)   chloride ion (2, 8, 8)

Ions formed from metal atoms are always created by the loss of electrons and are, therefore, positively charged. They are called cations. Ions formed from non-metal atoms by gaining electrons are negatively charged. They are called anions.

The oppositely charged ions are attracted to each other and this constitutes the ionic bond.

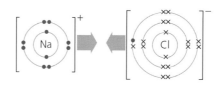

When sodium and chlorine react together, vast numbers of atoms react simultaneously. Each positive sodium ion produced is not attracted to just one chloride ion. It is attracted to other surrounding chloride ions; each chloride ion is attracted to other surrounding sodium ions. In fact, each sodium ion is attracted to six chloride ions and each chloride ion is attracted to six sodium ions (Figure 7.1).

In this way, the ions of sodium chloride build into a crystal structure. In sodium chloride, each ion is co-ordinated (bonded ionically) to six oppositely charged ions:

A giant ionic lattice is formed, which is said to be '6:6 coordinate'. In sodium chloride, the shape of the crystal is cubic.

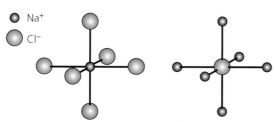

● Na⁺
● Cl⁻

**Figure 7.1** Ion attraction in sodium chloride.

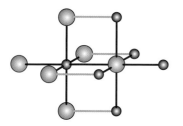

**Figure 7.2** Ionic bonding in sodium chloride.

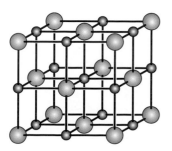

**Figure 7.3** Sodium chloride ionic lattice.

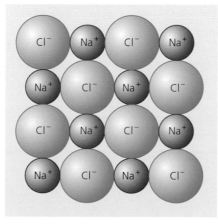

**Figure 7.4** The arrangement of ions in one layer of a sodium chloride crystal.

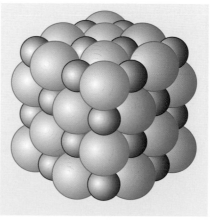

**Figure 7.5** The arrangement of the ions in several layers of a sodium chloride crystal.

Every ionic compound forms a giant ionic lattice that is held together by strong electrostatic attractions (ionic bonds) throughout the lattice. The ions are held in a fixed position within the lattice and are not free to move.

> **Tip**
>
> Ionic substances, like NaCl, do not exist as single molecules. Ionic substances form giant structures containing a very large number of positive and negative ions. The formula of sodium chloride might be better represented as $(Na^+Cl^-)_n$, where $n$ is a very large number.

## Activity

### Atoms, ions and the periodic table

Look carefully at Table 7.2. This shows the electron structures of the atoms and ions of the elements in Period 3 of the periodic table.

**Table 7.2** Electron structures of the atoms and ions of elements in Period 3.

| Elements in Period 3 | Na | Mg | Al | Si | P | S | Cl | Ar |
|---|---|---|---|---|---|---|---|---|
| Group | 1 | 2 | 13 | 14 | 15 | 16 | 17 | 18 |
| Electron structure | 2,8,1 | 2,8,2 | 2,8,3 | 2,8,4 | 2,8,5 | 2,8,6 | 2,8,7 | 2,8,8 |
| No. of electrons in outer shell | 1 | 2 | 3 | 4 | 5 | 6 | 7 | 8 |
| Common ion | $Na^+$ | $Mg^{2+}$ | $Al^{3+}$ | No ion | $P^{3-}$ | $S^{2-}$ | $Cl^-$ | No ion |
| Electron structure of ion | 2,8 | 2,8 | 2,8 | – | 2,8,8 | 2,8,8 | 2,8,8 | – |

1 What pattern can you see in the ions formed by elements in Groups 1, 2 and 13?
2 Predict the ions formed by the following atoms:
   a) Li
   b) Be
   c) K
   d) Ca.
3 We can write an equation for the formation of a sodium ion, $Na^+$, when a sodium atom, Na, loses an electron as:

$$Na \rightarrow Na^+ + e^-$$

   (2,8,1)  (2,8)  electron

   Write a similar equation for the formation of a magnesium ion, $Mg^{2+}$, from a magnesium atom, Mg.
4 What pattern can you see in the ions formed by elements in Groups 15, 16 and 17?
5 Predict the ions formed by the following atoms:
   a) N
   b) O
   c) F.
6 Why does argon not form an ion?
7 Use your answer to Question 6 to explain why argon is used to fill electric light bulbs.
8 Suggest why silicon does not form $Si^{4+}$ or $Si^{4-}$ ions.

## Typical properties of ionic compounds

Ionic compounds contain strong ionic bonds throughout the giant lattice. Typical properties of ionic substances are shown in Table 7.3.

**Table 7.3** Properties of ionic substances.

| Property | Explanation of property |
|---|---|
| High melting (and boiling) points | Ionic bonds are strong and occur throughout the giant lattice. In order to melt an ionic solid, all the ionic bonds in the lattice have to be broken. This requires a large amount of energy. Hence the melting (and boiling) point is high. |
| Poor conductor of electricity when solid | When solid, the ions are held in a fixed position within the lattice and cannot move. There are no mobile charge carriers, hence ionic solids are poor conductors. |
| Good conductor of electricity when molten or dissolved in water | When melted or dissolved in water the ions are mobile and free to move, hence molten and aqueous ionic compounds are good conductors. |
| Soluble in water and polar solvents | As a rough guide to solubility we can say 'like dissolves like'. Water has dipoles (charges), as do ionic solids (see below). |

### Solutions of ionic solids in water

It is not easy to see why a solid like sodium chloride readily dissolves in water. The explanation for the solubility of ionic salts, such as sodium chloride, in water is that the charged ions, both $Na^+$ and $Cl^-$ are strongly attracted to polar water molecules. The water molecules are attracted to the ions and bind to them. This binding of water molecules to ions is called hydration, which will be studied in detail in the second year of the A Level course. In the case of sodium chloride, the energy released as water molecules bind to the ions is enough to compensate for the energy needed to overcome the ionic bonding between the ions.

Some salts, like $Ca(OH)_2$ and $BaSO_4$, are sparingly soluble or insoluble in water because the energy that would be released when the ions are hydrated is not sufficient to balance the energy required to separate the ions.

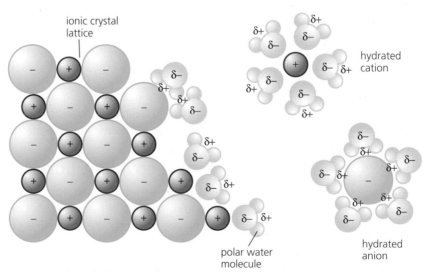

**Figure 7.6** Sodium and chloride ions leaving a crystal lattice and becoming hydrated as they dissolve in water. The bond between the ions and the polar water molecules is an electrostatic attraction.

## Test yourself

1 Draw 'dot-and-cross' diagrams for:
   a) lithium fluoride
   b) magnesium chloride
   c) lithium oxide
   d) calcium oxide.
2 With the help of a periodic table, predict the charges on ions of the following elements:
   a) caesium
   b) strontium
   c) gallium
   d) selenium
   e) astatine.
3 Why do metals form positive ions and non-metals form negative ions?

4 a) Write the electronic configurations for the atom and the ion formed by:
      i) magnesium
      ii) sulfur
      ii) rubidium
      iv) bromine.
   b) Predict the formula of magnesium bromide and rubidium sulfide.
5 The melting point of sodium fluoride is 993 °C but that of magnesium oxide is 2852 °C.
   a) Write the formulae of these two compounds showing charges on the ions.
   b) Suggest why the melting point of magnesium oxide is so much higher than that of sodium fluoride.

# Covalent bonding

### Key term

A single **covalent bond** is formed by the sharing of two electrons between two adjacent atoms. Each atom provides one electron. The electrostatic attraction between the shared pair of electrons and the nuclei of the two bonded atoms constitutes the covalent bond.

Covalent bonds are usually formed between two non-metal atoms.

Both non-metal atoms share electrons from their outer (valence) shell. Each atom achieves the electron configuration of a noble gas, which, apart from helium, is eight electrons in the outer shell.

Many compounds in everyday life are covalent. Almost all the compounds in food and most of the compounds in our bodies are bonded covalently. Covalent compounds usually exist as simple discrete molecules, but some form giant structures (page 110).

## Single covalent bonds

Some non-metallic elements exist as pairs of atoms linked together. The bonding is covalent and occurs because, by sharing electrons, each atom achieves the stability of a noble gas configuration. A **single covalent bond** is formed from the sharing of two electrons by adjacent atoms.

Each hydrogen atom has one electron in its outer orbit:

A hydrogen molecule is a simple diatomic molecule in which the atoms are held together by a single covalent bond:

The hydrogen molecule can be represented as H–H.

A fluorine molecule is also a diatomic molecule with the atoms held together by a single covalent bond. Each fluorine atom has seven electrons in its outer orbit and is, therefore, one electron short of attaining the octet.

By sharing electrons, each fluorine atom has eight electrons in its outer shell and the fluorine molecule can be represented as F–F (Figure 7.7).

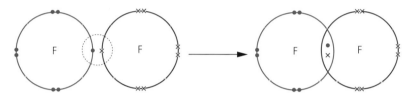

**Figure 7.7** Covalent bonding in a fluorine molecule.

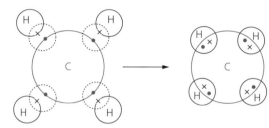

**Figure 7.8** Covalent bonding in methane.

Covalent bonds also occur between atoms of different non-metallic elements. For example, methane is a simple covalent compound. A molecule of methane consists of one carbon atom and four hydrogen atoms. A carbon atom requires four electrons to be isoelectronic (have the same number of electrons) with a noble gas; each hydrogen atom requires one electron. By sharing electrons, each atom in a molecule of methane achieves a stable electron configuration (Figure 7.8).

Dot-and-cross diagrams showing only the electrons in the outer shell provide a simple way of representing covalent bonding, as shown in Figure 7.9.

## Multiple covalent bonds

### Double bonds

If two atoms share four electrons, two covalent bonds are formed – a **double bond**. The simplest substance containing a double bond is an oxygen molecule. Each oxygen atom has six electrons:

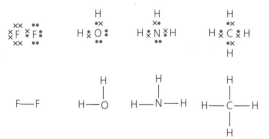

**Figure 7.9** Dot-and-cross diagrams representing covalent bonds.

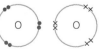

By sharing two of its electrons, each oxygen atom has eight electrons in its outer shell:

The oxygen molecule can be represented as O=O.

Carbon–carbon double bonds are often referred to as **unsaturated bonds** and are common in organic molecules. Ethene has a double bond, H₂C=CH₂. Carbon–oxygen double bonds are also common in organic

chemistry, for example methanal,
$$\begin{array}{c} H \\ | \\ H-C=O \end{array}$$

Figure 7.10 shows the single and double covalent bonds in these compounds.

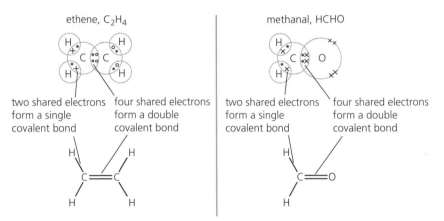

ethene, C₂H₄      methanal, HCHO

two shared electrons form a single covalent bond    four shared electrons form a double covalent bond    two shared electrons form a single covalent bond    four shared electrons form a double covalent bond

**Figure 7.10** Multiple covalent bonding in ethene and in methanal.

## Triple bonds

Nitrogen molecules contain a **triple** covalent bond, formed by sharing six electrons. Each nitrogen atom has five electrons in its outer shell and each, therefore shares three electrons with the other nitrogen atom:

By sharing three of its electrons, each nitrogen atom attains a stable outer octet:

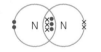

The nitrogen molecule can be represented as:

$$N\equiv N$$

## Lone pairs of electrons

In many molecules and ions, there are atoms with pairs of electrons in their outer shells which are not involved in the bonding between atoms. Chemists call these 'lone pairs' of electrons. Lone pairs of electrons:

- can form a variation on the covalent bond known as a dative covalent bond

- affect the shapes of molecules (see pages 99–103)

- are important in the chemical reactions of some compounds, including water and ammonia

- can behave as nucleophiles in organic reactions (see page 235).

A lone pair of electrons can also be described as a non-bonding pair of electrons.

## Dative covalent (coordinate) bonding

When adjacent atoms bond together and share two electrons, it is usual for each atom to provide one electron. Sometimes, however, one atom supplies *both* electrons. Two electrons are shared, so the bond is covalent, but, because both shared electrons come from one of the atoms, the bond is called a dative covalent bond (coordinate bond).

A simple example of a dative covalent bond also explains how ammonia ($NH_3$) forms the ammonium ion, $NH_4^+$. This is possible because the nitrogen atom in ammonia has eight electrons in its valence shell, two of which are not bonded to other atoms. This pair of electrons is called a **lone pair**.

If a hydrogen ion (which has no electrons) is provided by an acid, the lone pair of electrons on the nitrogen atom can form a dative covalent bond with the hydrogen ion to form an ammonium ion, which has a charge of 1+, is formed (Figure 7.11).

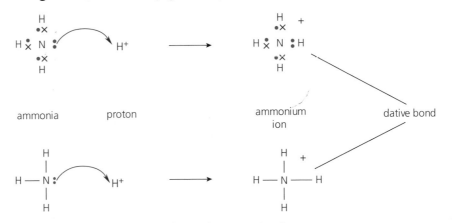

ammonia      proton      ammonium ion      dative bond

**Figure 7.11** Dative covalent bonding in the ammonium ion.

In an ammonium ion, all four bonds from nitrogen to hydrogen are equivalent. It is not possible to identify which is the dative covalent bond. It is simply the method of bond formation that is different. The electrical charge of the cation, $NH_4^+$, is spread over the whole ion.

Dative covalent bonding can sometimes occur between molecules – for example, when ammonia combines with boron trifluoride, $BF_3$. The boron atom has six electrons in its valence shell and, therefore, requires two electrons. These are provided by the lone pair of electrons on the nitrogen atom of the ammonia molecule:

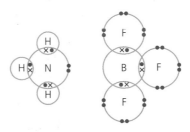

The lone pair of electrons donated by the nitrogen atom is shared with the boron atom. A dative covalent bond is formed and the nitrogen and boron atoms:

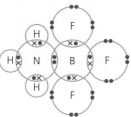

# Typical properties of covalent compounds

Many covalent compounds are liquids or gases with low melting points. If solid, they are not dense and tend to be soft and easily broken, unlike ionic crystals, which are hard and have a specific shape. Candle wax is a mixture of solid covalent compounds.

Most covalent compounds exist as small discrete molecules such as $H_2$, $P_4$, $H_2O$ and $CH_4$. There are some exceptions such as diamond, graphite, graphene and silicon. The covalent bonds within the molecules are strong but the bonds between individual molecules are much weaker and are easily broken.

General properties of covalent compounds are summarised in the table below.

**Tip**

The properties of covalent compounds listed in Table 7.4 apply only to simple molecular compounds and not to compounds with giant covalent structures.

**Table 7.4** Properties of covalent compounds.

| Property | Explanation of property |
|---|---|
| Low melting points and low boiling points | They exist as discrete (separate) molecules and the attraction between neighbouring covalent molecules is weak. When melted or boiled only the weak bonds between the molecules are broken so only a small amount of energy is required to break them, hence the melting and boiling points are low. |
| Poor conductor of electricity | There are no mobile charge carriers, ions or electrons, to carry the electric charge. Graphite and graphene are exceptions to this. |
| Soluble in non-polar solvents such as hexane | As a rough guide to solubility we can say 'like dissolves like' Covalent compounds are non-polar and hence tend to be soluble in non-polar solvents. |

**Test yourself**

6 Draw 'dot-and-cross' diagrams to show the covalent bonding in:
   a) hydrogen sulfide, $H_2S$
   b) ethane, $C_2H_6$
   c) carbon disulfide, $CS_2$
   d) nitrogen trifluoride, $NF_3$.
7 Identify the atoms with lone pairs of electrons in the following molecules and state the number of lone pairs:
   a) ammonia
   b) water
   c) hydrogen fluoride
   d) carbon dioxide.
8 Draw dot-and-cross diagrams for:
   a) $PH_3$
   b) $PH_4^+$
   c) $PH_2^-$.
9 In aqueous solutions acids donate protons, $H^+$, to water to form $H_3O^+$. Draw a dot-and-cross diagram of $H_3O^+$.

# Shapes of simple molecules and ions

Simple covalent compounds exist as discrete molecules with atoms held together by shared pairs of electrons. The shared pair of electrons is located between the two nuclei and this means that the atoms always stay in the same position relative to each other. The central atom may also be bonded to other atoms also held in fixed positions, which means that the covalent molecule will have a fixed shape.

## Electron pair repulsion theory

The shape of a molecule is determined by the numbers of bonded and lone (non-bonded) electron pairs around the central atom. A simple model of molecular shape is based on the fact that pairs of electrons (both bonded and non-bonded) repel each other to be as far apart in space as possible. This is known as the electron pair repulsion theory.

From strongest to weakest, the order of strength of repulsion is:

lone pair–lone pair > lone pair–bonded pair > bonded pair–bonded pair

The overall shape of a covalent molecule (or ion) is determined by the number and type of electron pairs around the central atom.

The way to determine the shape is to draw a dot-and-cross diagram and use this to deduce the number (and types) of electron pairs around the central atom.

## Two bonded pairs and no lone pairs

Covalent molecules containing two single covalent bonds are not common. The simplest example is beryllium chloride, $BeCl_2$, which, unusually for a metallic compound, is covalent. The beryllium atom has two bonded pairs and no lone pairs. The bonded pairs repel each other until they are as far apart as possible. The resultant shape is **linear**; the bond angle is 180°.

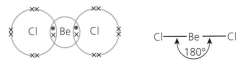

Linear shape is more common in molecules containing multiple – double or triple – bonds, such as carbon dioxide, $CO_2$. The central carbon atom and each oxygen atom in carbon dioxide are connected by two pairs of bonded electrons, each constituting a double covalent bond. The double-bonded pairs repel each other until they are as far apart as possible. The resultant shape is linear and the bond angle is 180°:

## Three bonded pairs and no lone pairs

Boron forms covalent molecules with the halogens. In boron trichloride, $BCl_3$, the central boron atom has three bonded pairs and no lone pairs.

### Key term

The **electron pair repulsion theory** states that the electron pairs repel each other and the overall shape of the molecule depends on the number and type of electron pairs around the central atom. Lone pairs (non-bonded pairs) repel more than bonded pairs of electrons.

### Tip

Lone pairs repel more because they are closer to the central atom than a bonded pair.

The bonded pairs repel each other so that they are as far apart as possible. The resultant shape is **trigonal planar**; the bond angle is 120°:

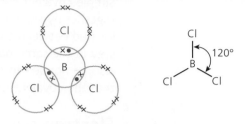

## Four bonded pairs and no lone pairs

Carbon forms many compounds containing four covalent bonds. The simplest is methane. The central carbon atom in methane has four bonded pairs and no lone pairs. The bonded pairs repel each other so that they are as far apart as possible. The resultant shape is **tetrahedral**; the bond angle is 109.5°:

**Tip**

The bond angle in a tetrahedron is actually 109° 28'.
28' represents 28 minutes – there are 60 minutes in 1°.

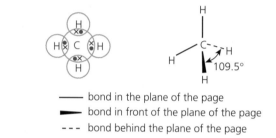

―――― bond in the plane of the page
▬▬▬ bond in front of the plane of the page
- - - bond behind the plane of the page

## Five bonded pairs and no lone pairs

The non-metallic elements in Period 3 can form compounds in which there are more than eight electrons around the central atom. An example is phosphorus pentafluoride, $PF_5$. The phosphorus atom has five bonded pairs and no lone pairs. The bonded pairs repel each other so that they are as far apart as possible. The resultant shape is **trigonal bipyramidal** (base-to-base triangular pyramids); the bond angles are 120° and 90°:

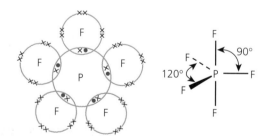

## Six bonded pairs and no lone pairs

Sulfur can also form compounds with more than eight electrons in the outer shell. Sulfur forms six covalent bonds in compounds such as sulfur hexafluoride. The sulfur atom has six bonded pairs and no lone pairs. The bonded pairs repel each other to the point of maximum separation to give an **octahedral** shape (base-to-base square pyramids); the bond angles are all 90°:

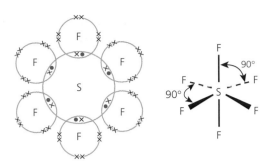

## Covalent molecules that contain lone pairs of electrons

Many covalent compounds contain lone pairs of electrons as well as bonded pairs. These lone pairs affect the shape of the molecule, because lone pairs repel more than bonded pairs.

### Ammonia

A dot-and-cross diagram of ammonia is shown below:

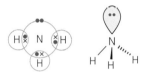

The central nitrogen atom in ammonia has four pairs of electrons around it and, therefore, you might expect them to be repelled to form tetrahedral bond angles of 109.5°. However, the four pairs of electrons are not identical; there are three bonded pairs and one lone pair. The lone pair repels more than the bonded pairs, which results in the bonded pairs being forced closer together. The angle between the N–H bonds contracts from 109.5° to about 107°. The shape of the ammonia molecule is **pyramidal**:

### Water

A dot-and-cross diagram for water is shown below:

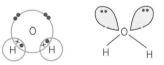

The central oxygen atom in a water molecule has four pairs of electrons around it and, therefore, you might expect the molecule to be tetrahedral with a bond angle of 109.5°. However, the four pairs of electrons surrounding the oxygen atom consist of two bonded pairs and two lone pairs. The lone pairs repel more than the bonded pairs, forcing the bonded pairs closer together. The bond angle contracts from 109.5° to about 104.5°. The shape of the water molecule is **angular** or **bent**.

The shapes of covalent compounds are summarised in Table 7.5.

**Table 7.5** Shapes of covalent compounds.

| Number of electron pairs around the central atom | Number of bonded pairs | Number of lone pairs | Shape | Bond angle | Example |
|---|---|---|---|---|---|
| 2 | 2 | 0 | Linear | 180° | $CO_2$  O=C=O |
| 3 | 3 | 0 | Trigonal planar | 120° | $BCl_3$ |
| 4 | 4 | 0 | Tetrahedral | 109.5° | $CH_4$ |
| 5 | 5 | 0 | Trigonal bipyramidal | 120° and 90° | $PF_5$ |
| 6 | 6 | 0 | Octahedral | 90° | $SF_6$ |
| 4 | 3 | 1 | Pyramidal | 107° | $NH_3$ |
| 4 | 2 | 2 | Angular (bent) | 104.5° | $H_2O$ |

## Shapes of ions

Ions such as the ammonium ion, $NH_4^+$, and the amide ion, $NH_2^-$, contain covalent bonds and, therefore, have a fixed shape and a defined bond angle.

When an ammonium ion is formed from ammonia, a proton is gained. The central nitrogen atom has four bonded pairs of electrons. The shape is tetrahedral and the bond angle is 109.5°.

When an amide ion is formed from ammonia, a proton is lost. The central nitrogen atom has two bonded pairs of electrons and two lone pairs. The shape is angular (bent) and the bond angle is about 104°.

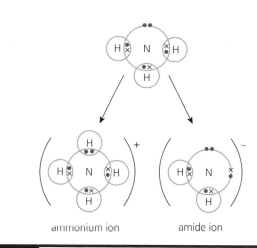

ammonium ion          amide ion

**Test yourself**

10  State the electron pair repulsion theory.
11  Determine the number, and type, of electrons pairs around the central
    atom(s) in each of the following. Predict the shape and bond angles of
    each. (Hint: it may help to draw dot-and-cross diagrams.)
    a) hydrogen sulfide, $H_2S$
    b) phosphine, $PH_3$
    c) sulfur dichloride, $SCl_2$
    d) dichloromethane, $CH_2Cl_2$
    e) $CH_3^+$
    f) $C_2H_4$
    g) $COCl_2$
    h) $XeF_4$

# Metallic bonding

**Key terms**

**Delocalised** electrons are bonding
electrons that are not fixed
between two atoms in a bond.
They are mobile and are shared
by several or many atoms.
The **metallic bond** is the electrostatic
attraction between the delocalised
electrons and the positive ions held
within the lattice.

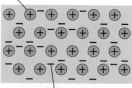

positive ion

delocalised electrons

**Figure 7.14** Metallic bonding results
from the strong attractions between the
metal ions in the lattice and the mobile
delocalised electrons.

Metal atoms pack together in a regular arrangement (Figures 7.12 and 7.13).

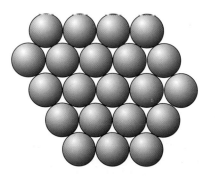

**Figure 7.12** The close packing of atoms
in one layer of a metal.

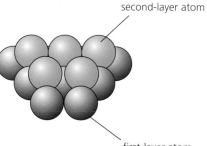

second-layer atom

first-layer atom

**Figure 7.13** Atoms in two layers of a metal
crystal.

The outer orbital electron clouds overlap and the electrons are free
to move from one atom to another – they are delocalised. The
attraction between these electrons and the positive ions forms the
metallic bond.

The electrons move through a giant lattice of positive ions (Figure 7.14). The
delocalised electrons are sometimes called 'a sea of mobile electrons'.

The delocalised electrons are able to move anywhere within the lattice, so the metallic bond is non-directional.

# Typical properties of metals

General properties of metallic compounds are summarised in Table 7.6.

**Table 7.6** Properties of metallic compounds.

| Property | Explanation of property |
|---|---|
| High melting points and high boiling points | High temperatures are needed to melt metals because the metallic bond is strong and exists throughout the giant lattice. In order to melt a solid metal, all the bonds in the lattice have to be broken, which requires a large amount of energy. |
| Good conductor of electricity | Metals are good conductors of electricity because they contain mobile charge carriers that are free to move, even in the solid. The mobile charge carriers are the delocalised electrons. |
| Malleable (re-shaped) and ductile (stretched into wires) | This is because the metallic bond is non-directional, is independent of shape and exists in all directions. |

## Activity

### Choosing metals for different uses

Various properties of six metals are shown in Table 7.7.

**Table 7.7** Various properties of six metals.

| Metal | Density/g cm$^{-3}$ | Tensile strength/ $10^7$ N m$^{-2}$ | Melting point/°C | Electrical resistivity/ $10^{-8}$ ohm m | Thermal conductivity/ J s$^{-1}$ cm$^{-1}$ K$^{-1}$ | Cost per tonne /£ |
|---|---|---|---|---|---|---|
| Aluminium | 2.7 | 8 | 660 | 2.5 | 2.4 | 960 |
| Copper | 8.9 | 33 | 1083 | 1.6 | 3.9 | 1200 |
| Iron | 7.9 | 21 | 1535 | 8.9 | 0.8 | 130 |
| Silver | 10.5 | 25 | 962 | 1.5 | 4.2 | 250000 |
| Titanium | 4.5 | 23 | 1660 | 43.0 | 0.2 | 27000 |
| Zinc | 7.1 | 14 | 420 | 5.5 | 1.1 | 750 |

1 Use the information in Table 7.7 to explain the following statements.
   a) Copper is used in most electrical wires and cables, but high-tension cables in the National Grid are made of aluminium.
   b) Bridges are built from steel, which is mainly iron, even though the tensile strength of iron is lower than that of some other metals.
   c) Metal gates and dustbins are made from steel, coated (galvanised) with zinc.
   d) Silver is no longer used to make coins in the UK.
   e) Aircraft are now constructed from an aluminium/titanium alloy rather than pure aluminium.
   f) The base of high-quality saucepans is copper rather than steel (iron).
2 If the atoms in a metal pack closer, the density should be higher, the bonds between atoms should be stronger and so the melting point should be higher. This suggests there should be a relationship between the density and melting point of a metal.
   a) Use the data in the table to check whether there is a relationship between density and melting point.
   b) Is there a relationship between the densities and melting points of the metals? Say 'yes' or 'no' and explain your answer.
3 The explanation of both electrical and thermal conductivity in metals uses the concept of delocalised electrons. This suggests there should be a relationship between the electrical and thermal conductivities of metals.
   a) Use the data in the table to check whether there is a relationship. (Hint: Electrical resistivity is the reciprocal of electrical conductivity.)
   b) Is there a relationship? Say 'yes' or 'no' and explain your answer.

# Electronegativity and bond polarity

It is rare for a compound to be 100% ionic. Only simple diatomic molecules like $H_2$ and $O_2$ in which the covalent bond exists between two identical atoms are 100% covalent. There is a range of intermediate types of bond between the extremes of ionic and covalent.

Many bonds are described as essentially:

- covalent with some ionic character
- ionic with some covalent character.

## Bond polarity

In a covalent molecule, ionic character is introduced when the pair of electrons making up the covalent bond is not shared evenly between the component atoms. This is a result of the difference in the electronegativity of the two atoms forming the covalent bond.

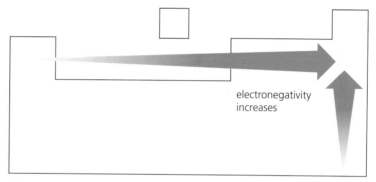

electronegativity increases

**Figure 7.15** Trends in electronegativity in the periodic table.

You should appreciate that:

- electronegativity increases across a period
- electronegativity decreases down a group.

Linus Pauling (1901–1994) developed, by comparing one element with another, a quantitative scale (Figure 7.16).

| H 2.1 |
|---|

| Li 1.0 | Be 1.5 | B 2.0 | C 2.5 | N 3.0 | O 3.5 | F 4.0 |
|---|---|---|---|---|---|---|
| Na 0.9 | Mg 1.2 | Al 1.5 | Si 1.8 | P 2.1 | S 2.5 | Cl 3.0 |
| K 0.8 | Ca 1.0 | | | | | Br 2.8 |
| Rb 0.8 | Sr 1.0 | | | | | I 2.5 |

**Figure 7.16**

In a covalent bond, the element with the greater electronegativity attracts the bonding electrons towards itself. Elements with high electronegativity such as fluorine and oxygen have a strong attraction for the shared covalent electrons.

In a molecule of hydrogen, $H_2$, both hydrogen atoms have the same attraction for the two bonded electrons. Hence, they are shared equally.

$$\underset{2.1 \quad 2.1}{H\!\!-\!\!H}$$

In a molecule of hydrogen chloride, HCl, the hydrogen atom and the chlorine atom have a different attraction for the two bonded electrons. Hence, the electrons are not shared equally. The molecules are not 100% covalent because the electrons are more concentrated towards the chlorine atom:

$$\underset{2.1 \quad 3.0}{H\!\rightarrow\!Cl}$$

The chlorine atom becomes slightly negatively charged with respect to the hydrogen; the hydrogen is left with a slight positive charge. A small degree of charge is indicated on a molecule by using the symbol $\delta+$ or $\delta-$:

$$\overset{\delta+ \quad \delta-}{H\!\!-\!\!Cl}$$

A covalent bond with some charge separation is said to have a **dipole**. A dipole indicates that the bond has a small amount of ionic charge; the bond is best described as essentially covalent, with some ionic character.

## Polar bonds and polar molecules

A polar bond is due to the difference in electronegativity of the atoms in a covalent bond. However, molecules with polar bonds are not necessarily polar molecules. If a molecule is symmetrical, the dipoles cancel out, even if the bonds are polar. For example, in a molecule of carbon dioxide, the electrons in the covalent bonds are pulled towards the oxygen atoms:

$$\overset{\delta-}{\underset{3.5}{O}}\!=\!\!\overset{\delta+}{\underset{2.5}{C}}\!=\!\!\overset{\delta-}{\underset{3.5}{O}}$$

However, because the molecule is symmetrical, the dipoles cancel each other out. The net result is that the C=O bonds are polar, but the carbon dioxide molecule is not.

Symmetrical shapes include linear, trigonal planar, tetrahedral and octahedral. If the atoms bonded to the central atom are the same, any dipoles in the bonds cancel each other out and the molecule is non-polar — for example, methane, $CH_4$.

Non-symmetrical shapes include pyramidal and angular (or bent). Molecules with these shapes are always polar, even when the central atom is attached to atoms that are the same. Ammonia molecules are pyramidal and polar; water molecules are angular and polar.

A compound consisting of two or more different non-metals is essentially covalent with some ionic character. In general, the molecules of such compounds are polar. However, if the molecule is symmetrical this is not the case, because the dipoles cancel each other out.

A compound consisting of a metal and a non-metal is essentially ionic with some covalent character.

**Tip**

If the central atom of a molecule has one or more lone pair of electrons the molecule will almost always be polar.

**Interpreting electronegativity values**

Electronegativity is defined as the attraction for electrons within a covalent bond.

Figure 7.16 on page 105 shows the electronegativity of some elements according to Pauling's electronegativity values. Use the data to answer the questions below.

1 Suggest why the electronegativity values for the noble gases are not included in Figure 7.16.
2 Explain the trend in electronegativity:
   **a)** across a period (Li to F)
   **b)** down a group (F to Br)
3 **a)** Draw dot-and-cross diagrams for $H_2S$, $SO_2$, $PH_3$ and $COCl_2$.
   **b)** Sketch molecules of $H_2S$, $SO_2$, $PH_3$ and $COCl_2$ showing shape, bond angles and any dipoles.
In a molecule of hydrogen chloride, HCl, the H atom and the chlorine atom have a different attraction for the two bonded electrons and hence they will **not** be equally shared. Molecules like HCl, cannot be regarded as 100% covalent.

the bonded electrons are pulled towards the Cl

H — Cl
2.3   3.0

this results in the Cl being slightly negatively charged, leaving the H with a slight positive charge

$\delta+$  $\delta-$
H — Cl

The dipoles show that the H–Cl bond has some charge and is hence a little bit ionic.

Table 7.8 tries to be quantitative about the effect of differences in electronegativity and uses the differences to estimate a % ionic character within a covalent bond.

**Table 7.8** Using electronegativity to estimate percentage ionic character in a covalent bond.

| Difference in electronegativity | 0.3 | 0.4 | 0.5 | 0.6 | 0.7 | 0.8 | 0.9 | 1.0 | 1.1 | 1.2 | 1.3 | 1.4 | 1.5 | 1.6 | 1.7 | 1.8 | 1.9 |
|---|---|---|---|---|---|---|---|---|---|---|---|---|---|---|---|---|---|
| % ionic character of the bond | 2 | 4 | 6 | 9 | 12 | 15 | 19 | 22 | 26 | 30 | 34 | 39 | 43 | 47 | 51 | 55 | 59 |

The difference in electronegativity between H and Cl is 0.9, which means that the bond has some ionic character. A difference in electronegativity of 0.9 suggests an ionic character of about 19%. The H–Cl bond is essentially covalent (81%) but with ionic (19%) character. This type of bond is described as a **polarised** bond and H–Cl has a **permanent dipole**.

4 **a)** Estimate the percentage ionic/covalent character of the:
   **i)** O–H bond in $H_2O$
   **ii)** S–H bond in $H_2S$
   **iii)** C–H bond in $CH_4$
   **iv)** C–Cl bond in $CH_3Cl$.
   **b)** Divide the molecules; $H_2O$, $H_2S$, $CH_4$ and $CH_3Cl$ into two groups – polar and non-polar.
   **c)** The boiling points of $H_2O$ and $H_2S$ are 100 °C and −61 °C respectively. $H_2O$ and $H_2S$ are both covalent molecules with similar shapes. Suggest an explanation for this large difference in boiling points.

12 Use the electronegativity values in Figure 7.16 on page 105 to help answer the following:
   **a)** Sketch HBr, $CO_2$, $NH_3$ and $NO_2$ and use $\delta+$ and $\delta-$ to show bonds that have charge separation.
   **b)** Divide HBr, $CO_2$ and $NH_3$ into polar and non-polar molecules.
   **c)** Draw a dot-and-cross diagram of $N_2H_4$.
   **d)** Is $N_2H_4$ polar or non-polar?

# Intermolecular forces

**Key term**

Intermolecular forces are forces of attraction that occur between molecules.

**Tip**

The prefix 'inter-' means between. A football match between two nations is an international match.

**Key terms**

Permanent dipole–dipole interactions are the weak electrostatic attractions between polar molecules that are essentially covalent but have some ionic character.

Hydrogen bonds are the relatively strong electrostatic attractions between polar molecules that contain hydrogen covalently bonded to elements with high electronegativity such as fluorine, oxygen and nitrogen.

Ionic, covalent and metallic bonds are all strong bonds; to break them generally requires between $200\,kJ\,mol^{-1}$ and $800\,kJ\,mol^{-1}$. There is a second set of bonds or types of interaction which form between molecules and are known as intermolecular forces. Compared to ionic, covalent and metallic bonds these are weak and only require between $2\,kJ\,mol^{-1}$ and $40\,kJ\,mol^{-1}$ to overcome them.

## Permanent dipole–dipole interactions

The covalent C–Cl bond in chloromethane, $CH_3Cl$ is has a permanent dipole and the molecules are also polar because chloromethane, $CH_3Cl$, is not symmetrical and the bond dipoles do not cancel each other out.

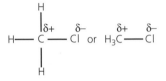

Oppositely charged dipoles attract:

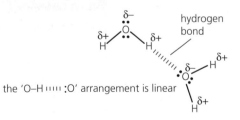

There is a weak electrostatic attraction between the dipoles of neighbouring molecules. This is called a permanent dipole–dipole interaction.

## Hydrogen bonding

**Hydrogen bonding** is a special form of permanent dipole–dipole interactions.

A hydrogen bond is a relatively strong intermolecular attraction between molecules that involves a lone pair of electrons from the nitrogen, oxygen or fluorine.

Hydrogen bonds exist between molecules in ammonia, $NH_3$, water, $H_2O$, and hydrogen fluoride, HF. They are also present in compounds that contain the –OH group (e.g. alcohols and carboxylic acids) and compounds that contain the $-NH_2$ group (e.g. amines). Hydrogen bonding is important in amino acids, peptides and proteins.

Water can be used to illustrate some features of hydrogen bonding. The hydrogen bond is formed between a lone pair of electrons on the oxygen atom in one water molecule and a hydrogen atom in an adjacent water molecule:

the 'O–H ⅢⅢ :O' arrangement is linear

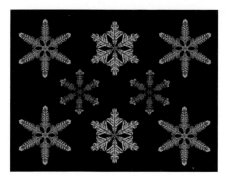

**Figure 7.17** Snowflakes have a characteristic six-fold symmetry and can come in an infinite range of shapes.

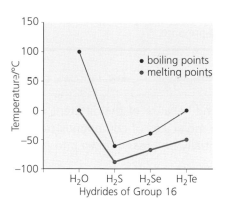

**Figure 7.18** Melting and boiling points of the hydrides of Group 16 elements.

### Tip

Induced dipole–dipole interactions can also be referred to as London (dispersion) forces.

### Key term

To define an **induced dipole–dipole interaction** there are three key features that have to be considered:

- The movement of electrons generates an instantaneous dipole.
- The instantaneous dipole induces other dipoles in neighbouring atoms/molecules.
- Two dipoles generate a weak temporary force of attraction between the atoms/molecules.

The presence of a lone pair of electrons on the electronegative atom is essential for hydrogen bonding to take place.

Hydrogen bonding in water causes some anomalous properties. The most important are that:

- the solid (ice) is less dense than the liquid (water)
- water has unexpectedly high melting and boiling points.

Ice floats on water because when water is frozen, the water molecules hydrogen-bond together to form interlocking hexagonal rings. This means that the molecules in any particular hexagon are held further apart, so that ice has a greater volume and thus lower density than water.

The structure of ice is complex, as can be seen in the patterns created by snowflakes (Figure 7.17).

Water is a hydride of oxygen. If water followed the trend shown by the hydrides of the elements in same group of the periodic table (Figure 7.18), the melting point of water would be approximately $-100\,^{\circ}C$ and the boiling point approximately $-80\,^{\circ}C$.

The unexpectedly high melting point of $0\,^{\circ}C$ and boiling point of $100\,^{\circ}C$ are due to the additional energy required to break the hydrogen bonds between adjacent water molecules. The intermolecular forces between the other Group 16 hydrides are smaller and, therefore, less energy is required to break them.

## Induced dipole–dipole interactions

The weakest of the intermolecular forces are induced dipole–dipole interactions **forces**. These act between all molecules, polar or non-polar. To understand these forces you must appreciate that the electrons are not static. Temporary movement of electrons leads to temporary induced dipoles. These temporary instantaneous dipoles can induce dipoles in neighbouring molecules. The attraction between the induced dipoles are called induced dipole–dipole interactions.

The strength of these forces depends on the number of electrons in the molecule. The greater the number of electrons in an atom or molecule, the greater the number of induced dipole–dipole forces.

Although these forces are weak, their presence explains some physical properties. For example, the boiling point of butane ($C_4H_{10}$) is higher than that of propane ($C_3H_8$). This is because a molecule of butane has more bonds, and hence more electrons, than a molecule of propane. Therefore, there are more intermolecular forces between butane molecules, and it is these intermolecular forces that have to be overcome for boiling to occur.

# Structure and physical properties

There are many types of solid compound because of the different ways in which the atoms of elements or the molecules of compounds can combine. The types of solids can essentially be broken down into four different sub-sets:

- giant ionic lattice – NaCl – this has been covered in the section on ionic bonding (see page 90)
- giant covalent lattices – graphite, graphene, diamond, silicon and silicon dioxide
- giant metallic lattices – this has already been covered in the section on metallic bonding (see page 103)
- simple molecular lattices – iodine and ice.

**Tip**

There are many other examples of the four different types of solids.

## Giant structures

For details of giant ionic lattices, giant metallic lattices and their properties, see pages 91 and 103.

## Giant covalent lattices

The group 14 elements carbon and silicon both have giant covalent lattices which are held together by strong covalent bonds between atoms. Carbon, in the form of diamond, and silicon have similar structures and they have high melting points because there are strong covalent bonds throughout the lattice. Carbon also exists as graphite which has a giant lattice. The covalent bonds within the graphite lattice are two dimensional whereas in diamond and in silicon they are three dimensional. Graphite like diamond has a very high melting point; in fact graphite is more stable than diamond, but unlike diamond graphite is a good conductor of electricity. This can be explained by the structure of graphite.

Graphite has a hexagonal layer structure:

Carbon can also exist in the form of graphene. Graphene is a pure form of carbon consisting of a very thin, almost transparent sheet, one

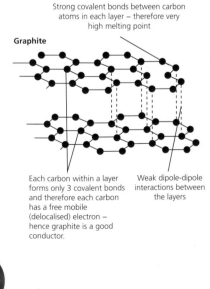

Strong covalent bonds between carbon atoms in each layer – therefore very high melting point

**Graphite**

Each carbon within a layer forms only 3 covalent bonds and therefore each carbon has a free mobile (delocalised) electron – hence graphite is a good conductor.

Weak dipole-dipole interactions between the layers

atom thick layer of carbon atoms. It is remarkably strong for its low weight (approximately 100 times stronger than steel) and it is a good conductor of both electricity and heat. Graphene was first isolated in the laboratory in 2004. In 2010 two scientists from the University of Manchester were awarded the Nobel Prize for their ground-breaking experimental work with the two-dimensional graphene.

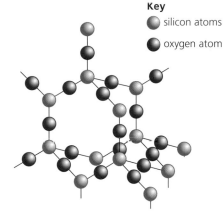

**Key**
● silicon atoms
● oxygen atom

**Figure 7.19** There are strong covalent bonds throughout the $SiO_2$ lattice.

The strength and the conductivity of graphene means it has may potential applications such as semiconductors, electronics, smart phones, tablets and battery energy.

## Simple molecular lattices

Simple molecules do form larger structures. However, the forces between the molecules are weak and the solids melt easily. An example is solid iodine, $I_2$, which has a regular structure built up by repeating the section shown below:

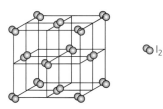

**Figure 7.20** The iodine solid in the test tube has been gently heated and purple fumes of iodine gas can be seen in the inverted round bottom flask surrounding the test tube.

The lattice is made up of $I_2$ molecules held in position by weak intermolecular forces. The I–I molecule is non-polar and the only forces that act between the separate $I_2$ molecules are induced dipole–dipole forces. When warmed the weak induced dipole–dipole forces are readily broken and individual molecules of $I_2$ are released such that iodine sublimes rather than melts (Figure 7.20).

> **Tip**
>
> Sublimation is when a solid changes directly to a gas on heating. $CO_2$ also sublimes and on heating $CO_2(s) \rightarrow CO_2(g)$, which is why $CO_2(s)$ is known as 'dry ice'.

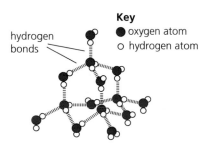

**Key**
● oxygen atom
○ hydrogen atom

hydrogen bonds

**Figure 7.21** The structure of ice.

Ice also forms a giant molecular lattice in which separate water molecules are held in position by a combination of intermolecular forces including hydrogen-bonding (Figure 7.21).

> **Test yourself**
>
> 17 Explain in terms of bonding and structure:
>    a) the melting point of $H_2O$ is 0 °C but the melting point of $H_2S$ is −85 °C
>    b) $CO_2$ sublimes at −56 °C but $SiO_2$ melts at 1610 °C.

# Practice questions

## Multiple choice questions 1–14

Answers to Questions 1 to 3 relate to the following types of bonding:

    **A**  ionic
    **B**  covalent
    **C**  metallic
    **D**  essentially covalent but with some ionic character (polar)

Which type of bonding is present in each of the following?

    **1** hydrogen sulfide, $H_2S$

    **2** Indium sulfide, $In_2S_3$

    **3** ammonium ion, $NH_4^+$

Answers to Questions 4 to 6 relate to the following types of intermolecular forces:

    **A**  permanent dipole–dipole interactions
    **B**  hydrogen bonding
    **C**  induced dipole–dipole interactions only
    **D**  a combination of hydrogen bonding and induced dipole–dipole interactions

Which type(s) of intermolecular forces are present in each of the following?

    **4** $N_2$

    **5** $NH_3$

    **6** Ne

Answers to Questions 7 to 10 relate to the following shapes:

    **A**  trigonal planar
    **B**  tetrahedral
    **C**  pyramidal
    **D**  angular

What is the shape of each of the following?

    **7** hydrogen selenide, $H_2Se$

    **8** phosphorus trichloride, $PCl_3$

    **9** carbonyl chloride, $COCl_2$

    **10** amide ion, $NH_2^-$

Use the key below to answer Questions 11 to 14.

| A | B | C | D |
|---|---|---|---|
| 1, 2 & 3 correct | 1, 2 correct | 2, 3 correct | 1 only correct |

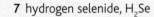

**11** Which of the following molecules would you expect to be linear?
    **1**  $H_2S$
    **2**  HCN
    **3**  $CO_2$

**12** Which of the following, in the solid state, has a crystal structure that contains simple discrete molecules?
    **1**  carbon dioxide
    **2**  silicon dioxide
    **3**  magnesium oxide

**13** Which of the following can accept a lone pair of electrons and form a dative covalent bond?
    **1**  $NH_3$
    **2**  $AlCl_3$
    **3**  $BF_3$

**14** The dot-and-cross diagram below represents the electronic structure of a sulfur hexafluoride molecule.

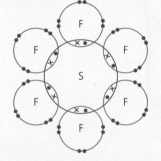

Correct statements about sulfur hexafluoride include:
    **1**  All S–F bonds are equivalent (the same).
    **2**  $SF_6$ is a polar molecule.
    **3**  $SF_6$ is a planar molecule.

**15** This question is about calcium and calcium oxide.
    **a)**  **i)**  Describe the bonding in calcium.
            **ii)**  Explain why calcium is a good conductor of electricity.
    **b)**  Draw 'dot-and-cross' diagrams for the ions in calcium oxide showing **all** the electrons and the ionic charges.
    **c)**  Under what conditions does calcium oxide conduct electricity?
        Explain your answer.        (8)

**16 a)** Use sodium chloride, hydrogen chloride and copper to explain what is meant by covalent, ionic and metallic bonding.

**b)** Compare and explain the conduction of electricity by sodium chloride and copper in terms of structure and bonding. (14)

**17 a)** Phosphorus forms the chloride $PCl_3$. Draw a 'dot-and-cross' diagram for $PCl_3$.

**b)** Draw and name the shape of the $PCl_3$ molecule.

**c)** Explain why $PCl_3$ has this shape.

**d)** How does $PCl_3$ form a stable compound with $BCl_3$? (7)

**18** The table shows the melting points of the elements in Period 3 of the periodic table.

| Element | Na | Mg | Al | Si | P | S | Cl | Ar |
|---|---|---|---|---|---|---|---|---|
| Melting point/°C | 98 | 649 | 660 | 1410 | 44 | 113 | −101 | −189 |

**a)** Explain why the melting point of sodium is much lower than that of magnesium.

**b)** Phosphorus and sulfur exist as molecules of $P_4$ and $S_8$ respectively. Explain their difference in melting points.

**c)** State the type of structure and the nature of the bonding in each of the following elements:

**i)** aluminium

**ii)** silicon

**iii)** chlorine. (12)

**19** When space travel was being pioneered, one of the first rocket fuels was hydrazine, $H_2NNH_2$.

**a)** Draw a 'dot-and-cross' diagram to show the electron structure of a hydrazine molecule.

**b)** Predict the value of the H–N–H bond angle in a hydrazine molecule and explain your reasoning.

**c)** Suggest a possible equation for the reaction that occurs when hydrazine vapour burns in oxygen.

**d)** When 1.00 g of hydrazine burns in excess oxygen, 18.3 kJ of thermal energy is released. Calculate the energy released when 1 mole of hydrazine burns in excess oxygen. (8)

**20** Three of the hydrides of Group 16 elements are $H_2O$, $H_2S$ and $H_2Se$.

**a)** Explain why all three hydrides have polar molecules.

**b)** State and explain the trend in electronegativity from O to Se down Group 16.

**c)** In which of the hydrides of Group 16 are the bonds most polar?

**d)** Which of the hydrides of Group 16 has the highest boiling point? Explain your answer. (7)

**21** Thin streams of some liquids are attracted towards a charged rod, but there is no such effect with other liquids.

**a)** Suggest why some liquids are attracted while others are not.

**b)** Which of the liquids, water, hexane, bromoethane and tetrachloromethane, are attracted towards a charged rod? Explain your answer.

**c)** Why are the affected liquids always attracted towards the charged rod and never repelled by it? (8)

**22** Aluminium chloride is a white covalent solid which sublimes at 180 °C.

**a)** Draw a dot-and-cross diagram of $AlCl_3$.

**b)** Predict the shape, bond angles and polarity of the bonds.

**c)** Determine whether or not a molecule of $AlCl_3$ would be polar or non-polar. Explain your answer.

**d)** Analysis of aluminium chloride suggests that the formula is in fact $Al_2Cl_6$. Draw two molecules of $AlCl_3$ side by side and suggest:

**i)** how $Al_2Cl_6$ could be formed

**ii)** the type of bonding in $Al_2Cl_6$

**iii)** the bond angles in $Al_2Cl_6$

**iv)** whether or not a molecule of $Al_2Cl_6$ would be polar or non-polar. Explain your answer.

**e)** Suggest an explanation for why aluminium chloride sublimes. (13)

## Challenge

**23** Electron pair repulsion theory can be used to predict the shapes of most molecules and ions. Sometimes, it requires careful consideration of how the electrons are distributed. An important point to note is that while atoms of the elements in the second period of the periodic table never have more than eight electrons in their outer orbits, those in the third and subsequent periods often do. Sulfur forms two oxides, $SO_2$ and $SO_3$, and a range of anions, including sulfite, $SO_3^{2-}$, and sulfate, $SO_4^{2-}$. In addition, it forms thio compounds. In a molecule of a thio compound, an oxygen atom is replaced by a sulfur atom. An example is thiosulfate, $S_2O_3^{2-}$.

**a)** In the structures in parts (i) and (ii), the sulfur atom has ten electrons in its outer shell. What is the likely shape of:

   **i)** $SO_2$

   **ii)** $SO_3^{2-}$ (the two electrons forming the '2–' charge must be included in the structure; two of the oxygen atoms each have one extra electron added)?

**b)** In the sulfate ion, the sulfur atom accommodates 12 electrons in its outer orbit. What is the shape of:

   **i)** $SO_4^{2-}$

   **ii)** $S_2O_3^{2-}$?

(In these two ions, two oxygen atoms each have one extra electron added.)

**c)** Another feature of sulfur chemistry is that sulfur atoms can link together. Naturally occurring sulfur consists of $S_8$ molecules, in which the sulfur atoms are linked by single bonds to create a ring.

Draw an $S_8$ ring, giving some indication of its likely shape and bond angles.

**d)** There are various anions containing linked sulfur atoms. An example is the dithionite ion, $S_2O_4^{2-}$. This ion can be thought of as two $SO_2^-$ ions joined by two sulfur atoms to make the $^-O_2S{-}SO_2^-$.

Draw the electron structure of the $S_2O_4^{2-}$ ion and suggest its shape. (15)

# Chapter 8

# The periodic table and periodicity

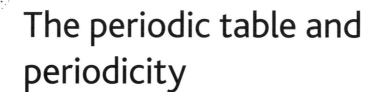

> ## Prior knowledge
>
> *In this chapter it is assumed that you are familiar with:*
> - atomic number, mass number and isotopes
> - structure of the atom
> - electronic configuration
> - bonding and structure.
>
> For example, you should be aware that the elements in the periodic table are arranged in order of increasing atomic number and that their chemical properties are related to the atomic number and the electronic configuration.

> ## Test yourself on prior knowledge
>
> 1 State the atomic number, the mass number and the number of electrons in each shell of a S atom.
> 2 Explain why elements 14 and 32 are in the same group of the periodic table.
> 3 Explain why $^{37}Cl$ and $^{35}Cl$ have the same chemical properties.
> 4 Give the number of electrons in each shell for:
>   a) P
>   b) $Br^-$
>   c) $Al^{3+}$
>   d) $Ni^{2+}$

## Introducing the periodic table and periodicity

In this chapter some features of the periodic table are considered. The periodic table is the arrangement of elements:

- by increasing atomic (proton) number

- in periods showing repeating trends in physical and chemical properties

- in groups having similar chemical properties.

**Key**
atomic number
symbol
relative atomic mass

| 1 | 2 | 3 | 4 | 5 | 6 | 7 | 8 | 9 | 10 | 11 | 12 | 13 | 14 | 15 | 16 | 17 | 18 |
|---|---|---|---|---|---|---|---|---|----|----|----|----|----|----|----|----|----|
| 1<br>H<br>1.0 | | | | | | | | | | | | | | | | | 2<br>He<br>4.0 |
| 3<br>Li<br>6.9 | 4<br>Be<br>9.0 | | | | | | | | | | | 5<br>B<br>10.8 | 6<br>C<br>12.0 | 7<br>N<br>14.0 | 8<br>O<br>16.0 | 9<br>F<br>19.0 | 10<br>Ne<br>20.2 |
| 11<br>Na<br>23.0 | 12<br>Mg<br>24.3 | | | | | | | | | | | 13<br>Al<br>27.0 | 14<br>Si<br>28.1 | 15<br>P<br>31.0 | 16<br>S<br>32.1 | 17<br>Cl<br>35.5 | 18<br>Ar<br>39.9 |
| 19<br>K<br>39.1 | 20<br>Ca<br>40.1 | 21<br>Sc<br>45.0 | 22<br>Ti<br>47.9 | 23<br>V<br>50.9 | 24<br>Cr<br>52.0 | 25<br>Mn<br>54.9 | 26<br>Fe<br>55.8 | 27<br>Co<br>58.9 | 28<br>Ni<br>58.7 | 29<br>Cu<br>63.5 | 30<br>Zn<br>65.4 | 31<br>Ga<br>69.7 | 32<br>Ge<br>72.6 | 33<br>As<br>74.9 | 34<br>Se<br>79.0 | 35<br>Br<br>79.9 | 36<br>Kr<br>83.8 |
| 37<br>Rb<br>85.5 | 38<br>Sr<br>87.6 | 39<br>Y<br>88.9 | 40<br>Zr<br>91.2 | 41<br>Nb<br>92.9 | 42<br>Mo<br>95.9 | 43<br>Tc | 44<br>Ru<br>101.1 | 45<br>Rh<br>102.9 | 46<br>Pd<br>106.4 | 47<br>Ag<br>107.9 | 48<br>Cd<br>112.4 | 49<br>In<br>114.8 | 50<br>Sn<br>118.7 | 51<br>Sb<br>121.8 | 52<br>Te<br>127.6 | 53<br>I<br>126.9 | 54<br>Xe<br>131.3 |
| 55<br>Cs<br>132.9 | 56<br>Ba<br>137.3 | 57–71 | 72<br>Hf<br>178.5 | 73<br>Ta<br>180.9 | 74<br>W<br>183.8 | 75<br>Re<br>186.2 | 76<br>Os<br>190.2 | 77<br>Ir<br>192.2 | 78<br>Pt<br>195.1 | 79<br>Au<br>197.0 | 80<br>Hg<br>200.6 | 81<br>Tl<br>204.4 | 82<br>Pb<br>207.2 | 83<br>Bi<br>209.0 | 84<br>Po | 85<br>At | 86<br>Rn |
| 87<br>Fr | 88<br>Ra | 89–103 | 104<br>Rf | 105<br>Db | 106<br>Sg | 107<br>Bh | 108<br>Hs | 109<br>Mt | 110<br>Ds | 111<br>Rg | 112<br>Cn | | 114<br>Fl | | 116<br>Lv | | |

| 57<br>La<br>138.9 | 58<br>Ce<br>140.1 | 59<br>Pr<br>140.9 | 60<br>Nd<br>144.2 | 61<br>Pm<br>144.9 | 62<br>Sm<br>150.4 | 63<br>Eu<br>152.0 | 64<br>Gd<br>157.2 | 65<br>Tb<br>158.9 | 66<br>Dy<br>162.5 | 67<br>Ho<br>164.9 | 68<br>Er<br>167.3 | 69<br>Tm<br>168.9 | 70<br>Yb<br>173.0 | 71<br>Lu<br>175.0 |
|---|---|---|---|---|---|---|---|---|---|---|---|---|---|---|
| 89<br>Ac | 90<br>Th<br>232.0 | 91<br>Pa | 92<br>U<br>238.1 | 93<br>Np | 94<br>Pu | 95<br>Am | 96<br>Cm | 97<br>Bk | 98<br>Cf | 99<br>Es | 100<br>Fm | 101<br>Md | 102<br>No | 103<br>Lr |

**Figure 8.1** The periodic table.

The International Union of Pure and Applied Chemistry, IUPAC, now recommends that the groups in the periodic table should be numbered 1 to 18.

- Groups 1 and 2 remain the same as before – classified as the s-block.

- The transition elements now become Group 3 to 12 – classified as the d-block.

- Groups 3 to 7 now become Groups 13 to 17 and the noble gases become Group 18 – classified as the p-block.

The blocks are shown in Figure 8.2.

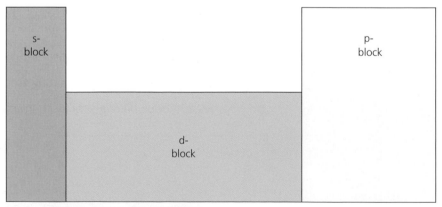

**Figure 8.2** Blocks of the periodic table.

**Key term**

Periodicity is a repeating pattern, in either physical or chemical properties, across different periods.

The vertical columns are called **groups** and the horizontal rows are called **periods**. Trends in the groups (vertical columns) and periods (horizontal rows) reflect the structures of the atoms of the elements within them, and these in turn affect the chemical properties of the elements. These repeating properties demonstrate periodicity.

Similarities within groups are illustrated by considering the properties and chemical reactions of the elements in Group 2 and Group 17 (the halogens).

Elements with the same outer-shell electron configuration are placed in the same group. For example, all elements with a single s electron in the outer shell are in Group 1; all elements with four outer-shell electrons ($s^2p^2$) are in Group 14 which until recently was called Group 4.

The elements within a group have similar but steadily changing properties.

Within a period, the elements go from metals on the left-hand side to non-metals on the right, with the noble gases on the extreme right.

## Trends across a period

Table 8.1 illustrates the trends across Period 3.

**Table 8.1** Trends across Period 3.

| Element | Sodium | Magnesium | Aluminium | Silicon | Phosphorus | Sulfur | Chlorine | Argon |
|---|---|---|---|---|---|---|---|---|
| Atomic radius/nm | 0.190 | 0.160 | 0.130 | 0.118 | 0.110 | 0.102 | 0.099 | 0.095 |
| Type of bonding | Metallic | Metallic | Metallic | Covalent | Covalent | Covalent | Covalent | / |
| Structure | Giant lattice | Giant lattice | Giant lattice | Giant lattice | Simple molecular | Simple molecular | Simple molecular | Simple atomic |
| Electrical conductivity | Good | Good | Good | Poor | Poor | Poor | Poor | Poor |
| Melting point/°C | 98 | 649 | 660 | 1410 | 44* | 113 | −101 | −189 |
| Boiling point/°C | 883 | 1107 | 2467 | 2355 | 281 | 445 | −35 | −186 |
| First ionisation energy/kJ mol⁻¹ | 496 | 738 | 578 | 789 | 1012 | 1000 | 1251 | 1521 |

*White phosphorus

## Trends down a group

The trends down Groups 2 and 17 are shown in Tables 8.2 and 8.3 respectively.

**Table 8.2** Trends down Group 2.

| Group 2 | Beryllium | Magnesium | Calcium | Strontium | Barium |
|---|---|---|---|---|---|
| Atomic radius/nm | 0.125 | 0.160 | 0.174 | 0.191 | 0.198 |
| Ionic radius/nm | 0.027 | 0.072 | 0.100 | 0.113 | 0.136 |
| Melting point/°C | 1278 | 649 | 839 | 769 | 725 |
| Boiling point/°C | 2970 | 1107 | 1484 | 1384 | 1640 |
| First ionisation energy/ kJ mol⁻¹ | 900 | 738 | 590 | 550 | 503 |

**Table 8.3** Trends down Group 17 (the halogens).

| Group 17 | Fluorine | Chlorine | Bromine | Iodine |
|---|---|---|---|---|
| Atomic radius/nm | 0.07 | 0.10 | 0.11 | 0.13 |
| Ionic radius/nm | 0.13 | 0.18 | 0.20 | 0.22 |
| Melting point/°C | −220 | −101 | −7 | 114 |
| Boiling point/°C | −188 | −35 | 59 | 184 |
| First ionisation energy/kJ mol⁻¹ | 1681 | 1251 | 1140 | 1008 |

# Explaining periodic patterns

## Atomic radius

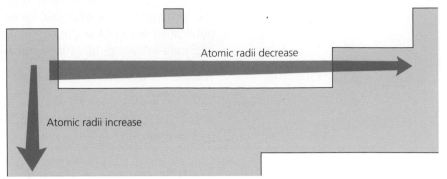

**Figure 8.3** Periodicity of atomic radii in the periodic table.

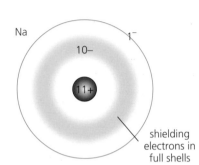

**Figure 8.4** Shielding effect of inner shell electrons reduces the pull of the nucleus in the outer shell.

**Across a period**, the atomic radius decreases from left to right (Figure 8.3).

From one atom of an element to the next across a period:

- the charge of the nucleus increases
- the shielding remains the same

Across the period the effect of the nuclear charge on the outer electrons increases and the atomic radii decreases.

**Down a group**, the atomic radius increases.

From one atom of an element to the next down a group:

- the charge of the nucleus increases
- the shielding increases

Down the group the effect of the nuclear charge on the outer electrons decreases and the atomic radii increases.

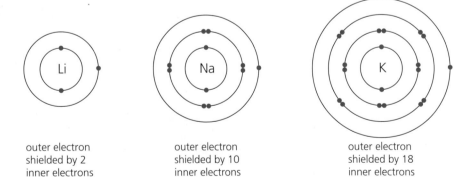

outer electron shielded by 2 inner electrons

outer electron shielded by 10 inner electrons

outer electron shielded by 18 inner electrons

**Figure 8.5** Atoms of the Group 1 elements lithium, sodium and potassium.

### Test yourself

1 Put the following elements in order of increasing atomic radius. Justify your answers:
   a) Mg, S, Si
   b) Mg, K, Al
   c) Si, Cl, K

# Bonding and structure

**Across a period**, bonding changes from metallic to covalent.

**Table 8.4** Trend across Period 3.

|  | Na | Mg | Al | Si | P | S | Cl | Ar |
|---|---|---|---|---|---|---|---|---|
| **Bonding** | metallic | | | covalent | | | | / |
| **Structure** | giant | | | | simple molecular | | | simple atomic |

**Down a group**, bonding changes from covalent to metallic. This is most evident in Group 14, formerly Group 4.

**Table 8.5** Trend down Group 14.

|  | C | Si | Ge | Sn | Pb |
|---|---|---|---|---|---|
| **Bonding** | non-metallic | Si and Ge show properties associated with metals and with non-metals and can be regarded as being **metalloid** | | metallic | |
| **Structure** | giant covalent lattice | | | giant metallic lattice | |

# Electrical conductivity

Electrical conductivity is related to the type of bonding. In the third period, sodium, magnesium and aluminium have metallic bonding. Metallic elements are good conductors because they possess mobile, outer-shell electrons. These electrons are delocalised, which allows metals to conduct heat and electricity, even in the solid state.

The remaining elements across Period 3 are poor conductors because they do not contain any mobile electrons.

Group 13 elements tend to be better conductors than elements in Groups 1 and 2 because Group 13 atoms have three electrons in their outer shells, compared with one and two outer-shell electrons for elements in Groups 1 and 2, respectively.

Graphite is also a good conductor of electricity as it also possesses mobile, outer-shell electrons. These electrons are delocalised, which allows graphite to conduct electricity, even in the solid state, see page 110.

### Test yourself

2 Put the following elements in order of increasing conductivity: Mg, Na, Al. Justify your answer.

# Melting point and boiling point

Melting points are related to the type of structure.

- Giant metallic and giant covalent lattices have strong bonds throughout the lattice, all of which have to be broken. This requires a great deal of energy and hence the melting points are high.

- Simple covalent molecules have weak intermolecular forces, such as induced dipole–dipole interactions forces, and little energy is needed to overcome them. Therefore, the melting points are low.

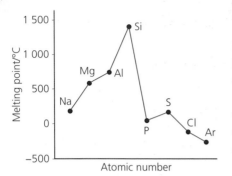

**Figure 8.6** Melting points for elements in Period 3.

Figure 8.6 shows how melting points change across Period 3. Boiling points follow the same pattern as melting points, but at a higher temperature.

Melting points increase from Group 1 to Group 13. These all have giant metallic lattices with strong bonds between all the atoms. Group 13 elements tend to have higher melting points than elements in Groups 1 and 2 because the metallic bond is stronger in Group 13 atoms compared to the elements in Groups 1 and 2 respectively. Silicon (and carbon) have very high melting points because they have very strong covalent bonds throughout the lattice.

In a non-metallic group, such as Group 17 (halogens), the only forces that have to be broken for both melting and boiling to occur are the intermolecular (induced dipole–dipole) forces (page 109), which depend on the number of electrons. As Group 17 (halogens) is descended, the number of electrons increases, so the melting and boiling points also increase.

> **Test yourself**
>
> 3 Explain why boiling points increase down Group 17.
> 4 Put the following elements in order of increasing melting point: Mg, Na, Al. Explain your answer.

# Ionisation energy

The ionisation energy of an element was introduced in Chapter 2 as evidence for the existence of electron subshells. It is possible to measure the energy required to remove electrons from an atom in order to create positive ions. This can be done stepwise. If an atom has seven electrons, seven successive ionisation energies can be measured, as each electron is removed to create ions with a charge of up to 7+.

This can be illustrated by an equation:

$$M(g) \rightarrow M^+(g) + e^-$$

## Trends in first ionisation energies

When removing an electron the attraction between the electron and the nucleus has to be overcome. There are three factors that influence the size of the first ionisation energy of an element.

Factor 1 **Atomic radius** – the further the electron is away from the nucleus the easier it is to remove an electron. This tends to **decrease** the ionisation energy.

Factor 2 **Nuclear charge** – as the number of protons in the nucleus increases its attraction for outer most electrons increases. This tends to **increase** the ionisation energy.

Factor 3 **Shielding effect** – electrons in inner shells exert a repelling effect on the electrons in an outer shell. This reduces the pull of the nucleus on the outer electrons. This reduces the effect of the nucleus and tends to **decrease** the ionisation energy.

### Down a group

Down a group, the ionisation energy **decreases**.

The first ionisation energies (kJ mol$^{-1}$) of the elements in Group 1 of the periodic table are shown in Table 8.6.

**Table 8.6** First ionisation energies of the Group 1 elements.

| Element | Li | Na | K | Rb | Cs |
|---|---|---|---|---|---|
| First ionisation energy/kJ mol$^{-1}$ | 520 | 496 | 419 | 403 | 376 |

Factor 1  **Atomic radius** – increases down the group and the outer electron is further from the nucleus. This tends to **decrease** the ionisation energy as we go down the group.

Factor 2  **Nuclear charge** – increases down the group. This tends to **increase** the ionisation energy as we go down the group.

Factor 3  **Shielding effect** – increases down the group. This reduces the effect of the nucleus. This tends to **decrease** the ionisation energy as we go down the group.

Factors 1 and 3 outweigh factor 2 as ionisation energy decreases down a group.

### Across a period

Across a period, ionisation energy **increases**. This is illustrated in Table 8.7.

**Table 8.7** First ionisation energies of elements in Period 2.

| Element | Li | Be | B | C | N | O | F | Ne |
|---|---|---|---|---|---|---|---|---|
| First ionisation energy / kJ mol$^{-1}$ | 520 | 900 | 801 | 1086 | 1402 | 1314 | 1681 | 2081 |
| Electron structure | $1s^22s^1$ | $1s^22s^2$ full sub shell | $1s^22s^22p^1$ | $1s^22s^22p^2$ | $1s^22s^22p^3$ ½ full sub-shell | $1s^22s^22p^4$ | $1s^22s^22p^5$ | $1s^22s^22p^6$ full shell |

We can also explain this in terms of the three factors.

Factor 1  **Atomic radius** – decreases across the period so the outer electrons are closer to the nucleus. This tends to **increase** the ionisation energy as we go across the period.

Factor 2  **Nuclear charge** – increases across the period. This tends to **increase** the ionisation energy as we go across the period.

Factor 3  **Shielding effect** – generally remains the same across a period and therefore has little effect on the nucleus.

Factors 1 and 2 both indicate that ionisation energy will **increase** across the period.

There is a general increase across a period but there are slight decreases after the second and the fifth element.

- Elements in Group 13 have an electron structure of $s^2p^1$ and the s electrons provide a slightly greater shielding of the p electron, which is therefore lost a little more readily.

- Elements in Group 15 have a half-filled set of p orbitals but in Group 16 there is a pair of p electrons and the repulsion between this pair of electrons is sufficient to make the elements in Group 16 ionise slightly more readily.

## Example

Explain why the first ionisation energy of barium is lower than that of calcium.

**Answer**

Calcium and barium have the same outer electronic structure ($s^2p^2$).

The barium atom has a larger radius than the calcium atom. Therefore, the outer electrons of barium are further away from the nucleus.

The outer electrons of barium are more shielded from the nucleus because there are more intermediate layers of electrons.

The force of attraction of the barium nucleus on the outermost electrons is lower than that in calcium.

Therefore, the first ionisation energy of barium is lower than that of calcium.

## Test yourself

5 For each of the following pairs, state which element has the higher first ionisation energy and explain your answer.
  a) Mg and Na
  b) Mg and Ca
  c) Ne and Na
6 Explain why the first ionisation energy of aluminium, $578\,kJ\,mol^{-1}$, is less than that of magnesium, $738\,kJ\,mol^{-1}$.

# Practice questions

## Multiple choice questions 1–9

Questions 1 to 4 relate to the elements shown on the outline of the periodic table below as A to D. (Note: the letters used A to D are **not** the chemical symbols for the elements concerned.)

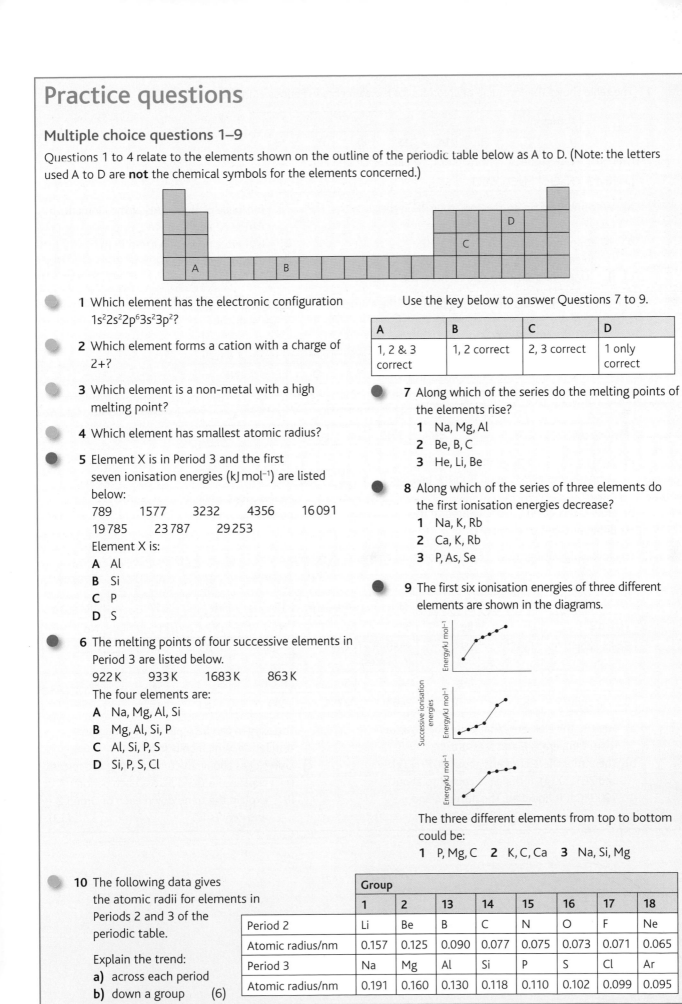

**1** Which element has the electronic configuration $1s^2 2s^2 2p^6 3s^2 3p^2$?

**2** Which element forms a cation with a charge of 2+?

**3** Which element is a non-metal with a high melting point?

**4** Which element has smallest atomic radius?

**5** Element X is in Period 3 and the first seven ionisation energies (kJ mol⁻¹) are listed below:

789    1577    3232    4356    16 091
19 785    23 787    29 253

Element X is:

A  Al
B  Si
C  P
D  S

**6** The melting points of four successive elements in Period 3 are listed below.

922 K    933 K    1683 K    863 K

The four elements are:

A  Na, Mg, Al, Si
B  Mg, Al, Si, P
C  Al, Si, P, S
D  Si, P, S, Cl

Use the key below to answer Questions 7 to 9.

| A | B | C | D |
|---|---|---|---|
| 1, 2 & 3 correct | 1, 2 correct | 2, 3 correct | 1 only correct |

**7** Along which of the series do the melting points of the elements rise?

1  Na, Mg, Al
2  Be, B, C
3  He, Li, Be

**8** Along which of the series of three elements do the first ionisation energies decrease?

1  Na, K, Rb
2  Ca, K, Rb
3  P, As, Se

**9** The first six ionisation energies of three different elements are shown in the diagrams.

The three different elements from top to bottom could be:

1  P, Mg, C    2  K, C, Ca    3  Na, Si, Mg

**10** The following data gives the atomic radii for elements in Periods 2 and 3 of the periodic table.

Explain the trend:
a) across each period
b) down a group    (6)

| Group | | | | | | | | |
|---|---|---|---|---|---|---|---|---|
| | 1 | 2 | 13 | 14 | 15 | 16 | 17 | 18 |
| Period 2 | Li | Be | B | C | N | O | F | Ne |
| Atomic radius/nm | 0.157 | 0.125 | 0.090 | 0.077 | 0.075 | 0.073 | 0.071 | 0.065 |
| Period 3 | Na | Mg | Al | Si | P | S | Cl | Ar |
| Atomic radius/nm | 0.191 | 0.160 | 0.130 | 0.118 | 0.110 | 0.102 | 0.099 | 0.095 |

11 The table shows the formulae of chlorides for the elements in Periods 2 and 3.

| Group | | | | | | | | |
|---|---|---|---|---|---|---|---|---|
| | 1 | 2 | 13 | 14 | 15 | 16 | 17 | 18 |
| Period 2 | LiCl | $BeCl_2$ | $BCl_3$ | $CCl_4$ | $NCl_3$ | $OCl_2$ | FCl | – |
| Period 3 | NaCl | $MgCl_2$ | $AlCl_3$ | $SiCl_4$ | $PCl_3$ | $SCl_2$ | $Cl_2$ | – |

a) Why are there no entries for Group 18 in the table?

b) What periodic pattern is shown by the formulae of these chlorides?

c) i) Draw up a similar table to show the formulae of the oxides of the elements in Periods 2 and 3.

ii) Is there a periodic pattern in the formulae of these oxides? (5)

12 The table below shows the melting points of the elements in Periods 2 and 3 of the periodic table.

| Group | | | | | | | | |
|---|---|---|---|---|---|---|---|---|
| | 1 | 2 | 13 | 14 | 15 | 16 | 17 | 18 |
| Period 2 | Li | Be | B | C | N | O | F | Ne |
| Melting point/°C | 181 | 1278 | 2300 | 3655 | −210 | −218 | −220 | −248 |
| Period 2 | Na | Mg | Al | Si | P | S | Cl | Ar |
| Melting point/°C | 98 | 649 | 660 | 1410 | 44 | 119 | −101 | −189 |

a) What is the general pattern in melting temperatures across periods in the periodic table?

b) How is this general trend related to the different types of elements?

c) What is meant by the term *periodicity*?

d) State two other physical properties that can be described as periodic in relation to the periodic table. (7)

13 The following data gives the **boiling point** for elements in Periods 2 and 13 of the periodic table.

| Group | | | | | | | | |
|---|---|---|---|---|---|---|---|---|
| | 1 | 2 | 13 | 14 | 15 | 16 | 17 | 18 |
| Period 2 | Li | Be | B | C | N | O | F | Ne |
| Boiling point/°C | 1342 | 2970 | 2550 (sublimes) | 4554 | −196 | −183 | −188 | −246 |
| Period 3 | Na | Mg | Al | Si | P | S | Cl | Ar |
| Boiling point/°C | 883 | 1107 | 2467 | 2355 | 280 | 445 | −35 | −186 |

a) Identify the elements that would be a gas at room temperature and pressure.

b) The atmosphere contains about 80% $N_2(g)$ and 20% $O_2(g)$. If the air is cooled to about −200 °C it is liquefied. The nitrogen and the oxygen can be separated by fractional distillation. Which boils first?

c) Using ideas about structure, bonding and forces:

i) explain the trend across each period

ii) explain the trend down each of Group 2 and Group 17. (13)

**14** The graph shows the boiling points of eight elements in Periods 2 and 3. The sequence of the eight elements follows the order in which they appear in the periodic table.

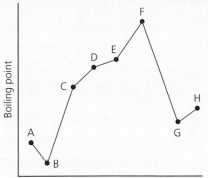

Elements in sequence from Periods 2 and 3

  a) Identify element F. Justify your answer.
  b) Explain the difference in boiling points between C and D.
  c) Identify B and explain why its boiling point is so low. (7)

**15** The graph shows the first and second ionisation energies of the elements lithium to sodium.

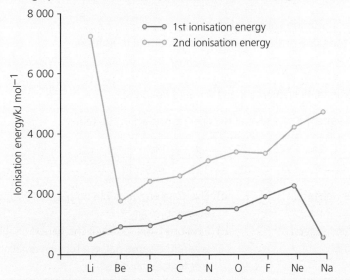

  a) Write an equation to illustrate the first ionisation energy of carbon.
  b) Write an equation to illustrate the second ionisation energy of nitrogen.
  c) Explain why the second ionisation energy is greater than the first ionisation energy for all elements.
  d) Explain why the difference between first and second ionisation energies is much larger for Li and for Na than it is for any of the other elements.
  e) Explain the general trend in first ionisation energies from Li to Ne.
  f) Explain why the trend in second ionisation energies from Be to Na is similar to the trend in first ionisation energies from Li to Ne.
  g) Why is the second ionisation energy of Li much greater than the second ionisation energy of Na? (12)

## Challenge

16 Two elements are adjacent to each other in a row of the periodic table.
Their successive ionisation energies are shown in the graph below.

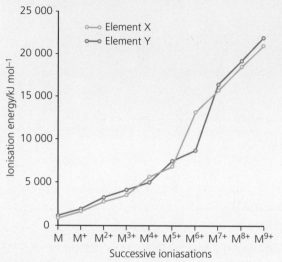

On the $x$-axis, each point represents the ionisation energy for that ion.
So, for example, the point at $M^{4+}$ gives the value of the ionisation energy for the ionisation:

$$M^{4+}(g) \rightarrow M^{5+}(g) + e^-$$

a) Decide in which groups of the periodic table elements X and Y would be found.

Element Y has an atomic mass of 127. Element X has several isotopes, as shown in the table.

| Mass number | 120 | 122 | 123 | 124 | 125 | 126 | 128 | 130 |
|---|---|---|---|---|---|---|---|---|
| % abundance | 0.1 | 2.5 | 0.9 | 4.6 | 7.0 | 18.7 | 31.7 | 34.5 |

b) Use this information to calculate the weighted mean atomic mass of element X.

c) Compare your answers to (a) and (b) and suggest what is unusual about elements X and Y.

d) Use the periodic table on page 116 to identify elements X and Y.

e) Find one other place in the periodic table where the same pattern for two adjacent elements is observed.

(8)

# Chapter 9

# Group 2 and the halogens, qualitative analysis

**Prior knowledge**

*In this chapter it is assumed that you are familiar with:*
- the periodic table and periodicity
- electronic configuration
- redox reactions.

The elements in the periodic table are arranged in order of their atomic number and they are grouped together in terms of their chemically properties. Metallic groups tend to occur on the left-hand side of the periodic table and groups on the right-hand side tend to show properties of non-metals.

**Test yourself on prior knowledge**

1 What is the electron structure for these elements?
   **a)** C    **b)** V    **c)** As
2 Identify the following atoms or ions:
   **a)** X has electron structure $1s^22s^22p^63s^23p^5$
   **b)** $Y^{2-}$ has electron structure $1s^22s^22p^63s^23p^6$
   **c)** $Z^{2+}$ has electron structure $1s^22s^22p^63s^23p^64s^23d^{10}4p^6$
3 What is the oxidation number of:
   **a)** Cr in $K_2CrO_4$
   **b)** O in $K_2O_2$
   **c)** H in $LiAlH_4$?

**Figure 9.1** Relative sizes of the atoms and ions of Group 2 elements.

## Group 2

### Electron configuration

Atoms of Group 2 elements have two electrons in their outer shell and readily form $2^+$ ions (cations) that have the same electron configuration as a noble gas.

The electron configurations of the elements are shown in Table 9.1.

**Table 9.1** Electron configurations of the Group 2 elements.

| Group 2 element | Full electron configuration | Shorthand configuration |
|---|---|---|
| Beryllium, $_4$Be | $1s^22s^2$ | $[He]2s^2$ |
| Magnesium, $_{12}$Mg | $1s^22s^22p^63s^2$ | $[Ne]2s^2$ |
| Calcium, $_{20}$Ca | $1s^22s^22p^63s^23p^64s^2$ | $[Ar]2s^2$ |
| Strontium, $_{38}$Sr | $1s^22s^22p^63s^23p^63d^{10}4s^24p^6\,5s^2$ | $[Kr]2s^2$ |
| Barium, $_{56}$Ba | $1s^22s^22p^63s^23p^63d^{10}4s^24p^64d^{10}5s^25p^66s^2$ | $[Xe]2s^2$ |

**Test yourself**

1 Explain why the ionic radius of a Group 2 ion is smaller than the atomic radius of the corresponding Group 2 atom.
2 The graph shows the sum of the first and the second ionisation energies of the Group 2 metals.

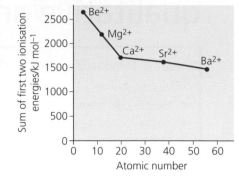

Explain the trend shown by the graph.
3 Explain why Group 2 ions in any period are smaller than the Group 1 ion in the same period.

**Tip**

Questions are often set about conductivity. Remember that electricity is conducted in metals by mobile electrons; in ionic compounds that are molten or in aqueous solution, it is conducted by mobile ions.

**Tip**

Isoelectronic with a noble gas means that the ions have the same number of electrons as a noble gas.

# Physical properties

The elements in Group 2 are all metals. Therefore, they are good conductors of electricity and have high melting and boiling points. With the exception of beryllium, generally they form colourless ionic compounds that also have high melting and boiling points. Their compounds are good conductors of electricity when molten or in aqueous solution, but poor conductors when solid.

# Chemical properties

Atoms of Group 2 elements react by losing two electrons. They give up their two s electrons to form $M^{2+}$ ions (where M represents Mg, Ca, Sr or Ba) which are isoelectronic with a noble gas.

$$M \rightarrow M^{2+} + 2e^-$$

It follows that Group 2 elements are reducing agents.

## Trend in reactivity

Reactivity increases down the group such that Ba is the best reducing agent. The reactivity can be explained in terms of the ease with which they lose electrons. This has been covered in Chapter 8 on page 120.

## Reaction with oxygen

Group 2 elements undergo redox reactions with $O_2$. Apart from beryllium the Group 2 metals burn brightly in oxygen to form ionic oxides, $M^{2+}O^{2-}$.

● Magnesium burns, emitting a bright white light:

$$Mg(s) + \frac{1}{2}O_2(g) \rightarrow MgO(s)$$

- Calcium burns with a brick-red flame:

$$Ca(s) + \tfrac{1}{2}O_2(g) \rightarrow CaO(s)$$

- Strontium burns with a crimson flame:

$$Sr(s) + \tfrac{1}{2}O_2(g) \rightarrow SrO(s)$$

- Barium burns with a green flame:

$$Ba(s) + \tfrac{1}{2}O_2(g) \rightarrow BaO(s)$$

The oxidation number of the Group 2 element increases from 0 to +2, hence it is oxidised; the oxidation number of oxygen decreases from 0 to −2, hence it is reduced:

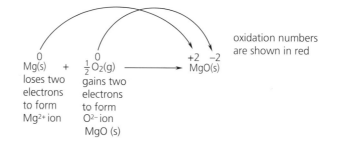

The Group 2 element is a reducing agent.

The oxides of Group 2 elements, apart from BeO, are all white solids with high melting points that react with water to form solutions of the corresponding hydroxide. The oxides are all basic oxides and react with acids to produce salts.

Table 9.2 shows the melting points of the oxides.

**Table 9.2** Melting points of the Group 2 oxides.

| Group 2 oxide | MgO | CaO | SrO | BaO |
|---|---|---|---|---|
| Melting point/°C | 2852 | 2614 | 2430 | 1918 |

MgO has a very high melting point and is used as a refractory ceramic to line furnaces.

CaO is known as 'quicklime'; it reacts rapidly with water to produce calcium hydroxide and is used to reduce acidity in soil.

## Reaction with water

Group 2 elements undergo redox reactions with water to produce the hydroxide and hydrogen. The rate of reaction increases down the group.

The general equation for Group 2 metals reacting with water is:

$$M(s) + 2H_2O(l) \rightarrow M(OH)_2(aq) + H_2(g) \quad \text{where M is Ca, Sr or Ba.}$$

Magnesium reacts very slowly with water and the $Mg(OH)_2$ is formed, but it is formed as a white precipitate.

$$Mg(s) + 2H_2O(l) \rightarrow Mg(OH)_2(s) + H_2(g)$$

Magnesium reacts more readily on heating with steam to form the oxide:

$$Mg(s) + 2H_2O(g) \rightarrow MgO(s) + H_2(g)$$

The oxidation number of the Group 2 element increases from 0 to +2; the oxidation number of hydrogen changes from +1 to 0:

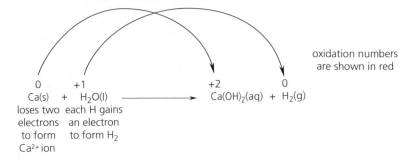

oxidation numbers are shown in red

$$\underset{\substack{0 \\ \text{loses two} \\ \text{electrons} \\ \text{to form} \\ Ca^{2+}\text{ion}}}{Ca(s)} + \underset{\substack{+1 \\ \text{each H gains} \\ \text{an electron} \\ \text{to form } H_2}}{H_2O(l)} \longrightarrow \underset{+2}{Ca(OH)_2(aq)} + \underset{0}{H_2(g)}$$

## Reaction with chlorine

All Group 2 metals react with chlorine to form white chlorides.

$M(s) + Cl_2(g) \rightarrow MCl_2(s)$ where M is Mg, Ca, Sr or Ba.

## Reaction with dilute acids

All Group 2 metals react with the dilute acids to produce salts and hydrogen gas.

$M(s) + 2HCl(g) \rightarrow MCl_2(aq) + H_2(g)$  where M is Mg, Ca, Sr or Ba.

> **Test yourself**
>
> 4 Write balanced equations for the reactions of:
>   a) calcium with oxygen
>   b) barium with oxygen
>   c) strontium with chlorine
>   d) barium with water.
> 5 Show that calcium acts as a reducing agent when it reacts with:
>   a) oxygen
>   b) water.

# Reactions of Group 2 compounds

## Oxides

All Group 2 metal oxides , except beryllium oxide, BeO, are basic oxides and react with:

- water to form hydroxides

- acids to form salts.

The general equation for the reaction of the oxide with water is:

$MO(s) + H_2O(l) \rightarrow M(OH)_2(aq)$  where M is Ca, Sr or Ba.

Magnesium hydroxide is sparingly soluble and forms a white suspension known as 'milk of magnesia' which can be used as an antacid to treat indigestion.

Calcium oxide reacts vigorously with water to form a solution of calcium hydroxide which is known as limewater.

The reactions of Group 2 oxides with water are *not* redox reactions. The oxidation numbers of all the elements remain the same, for example:

$$\underset{CaO(s)}{\overset{+2\ -2}{}} + \underset{H_2O(l)}{\overset{+1\ -2}{}} \longrightarrow \underset{Ca(OH)_2(aq)}{\overset{+2\ -2+1}{}}$$

each H gains an electron to form $H_2$

The resulting hydroxide solutions are alkaline and have pH values in the range 8–12. The actual pH value depends on the concentration of the solution and the solubility of the Group 2 hydroxide. Solubility of the hydroxides increases down the group and it follows that for saturated solutions the alkalinity will also increases down the group.

The general equation for the reaction of the oxide with an acid is:

$$MO(s) + 2HCl(aq) \rightarrow MCl_2(aq) + H_2O(l) \quad \text{where M is Mg, Ca, Sr or Ba.}$$

## Carbonates

The carbonates of Group 2 all have the general formula $MCO_3$. They are insoluble in water and all Group 2 carbonates decompose on heating to form oxides and carbon dioxide. The general equation for the reaction is:

$$MCO_3(s) \xrightarrow{\text{heat}} MO(s) + CO_2(g) \quad \text{where M is Mg, Ca, Sr or Ba.}$$

The ease with which the carbonates decompose decreases down the group. Beryllium carbonate, $BeCO_3$, is so unstable that it does not exist at room temperature. Barium carbonate, $BaCO_3$, requires strong heating to bring about the decomposition. The thermal decomposition of Group 2 carbonates is not a redox reaction. The oxidation number of all the elements remains the same, for example:

$$\underset{MgCO_3(s)}{\overset{+2\ +4\ -2}{}} \longrightarrow \underset{MgO(s)}{\overset{+2-2}{}}$$

Calcium carbonate is used as an antacid to treat indigestion.

## Sulfates

The sulfates have the general formula $MSO_4$. They are all white solids whose solubility in water decreases down the group.

Epsom salts consist of hydrated magnesium sulfate, $MgSO_4.7H_2O$, which is a laxative.

Barium sulfate, $BaSO_4$, is insoluble in water and can be used as a test for a sulfate.

$$BaCl_2(aq) + Na_2SO_4(aq) \rightarrow BaSO_4(s) + 2NaCl(aq)$$
white
precipitate

Barium sulfate also absorbs X-rays strongly so it is used in 'barium meals' to help diagnose disorders of the stomach and intestines. Soluble barium compounds are toxic, but barium sulfate is insoluble so it is not absorbed. X-rays cannot pass through the 'barium meal', which therefore creates a shadow on the X-ray film.

## Activity

### Limestone and its uses

Calcium carbonate occurs naturally as limestone (Figure 9.2). Limestone is an important mineral. Some of the rock is quarried for road building and construction. Pure limestone is used in the chemical industry.

Heating limestone in a furnace at 1200 K converts it to calcium oxide (quicklime). The reaction of quicklime with water produces calcium hydroxide (slaked lime).

1 Name two other minerals that consist largely of calcium carbonate.
2 Suggest a reason for grinding up limestone lumps before heating them with sodium carbonate and sand to make glass.
3 Write an equation for the decomposition of calcium carbonate on heating in a furnace.
4 With the help of equations, show that both calcium oxide and calcium hydroxide can be used to increase the pH in soils that are too acidic.
5 Lime mortar was used in older buildings. It is a mixture of slaked lime, sand and water. It sets slowly by reacting with carbon dioxide in the air. Identify the main product of the reaction of slaked lime with carbon dioxide.
6 A suspension of calcium hydroxide in water is used as an industrial alkali. Suggest why a suspension and not just a solution of the hydroxide is used.
7 What is the laboratory use of a solution of calcium hydroxide?

**Figure 9.2** Limestone cliff at Malham Cove in Yorkshire. Limestone gives rise to attractive scenery.

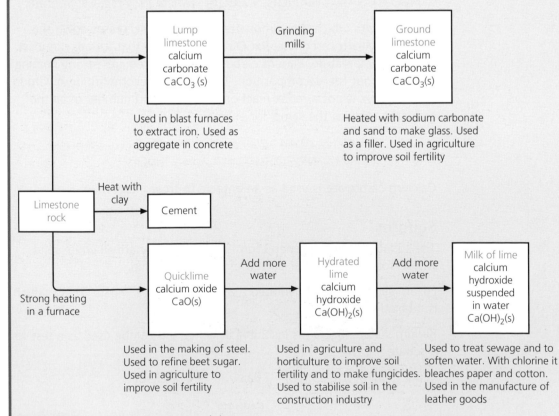

**Figure 9.3** Products from limestone and their uses.

# The halogens (Group 17)

Fluorine, chlorine, bromine and iodine belong to the family of halogens. All four are reactive non-metals readily forming halide anions, $X^-$. The reactivity increases up the group such that fluorine and chlorine are particularly hazardous.

The name 'halogen' means 'salt-former' and halides occur naturally as salts with metals. The halogens are important economically as the ingredients of plastics, pharmaceuticals, anaesthetics, dyestuffs and chemicals for water treatment.

All the halogens consist of diatomic molecules, $X_2$, linked by a single covalent bond. The halogen molecules are non-polar.

The halogens have similar chemical properties because they all have seven electrons in the outer shell – one less than the next noble gas in Group 18.

The halogens (Group 17), like all groups in the periodic table, show a range of trends including atomic radius (increases down the group) and ionisation energy (decreases down the group). Reactivity, however, increases up the group which is in contrast to Group 2. When halogens react they gain an electron whereas Group 2 elements lose electrons.

## Electron configuration

Atoms of Group 17 elements have seven electrons in their outer shell and readily form a 1– ion (anion) that has the same electron configuration as a noble gas. The electron configurations of the halogens are shown in Table 9.2.

**Table 9.2** Electron configuration of the Group 17 elements.

| Halogen | Full electron configuration |
| --- | --- |
| Fluorine, $_9$F | $1s^2 2s^2 2p^5$ |
| Chlorine, $_{17}$Cl | $1s^2 2s^2 2p^6 3s^2 3p^5$ |
| Bromine, $_{35}$Br | $1s^2 2s^2 2p^6 3s^2 3p^6 4s^2 4p^5$ |
| Iodine, $_{53}$I | $1s^2 2s^2 2p^6 3s^2 3p^6 3d^{10} 4s^2 4p^6 4d^{10} 5s^2 5p^5$ |

# Characteristic physical properties

The elements in Group 17 are non-metals and, hence, are poor conductors.

At room temperature fluorine, $F_2$, is a yellow gas, chlorine, $Cl_2$, is a green gas, bromine, $Br_2$, is a dark red/brown liquid and iodine, $I_2$, is a grey/black solid which sublimes to form a purple/violet gas.

**Figure 9.4** Chlorine is a toxic green gas.

**Figure 9.5** Bromine is a dark red/brown liquid – it is very volatile and gives off a toxic brown/orange vapour.

**Figure 9.6** Iodine is a shiny grey/black solid but when warmed gently sublimes to give a toxic purple vapour.

Descending the group from fluorine to iodine, there is an increase in induced dipole-dipole interactions corresponding to the increased number of electrons in the molecules. This increase in induced dipole-dipole interactions reduces volatility. Therefore, the boiling points (and melting points) increase down the group as shown below (Figure 9.7).

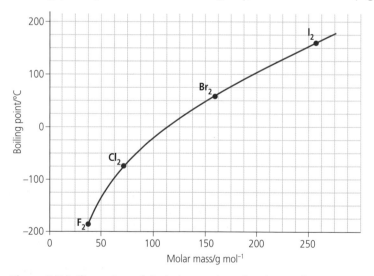

**Figure 9.7** Boiling points of the halogens plotted against molar mass.

Fluorine is the most electronegative of all elements (see page 105) and it is a powerful oxidising agent. Its oxidation state is −1 in all its compounds. Uses of fluorine include the manufacture of a wide range of compounds consisting of only carbon and fluorine (fluorocarbons). The most familiar of these is the very slippery, non-stick polymer poly(tetrafluorethene) or Teflon®.

Chlorine reacts directly with most elements. In its compounds, chlorine is usually present in the −1 oxidation state, but it can be oxidised to positive oxidation states by oxygen and by fluorine. Chlorine is used in the production of polymers such as polyvinylchloride (PVC). Water companies use chlorine to kill bacteria in drinking water. It is used to bleach paper and textiles.

Bromine, like the other halogens, is a reactive element, but it is a less powerful oxidising agent than chlorine. Bromine is used to make a range of products including flame retardants and medicines.

Iodine is also an oxidising agent but less powerful than the other halogens. Iodine is also used in a range of products including medicines, dyes and catalysts. Iodine is also needed in our diet so that the thyroid gland can make thyroxine, a hormone which regulates growth and metabolism. Foods such as sea-weed, yogurt, milk and eggs provide a source of iodine but in many regions sodium iodide, NaI, is added to table salt to supplement the iodine in the diet.

### Test yourself

9 Explain the variation in states of the halogens. Predict the state of astatine, give your reasons.
10 Write the full electronic configuration of:
   a) a chlorine atom
   b) a bromide ion.
11 Explain why:
   a) atomic radii of halogens increase down the group
   b) ionic radii of halides are larger than atomic radii of the corresponding halogen.
12 Draw dot-and-cross diagrams of the following molecules:
   a) fluorine
   b) hydrogen bromide
   c) iodine monochloride, ICl.
13 Explain why $Cl_2(g)$ is a non-polar molecule but $HCl(g)$ is a polar molecule.

## Chemical properties

The halogens in Group 17 react by **gaining** electrons to form halide anions. The ease with which the electron is gained by a halogen atom decreases down the group. This is because atomic radius and shielding increase down the group and this reduces the attraction of the nucleus for electrons.

Fluorine is the most reactive halogen. It readily gains electrons and is a powerful oxidising agent. Iodine is the least reactive halogen. In order of decreasing activity, halide ions are formed thus:

$$F_2 + 2e^- \rightarrow 2F^-$$
$$Cl_2 + 2e^- \rightarrow 2Cl^-$$
$$Br_2 + 2e^- \rightarrow 2Br^-$$
$$I_2 + 2e^- \rightarrow 2I^-$$

### Tip

Make sure that you know the difference between a halogen and a halide. Many students confuse chlorine with chloride.

135

Halogens react with most other elements to form halides.

They react directly with:

metals:
$$2Na(s) + Cl_2(g) \rightarrow 2NaCl(s)$$
$$Fe(s) + Cl_2(g) \rightarrow FeCl_2(s)$$

non-metals:
$$Si(s) + 2Cl_2(g) \rightarrow SiCl_4(l)$$
$$P_4(s) + 10Cl_2(g) \rightarrow 4PCl_5(l)$$

Chlorine, however, does not react directly with carbon, oxygen or nitrogen.

## Displacement reactions

A halogen ($F_2$, $Cl_2$ or $Br_2$) displaces a heavier halide ($Cl^-$, $Br^-$ or $I^-$) from one of its salts (Table 9.3).

**Table 9.3** Displacement reactions.

| | Fluoride, $F^-$ | Chloride, $Cl^-$ | Bromide, $Br^-$ | Iodide, $I^-$ |
|---|---|---|---|---|
| Fluorine, $F_2$ | | ✓ | ✓ | ✓ |
| Chlorine, $Cl_2$ | ✗ | | ✓ | ✓ |
| Bromine, $Br_2$ | ✗ | ✗ | | ✓ |
| Iodine, $I_2$ | ✗ | ✗ | ✗ | |

- Fluorine displaces $Cl^-$, $Br^-$ and $I^-$ from solution.
- Chlorine displaces $Br^-$ and $I^-$ from solution.
- Bromine displaces $I^-$ from solution.
- Iodine cannot displace any of the above halides.

These displacement reactions illustrate the decrease in oxidising power down the group. Chlorine oxidises both bromide and iodide ions to form orange-brown bromine and in aqueous solution iodine which has a brown-black colour, respectively. Addition of an organic solvent produces a distinctive violet colour with iodine.

The halogens dissolve readily in non-polar solvents such as hexane, which is immiscible with water and forms two separate layers. When dissolved in hexane the solutions have the same colour as the halogen vapour. Iodine in hexane has a very distinctive purple/violet colour (Figure 9.8).

The oxidation reactions of chlorine are represented by the equations:

$$Cl_2(g) + 2Br^-(aq) \rightarrow 2Cl^-(aq) + Br_2(l)$$
$$Cl_2(g) + 2I^-(aq) \rightarrow 2Cl^-(aq) + I_2(s)$$

**Figure 9.8** Bromine (left) and iodine (right) have distinctive colours when dissolved in the upper organic layer – the colours of the halogens are used to identify if a reaction has occurred or not.

Bromine oxidises I⁻ only.

The oxidation of iodide ions by bromine is represented by the equation:

$$Br_2 + 2I^- \rightarrow 2Br^- + I_2$$

Iodine does not oxidise either chloride or bromide ions.

Each displacement reaction is a redox reaction. The halogen higher in the group gains electrons (is reduced) to form the corresponding halide.

## Uses of chlorine

Chlorine is used in water treatment. It reacts reversibly with water and the resultant mixture kills bacteria.

$$Cl_2(aq) + H_2O(l) \rightleftharpoons HCl(aq) + HClO(aq)$$

This reaction is a redox reaction, but it is unusual in that chlorine undergoes both oxidation and reduction:

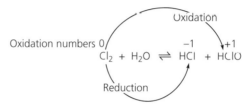

**Key term**

**Disproportionation** occurs when the same element both increases and decreases its oxidation numbers so that the element is simultaneously oxidised and reduced.

One chlorine atom in the $Cl_2$ molecule is oxidised. Its oxidation number changes from 0 to +1. The other chlorine atom is reduced. Its oxidation number changes from 0 to −1. This type of reaction is called disproportionation.

## Benefits and risks of water treatment

Chlorine is added to drinking water to kill micro-organisms such as bacteria and viruses. This greatly reduces the risk of waterborne diseases such as cholera and typhoid fever. The chloric(I) acid molecules are able to pass through the cell walls of bacteria and once inside the bacterial cells, the HClO molecules break up and kill the organism by oxidising and chlorinating molecules that make up the structure of the cells.

Disinfection with chlorine can produce by-products that may be harmful. Chlorine can also react with organic matter in the water. This produces traces of chlorinated hydrocarbons such as trichloromethane, $CHCl_3$. The amount produced is very small but some studies suggest an increased risk of bladder cancer and possibly colon cancer in people who drank chlorinated water for 35 years or more.

However, the interpretation of the evidence remains controversial, and the links between trichloromethanes and cancer are not firmly established. Whatever these risks, the benefits of chlorination to prevent waterborne disease are widely accepted to be much greater than the small health risks from trihalomethanes.

**Water treatment in swimming pools**

Swimming pools can be sterilised with chlorine compounds that produce chloric(I) acid when they dissolve in water. Chloric(I) acid is a weak acid so it does not easily ionise. The concentration of un-ionised HClO in a solution depends on the pH, as shown in the graph in Figure 9.9.

Swimming pool managers have to check the pH of the water carefully. They aim to keep the pH in the range 7.2–7.8.

1 Write an equation to show the formation of chloric(I) acid when chlorine reacts with water.

2 Swimming pools were once treated with chlorine gas from cylinders that contained the liquefied gas under pressure. Nowadays they are usually treated with chemicals that produce chloric(I) acid when they are added to water. Suggest reasons for this change.

3 Explain why sodium chlorate(I) produces chloric(I) acid when it is added to water at pH 7–8.

4 Explain why the pH of water in swimming pools must not be allowed to increase above 7.8.

5 Suggest why pool water must not become acidic, even though this would increase the concentration of un-ionised HClO.

6 Nitrogen compounds, including ammonia, urea and proteins, react with HClO to form chloramines, which are irritating to skin and eyes. Chloramines formed from ammonia can react with themselves to form nitrogen and hydrogen chloride, which gets rid of the problem. However, if there is excess HClO, another reaction produces nitrogen trichloride, which is responsible for the so-called 'swimming pool smell' and is very irritating to the eyes.

Write equations to show:
a) the formation of the chloramine, $NH_2Cl$, from ammonia and chloric(I) acid
b) the removal of chloramines by the reaction of $NH_2Cl$ with $NHCl_2$
c) the formation of nitrogen trichloride from chloramine, $NH_2Cl$.

7 Explain why swimming pools do not smell of chlorine if properly maintained.

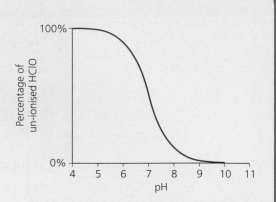

**Figure 9.9** Concentration of chloric(I) acid, HClO, varies over a range of pH values at 20 °C.

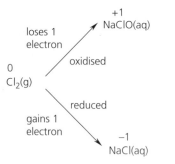

**Figure 9.10** Chlorine compounds treat the water in swimming pools.

Chlorine also reacts with sodium hydroxide to form bleach, which is a mixture of sodium chloride and sodium chlorate(I). This is also a disproportionation reaction of chlorine:

$$Cl_2(g) + 2NaOH(aq) \rightarrow NaCl(aq) + NaClO(aq) + H_2O$$

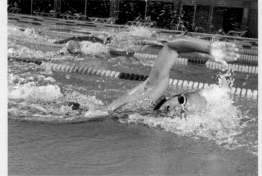

The changes in oxidation numbers of chlorine are shown below:

Chlorine reacts with NaOH(aq) to form a number of different chlorates including sodium chlorate(I), NaClO, sodium chlorate(III), $NaClO_2$, and sodium chlorate(V), $NaClO_3$. In each of these reactions NaCl and water are also formed. Each reaction is a disproportionation reaction.

## Example

When chlorine reacts with a hot concentrated solution of NaOH(aq), sodium chlorate(V), $NaClO_3$(aq) is formed. Construct an equation for this reaction.

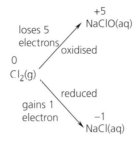

The 5 electrons lost must be counter balanced by 5 electrons being gained.

We, therefore need 5 NaCl where each Cl gains 1 electron

### Answer

We know that NaCl(aq) will be formed along with $NaClO_3$(aq). The oxidation changes in Cl are −1 in NaCl and +5 in $NaClO_3$, as shown below:

We now know that the products will contain $1NaClO_3$(aq), 5NaCl(aq) and $H_2O$(l) such that the balanced equation can be constructed to give:

$$3Cl_2(g) + 6NaOH(aq) \rightarrow NaClO_3(aq) + 5NaCl(aq) + 3H_2O(l)$$

It is important to remember that oxidation number changes have to be balanced as well as balancing symbols.

Balancing redox equations will play an important part of Module 5 in the A level course.

**Figure 9.11** Precipitates (from left to right) of silver chloride, silver bromide and silver iodide.

## Testing for halide ions

Silver chloride, silver bromide and silver iodide are insoluble in water. Therefore the presence of chloride, bromide and iodide ions can be detected by the addition of an aqueous solution of silver nitrate, $AgNO_3$(aq). The test solution is first acidified with dilute nitric acid and silver nitrate is then added. Each of the silver halides forms a different coloured precipitate (Figure 9.11). The identity of the precipitate is confirmed by its solubility in ammonia.

- Silver chloride is a white precipitate, which quickly turns grey in sunlight:

  $Ag^+$ (aq) + $Cl^-$ (aq) $\rightarrow$ AgCl(s)

- Silver bromide is a cream precipitate, which rapidly darkens in sunlight:

  $Ag^+$ (aq) + $Br^-$ (aq) $\rightarrow$ AgBr(s)

- Silver iodide is a yellow precipitate:

  $Ag^+$ (aq) + $I^-$(aq) $\rightarrow$ AgI(s)

The colour changes are not very distinctive but a further test with ammonia helps to distinguish the precipitates. Silver chloride dissolves in dilute ammonia, silver bromide dissolves in concentrated ammonia and silver iodide doesn't dissolve in ammonia at all (Table 9.4).

**Table 9.4** Tests for halides.

| Halide | Effect of adding $AgNO_3$(aq) | Effect of adding ammonia |
|---|---|---|
| Chloride | White precipitate | Precipitate dissolves in dilute $NH_3$ to give a colourless solution |
| Bromide | Cream precipitate | Precipitate dissolves in concentrated $NH_3$ to give a colourless solution |
| Iodide | Yellow precipitate | Precipitate does not dissolves in $NH_3$ (dilute or concentrated) |

**Test yourself**

14 Describe the colour changes on adding:
   a) a solution of chlorine in water to aqueous sodium bromide
   b) a solution of bromine in water to aqueous potassium iodide.
15 Write ionic equations for the reactions of aqueous chlorine with:
   a) bromide ions
   b) iodide ions.
16 Put the chloride, bromide, and iodide ions in order of their strength as reducing agents, with the strongest reducing agent first. Explain your answer.
17 Write ionic equations for the reactions of silver nitrate solution with:
   a) potassium iodide solution
   b) sodium bromide solution.
18 a) Use oxidation numbers to write a balanced equation for the reaction of iodine with hot aqueous hydroxide ions to form $IO_3^-$ and $I^-$ ions.
   b) Show that this is a disproportionation reaction.

# Practice questions

## Multiple choice questions 1–10

Questions 1 to 5 concern the graphs shown below.

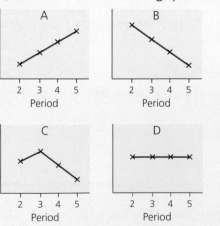

The graphs represent the approximate variations in properties of the elements down Groups 2 and/or 17.

1 Which one represents the oxidation state of all Group 2 elements when they form compounds?

2 Which one represents the ionic radii of Group 17 elements?

3 Which one represents the thermal stability of Group 2 carbonates ?

4 Which one represents the oxidising power of the halogens ?

5 Which one represents the strength of the hydride bond in the hydrogen halides of Group 17?

The answers to Questions 6 to 8 refer to one of the following compounds.

   A  sodium fluoride
   B  magnesium chloride
   C  potassium bromide
   D  barium iodide

6 Which one of the substances when dissolved in water would give a yellow precipitate when a few drops of aqueous silver nitrate was added to it?

7 Which one of the substances when dissolved in water would give a white precipitate when dilute sulfuric acid was added to it?

**8** Which one of the substances when dissolved in water reacts with an aqueous solution of bromine?

Use the key below to answer Questions 9 and 10.

| A | B | C | D |
|---|---|---|---|
| 1, 2 & 3 correct | 1, 2 correct | 2, 3 correct | 1 only correct |

**9** The highest occupied energy level of the Group 17 element astatine, At, is $6s^2 6p^5$. Which of the following statements are likely to be correct?

1 Silver astatide, AgAt, will be soluble in concentrated ammonia.
2 If astatine, $At_2$, is added to a solution of sodium bromide, the solution turns orange/brown.
3 Astatine, $At_2$, is a solid at room temperature and pressure.

**10** In which of the following reactions would you expect to see an effervescence?

1 a solution of sulfuric acid is added to a solution of magnesium carbonate
2 calcium is added to a solution of hydrochloric acid
3 a solution of potassium chloride is added to a solution of fluorine.

**11** The elements Mg to Ba in Group 2, and their compounds, can be used to show the trends in properties down a group of the periodic table. State and explain the trend down the group in:
**a)** atomic radius
**b)** first ionisation energy
**c)** reactivity of the metals. (8)

**12 a)** Explain, with the aid of an equation, the use of calcium hydroxide in agriculture.
**b)** Explain, with the aid of an equation, the use of magnesium hydroxide in some indigestion tablets as an antacid. (4)

**13** Radium is a highly radioactive element. Use your knowledge of the chemistry of the elements Mg to Ba in Group 2 to predict properties of radium and its compounds in terms of the following. Include descriptions of any chemicals changes, equations and suggest the appearance of the products in your predictions.
**a)** Reaction of radium with oxygen
**b)** Reaction of radium with water
**c)** Reaction of radium oxide with water
**d)** Reaction of radium hydroxide with dilute hydrochloric acid

**e)** Solubility of radium sulfate in water
**f)** Effect of heating radium carbonate (17)

**14** Identify the element that disproportionates in each of the following reactions by giving the oxidation states of the element before and after the reaction.
**a)** $2H_2O_2(aq) \rightarrow 2H_2O(l) + O_2(g)$
**b)** $Cl_2(aq) + 2NaOH(aq) \rightarrow NaCl(aq) + NaClO(aq) + H_2O(l)$
**c)** $Cu_2O(s) + H_2SO_4(aq) \rightarrow CuSO_4(aq) + Cu(s) + H_2O(l)$
**d)** $3MnO_4^{2-}(aq) + 4H+(aq) \rightarrow 2MnO_4^-(aq) + MnO_2(s) + 2H_2O(l)$ (4)

**15** Two students were asked to design an experiment to identify an unknown Group 2 carbonate. Student 1 decided to heat the carbonate to constant mass and use the results to determine the identity of the carbonate. Her results are shown below.

| | |
|---|---|
| Mass of crucible | = 19.58 g |
| Mass of crucible + carbonate | = 23.27 g |
| Mass of crucible + solid after heating for 3 minutes | = 22.54 g |
| Mass of crucible + solid after heating for 6 minutes | = 22.17 g |
| Mass of crucible + solid after heating for 9 minutes | = 22.17 g |

Use the results to identify the Group 2 carbonate. A second student decided to identify the carbonate by reacting it with HCl(aq) and collected the carbon dioxide by the displacement of water as shown below.

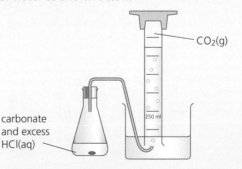

The student removed the bung and added 1.23 g of the carbonate to the conical flask followed by 100 cm³ of 1.0 mol dm⁻³ HCl(aq) and quickly replaced the bung. He collected 166 cm³ of $CO_2(g)$.
Use the results to deduce the molar mass of the carbonate. Explain why the results are inconclusive. Identify two possible sources of error in the experiment. (16)

**16 a)** Write an equation for the reaction of chlorine with aqueous sodium hydroxide and use this example to explain what is meant by disproportionation.

**b)** On heating, chlorate(I) ions in solution disproportionate to chlorate(V) ions and chloride ions. Write an ionic equation for this reaction.

**c)** On heating to just above its melting temperature, $KClO_3$ reacts to form $KClO_4$ and $KCl$. Write a balanced equation for the reaction and show that it is a disproportionation reaction. (5)

**17** State the trend in the power of chloride, bromide and iodide ions as reducing agents. Describe how you could demonstrate the trend by carrying out experiments in the laboratory. Include in your description the main observations that illustrate the differences between the ions and write equations for the reactions. (12)

**18** There are four known oxides of chlorine; $Cl_2O$, $ClO_2$, $Cl_2O_6$ and $Cl_2O_7$.

**a)** Determine the oxidation number of Cl in each of the oxides.

**b)** Draw dot-and-cross diagrams of $Cl_2O$ and predict the shape and bond angle.

Each of the oxides of chlorine is unstable and dangerously explosive. They all react with alkalis, $OH^-(aq)$ to produce a mixture of chlorate ions and water.

Construct equations for each of the following oxides of chlorine when it reacts with $OH^-(aq)$:

**c)** $Cl_2O$ reacts with $OH^-(aq)$ to form chlorate(I) ions and water

**d)** $ClO_2$ reacts with $OH^-(aq)$ to form chlorate(III) ions, chlorate(V) ions and water

**e)** $Cl_2O_6$ reacts with $OH^-(aq)$ to form chlorate(V) ions, chlorate(VII) ions and water

**f)** $Cl_2O_7$ reacts with $OH^-(aq)$ to form chlorate(VII) ions and water. (10)

## Challenge

**19** Chlorine can be prepared by the reaction between $KMnO_4(s)$ and hydrochloric acid. The equation for the reaction is:

$$2KMnO_4(s) + 16HCl(aq) \rightarrow 2KCl(aq) + 2MnCl_2(s) + 5Cl_2(g) + 8H_2O(l)$$

**a)** Use oxidation numbers to show that this is a redox reaction.

**b)** Identify what has been:
   **i)** oxidised
   **ii)** reduced.

**c)** Calculate the volume of $Cl_2(g)$ that could be made by reacting 3.95 g $KMnO_4(s)$ with excess $HCl(aq)$.

**d)** When 3.95 g $KMnO_4(s)$ was reacted with excess $HCl(aq)$, less $Cl_2(g)$ than the theoretical volume calculated in part (c) was collected. Suggest why. (11)

# Chapter 10

# Enthalpy changes

## Prior knowledge

*In this chapter it is assumed that you will be able to:*
- balance equations for the combustion of organic compounds
- calculate amounts of substance in moles.

## Test yourself on prior knowledge

1 Write equations for the combustion reactions that occur when each of the following is burned in a plentiful supply of oxygen:
   **a)** ethene ($C_2H_4$)
   **b)** ethane ($C_2H_6$)
   **c)** ethanol ($C_2H_5OH$).
2 Calculate the amounts (in mol) of carbon dioxide molecules and steam molecules produced when each of the following are burned in a plentiful supply of oxygen:
   **a)** 1 mol of ethane molecules
   **b)** 6 g of ethane molecules
   **c)** 2.3 g of ethanol molecules.

# Energy changes

## Tip

The symbol $\Delta$ is used to indicate 'change in':

- $\Delta T$ = change in temperature
- $\Delta V$ = change in volume
- $\Delta P$ = change in pressure
- $\Delta H$ = change in enthalpy.

## Key term

**Enthalpy change, $\Delta H$,** is the difference between the enthalpy of the reactants and the enthalpy of the products.

$\Delta H$ = enthalpy of products − enthalpy of reactants

Why do some chemicals react together and others do not? This is a fundamental question central to the study of chemistry. All chemical reactions involve a change in which energy is given out or energy is taken in. One way that chemists have tried to address this issue is to analyse the energy changes that occur when reactions take place. It is usually true that reactions in which the products have lower energy than the reactants are able to take place, whereas reactions in which the products have higher energy cannot. An important point, however, is that the rate at which the reaction occurs may be extremely slow. Thus, a reaction might potentially proceed, but its progress may not be readily apparent.

## Enthalpy change

All chemical reactions involve a change in energy. The exchange of energy between a reaction mixture and its surroundings is the enthalpy change. It is given the symbol $\Delta H$ and its units are joules, J, or kilojoules, kJ.

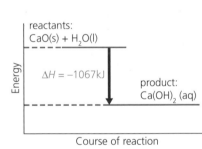

**Tip**

The sign given to an energy change is always from the point of view of the reaction. This means if the reaction loses heat to the surroundings then the sign will be '–' but if it takes heat from the surroundings it will be '+'.

Enthalpy is an energy change measured under constant pressure, which is normally the condition that applies when a laboratory experiment is carried out.

When an **exothermic reaction** takes place, energy is transferred from the reaction mixture to the surroundings. Chemical energy is released by the reactants and the temperature of the surroundings increases. For example, hot packs in self-warming drinks (Figure 10.1, Figure 10.2 and Figure 10.3) or for treating painful rheumatic conditions involve exothermic reactions.

In an **exothermic** reaction, $\Delta H$ is **negative** because the reactants have lost energy and given it to the surroundings.

**Figure 10.1** A self-warming can of coffee – pressing the bulb on the bottom of the can starts an exothermic reaction in a sealed compartment. The energy released heats the coffee.

reactants:
CaO(s) + H₂O(l)

$\Delta H = -1067\,kJ$

product:
Ca(OH)₂ (aq)

Energy

Course of reaction

**Figure 10.2** An enthalpy profile diagram for the reaction of calcium oxide with water.

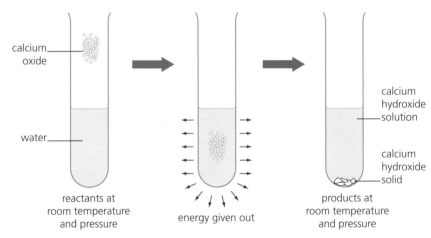

calcium oxide

water

reactants at room temperature and pressure

energy given out

calcium hydroxide solution

calcium hydroxide solid

products at room temperature and pressure

**Figure 10.3** The exothermic reaction that takes place in some hot packs.

**Figure 10.4** Twist a cold pack and it gets cold enough to reduce the pain of a sports injury. When chemicals in the cold pack react, they take in energy as heat and the pack gets cold. This, in turn, cools the sprained or bruised area and helps to reduce painful swelling.

Reactions can sometimes be **endothermic**. In these cases $\Delta H$ is **positive** because the products have a higher energy than the reactants and heat is taken from the surroundings. A cold pack to reduce the swelling that occurs from a bruise utilises an endothermic reaction to draw heat from its surroundings (Figure 10.4) and the absorption of carbon dioxide by green leaves during photosynthesis is another important example of an endothermic reaction.

## Enthalpy profile diagrams and activation energy

Enthalpy changes can be represented by simple enthalpy profile diagrams as shown in Figure 10.5.

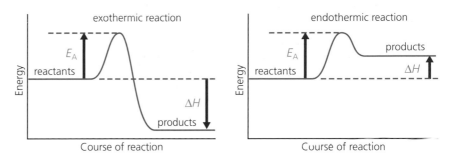

**Figure 10.5** Enthalpy profile diagrams showing the activation energy for both exothermic and endothermic reactions.

The enthalpy profiles show that, initially, energy is taken from the surroundings. Reactions need an input of energy to get them started because the bonds of the reactants have to be broken before the new bonds of the products are formed. The energy required to do this is called the activation energy, $E_A$, of the reaction

## Calculating the enthalpy changes of gaseous reactions using bond enthalpies

If the enthalpies required to break and to make bonds are known it is straightforward to calculate the overall enthalpy change that occurs as a result of a reaction. These enthalpies, called bond enthalpies, are values that chemists have collected together using the results of a wide range of experiments. They are expressed in $kJ\,mol^{-1}$.

Some average bond enthalpies are shown in Table 10.1.

**Table 10.1** Average bond enthalpies.

| Bond | $\Delta H/kJ\,mol^{-1}$ |
|---|---|
| C–C | +347 |
| C=C | +612 |
| C≡C | +838 |
| C–H | +413 |
| C=O | +805 |
| O=O | +498 |
| O–H | +464 |
| C–N | +286 |
| N≡N | +945 |
| N–H | +391 |
| H–H | +436 |

Bond enthalpies can be used to calculate the overall enthalpy change that occurs in a simple reaction involving gaseous covalent molecules.

For these reactions the correct sign to $\Delta H$ is given by using:

$\Delta H$ = sum of bond enthalpies of reactants − sum of bond enthalpies of products

For example, Figure 10.6 shows the bonds that are broken and formed when 1 mol of methane is burned. The reaction occurring in this case is:

$$CH_4(g) + 2O_2(g) \rightarrow CO_2(g) + 2H_2O(l)$$

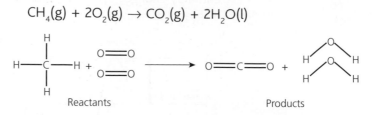

Reactants                                          Products

**Figure 10.6** The enthalpy of combustion of methane using bond enthalpies.

The relevant bond enthalpies from Table 10.1 are:

| | | | |
|---|---|---|---|
| C–H | 413 kJ mol⁻¹ | O=O | 498 kJ mol⁻¹ |
| C=O | 805 kJ mol⁻¹ | O–H | 464 kJ mol⁻¹ |

Bond enthalpies of reactants:

$4 \times (C–H) = 4 \times 413 = 1652\,kJ\,mol^{-1}$

$2 \times (O=O) = 2 \times 498 = 996\,kJ\,mol^{-1}$

sum of bond enthalpies of reactants = 1652 + 996 = 2648 kJ mol⁻¹

Bond enthalpies of products:

$2 \times (C=O) = 2 \times 805 = 1610\,kJ\,mol^{-1}$

$4 \times (O–H) = 4 \times 464 = 1856\,kJ\,mol^{-1}$

sum of bond enthalpies of products = 1610 + 1856 = 3466 kJ mol⁻¹

$\Delta H$ = sum of bond enthalpies of reactants − sum of bond enthalpies of products

= 2648 − 3466 = −818 kJ mol⁻¹

It is important to note that the figures given for bond enthalpies are average figures for that type of bond and the true figure will be slightly different depending on the compound used. The strength of a particular bond is affected by the neighbouring bonds in the molecule.

- Bond enthalpies are those specific to a bond in a real example (e.g. the C–O bond in $CH_3OH$).

- An average bond enthalpy is the figure obtained by averaging the bond enthalpies in all compounds containing that particular bond (e.g. all C–O bonds).

The difference can be shown by considering a molecule such as water. As shown in Figure 10.7, the O–H bonds can be broken successively but the energy required to do this is different for each bond.

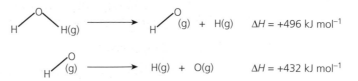

**Figure 10.7** The different enthalpies required to break successive OH bonds in water.

The O–H bond enthalpy in water is quoted as $\frac{496 + 432}{2} = +464\,\text{kJ mol}^{-1}$

This is the average (mean) bond enthalpy.

The enthalpy change for the combustion of methane is quoted in data lists as $-890\,\text{kJ mol}^{-1}$, which differs from the value $-818\,\text{kJ mol}^{-1}$ calculated above. This is largely explained by the use of average bond enthalpies for the C–H, C=O and O–H bonds in the calculation, rather than specific bond enthalpies.

> ### Test yourself
>
> 1 Draw an enthalpy profile to illustrate the reaction of methane and oxygen as calculated in example above.
> 2 a) Explain, with the aid of an equation, what is meant by the bond enthalpy of HCl(g).
>    b) Calculate the enthalpy change that occurs when hydrogen reacts with chlorine to form hydrogen chloride.
>
>    $E(\text{H–H}) = 436\,\text{kJ mol}^{-1}$  $E(\text{Cl–Cl}) = 243\,\text{kJ mol}^{-1}$  $E(\text{H–Cl}) = 432\,\text{kJ mol}^{-1}$
>
> 3 Calculate the enthalpy change that occurs in the reaction between ethene and hydrogen to form ethane.
>
>    $E(\text{H–H}) = 436\,\text{kJ mol}^{-1}$  $E(\text{C=C}) = 612\,\text{kJ mol}^{-1}$
>
>    $E(\text{C–H}) = 413\,\text{kJ mol}^{-1}$  $E(\text{C–C}) = 347\,\text{kJ mol}^{-1}$
>
> 4 a) Use bond enthalpies to calculate the enthalpy change that would occur if nitrogen could react with hydrogen to make hydrazine.
>    $N_2 + 2H_2 \rightarrow N_2H_4$
>    $E(\text{N}\equiv\text{N}) = 945\,\text{kJ mol}^{-1}$
>    $E(\text{H–H}) = 436\,\text{kJ mol}^{-1}$
>    $E(\text{N–N}) = 158\,\text{kJ mol}^{-1}$
>    $E(\text{N–H}) = 391\,\text{kJ mol}^{-1}$
>    b) Use your answer to part (a) and Question 3 to suggest why ethene reacts quite readily hydrogen but nitrogen does not.

### Key term

The **specific heat capacity** of a substance is the heat required to increase the temperature of 1.0 g of the substance by 1 °C (1 K).

## Measuring enthalpy changes experimentally

It is relatively easy to measure the temperature changes for some of the reactions that take place in solution and also for some combustion reactions.

### Reactions in solution

The enthalpy change, $\Delta_r H$, for reactions that take place in solution can usually be measured directly using the apparatus shown in Figure 10.8 although the result obtained is approximate because of heat loss to the surroundings.

The energy released or absorbed when the reaction takes place, $q$, can be calculated using the equation:

$q = mc\Delta T$

- $m$ is the mass of the reaction mixture
- $c$ is the specific heat capacity of the reaction mixture
- $\Delta T$ is the change in temperature.

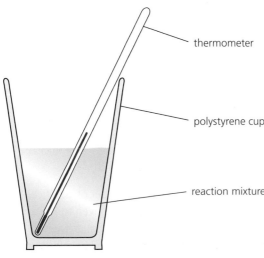

thermometer

polystyrene cup

reaction mixture

**Figure 10.8** Apparatus to measure the enthalpy of reaction.

The solvent used for most reactions is water. The specific heat capacity of water is $4.18\,J\,g^{-1}\,K^{-1}$.

Using this value for the specific heat capacity, the enthalpy change for the reaction mixture, $q$, has a value in joules (J). It is usual to adjust this value so that $\Delta H$ can be quoted for 1 mol of reactant with the units of $kJ\,mol^{-1}$. This means dividing $q$ by 1000 to convert to kJ and then dividing this by the amount, in moles, $n$, used.

$$\Delta H = \frac{q}{1000n} = \frac{(mc\Delta T)}{1000n}$$

Using the formula above the value of $\Delta H$ is always positive. It is essential to take the temperature change into consideration:

- If $\Delta T$ increases the reaction is exothermic so $\Delta H$ is negative.
- If $\Delta T$ deceases the reaction is endothermic so $\Delta H$ is positive.

**Example 1**

When 1.50 g of sodium carbonate is added to $40.0\,cm^3$ of $2.00\,mol\,dm^{-3}$ hydrochloric acid (an excess) the temperature rises from 19.0 °C to 24.6 °C.

Calculate $\Delta H$ for the reaction assuming the density of the acid is $1.00\,g\,cm^{-3}$ and the specific heat capacity is $4.18\,J\,g^{-1}\,K^{-1}$.

**Answer**

**Step 1** Calculate the energy change in the reaction by using $q = mc\Delta T$.

The heat that is released is given to the $40.0\,cm^3$ or 40.0 g of hydrochloric acid. You can ignore the contribution of the 1.50 g of sodium carbonate. The temperature rise is 24.6 − 19.0 = 5.6 °C.

energy change = 40 × 4.18 × 5.6 = 936.32 J

**Step 2** Convert the answer to $kJ\,mol^{-1}$ by dividing by 1000.

$$\text{energy change} = \frac{940.8}{1000} = 0.936\,32\,kJ$$

**Step 3** Convert the answer to $kJ\,mol^{-1}$ by dividing by the amount (in moles) used.

Since the hydrochloric acid was in excess it is the sodium carbonate which is the cause of the rise in temperature that has been measured.

molar mass of sodium carbonate is 46.0 + 12.0 + 48.0 = $106.0\,g\,mol^{-1}$

amount of sodium carbonate used is $\dfrac{1.50}{106.0}$ = 0.01415 mol

$$\Delta H = \frac{0.936\,32}{0.014\,15} = -66.171\,0247\,kJ\,mol^{-1}$$

When giving a value of $\Delta H$ it is essential that you indicate whether the reaction is endothermic or exothermic. In this case the reaction is exothermic so a '−' sign must be included.

$$\Delta H = -66.2\,kJ\,mol^{-1}$$

The final answer should be given to a sensible number of significant figures. In this case since all the quantities used are given to 3 significant figures it is best to do the same for the final answer.

Notice in Example 1, it was specified that the hydrochloric acid used was 'in excess'. This was necessary because the calculation assumed that all of the 1.50 g of sodium carbonate had reacted. In the next example, the two solutions are in the correct reacting quantities so it is not necessary to specify which solution was in excess.

## Example 2

When 50.0 cm³ of 2.00 mol dm⁻³ hydrochloric acid is mixed with 50.0 cm³ of 2.00 mol dm⁻³ sodium hydroxide solution, the temperature increases by 13.7 °C.

Calculate the enthalpy change of neutralisation of hydrochloric acid. Assume that the specific heat capacity, $c = 4.18\,J\,g^{-1}\,K^{-1}$ and that the density of both the hydrochloric acid and sodium hydroxide solution is $1.00\,g\,cm^{-3}$.

### Answer

**Step 1** Calculate the energy transferred in the reaction by using:

$$q = mc\Delta T$$

$m$ = mass of the two solutions = 50.0 + 50.0 = 100.0 g

$q = 100 \times 4.18 \times 13.7 = 5726.6\,J = 5.7266\,kJ$

**Step 2** Convert the answer to kJ mol⁻¹ by dividing by the number of moles used.

amount, $n$ in moles of HCl $= cV = 2.00 \times \dfrac{50.0}{1000} = 0.100\,mol$

$\Delta H = \dfrac{q}{n} = \dfrac{5.7266}{n} = \dfrac{5.72266}{0.100} = -57.2266\,kJ\,mol^{-1}$

(Remember that because the temperature rises, this is an exothermic reaction so $\Delta H$ is negative.)

$\Delta H = -57.2\,kJ\,mol^{-1}$ to 3 significant figures

## Combustion

The enthalpy change of combustion, $\Delta_c H^\ominus$ for a volatile liquid (e.g. ethanol) can be measured directly using the apparatus shown in Figure 10.9. The result will be approximate because of heat loss.

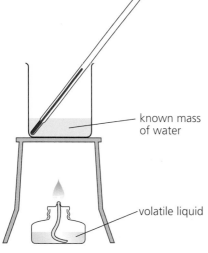

known mass of water

volatile liquid

**Figure 10.9** Apparatus used to determine the enthalpy change of combustion of a volatile liquid, such as ethanol.

## Example 3

When 0.230 g of ethanol, $CH_3CH_2OH$, is burned, it raises the temperature of 200 cm³ of water by 7.80 °C. Calculate the enthalpy change of combustion of ethanol. Assume that the specific heat capacity, $c$, is $4.18\,J\,g^{-1}\,K^{-1}$ and that the density of water is $1.00\,g\,cm^{-3}$.

### Answer

**Step 1** Calculate the energy transferred in the reaction by using:

$$q = mc\Delta T$$

$m$ = mass of water = 200 g

$q = 200 \times 4.18 \times 7.80 = 6520.8\,J = 6.5208\,kJ$

▶▶▶

## Activity

**Measuring the enthalpy change for the reaction of zinc with copper sulfate solution**

Two students decided to measure the enthalpy change for the reaction between zinc and copper(II) sulfate solution.

$$Zn(s) + CuSO_4(aq) \rightarrow ZnSO_4(aq) + Cu(s)$$

The method they used is shown in Figure 10.10 and their results are shown in Table 10.2. After adding the zinc, it took a little while for the temperature to reach a peak and then the mixture began to cool.

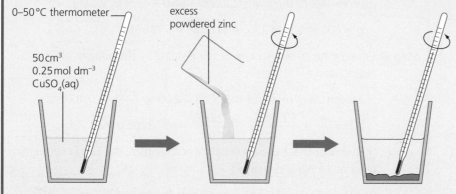

Measure the temperature every 30 s for 2.5 minutes.

At 3.0 minutes add excess powdered zinc and stir.

Continue stirring and record the temperature every 30 s for a further 6 minutes.

**Figure 10.10** Measuring the enthalpy change for the reaction of zinc with copper sulfate solution.

**Table 10.2**

| Time/min | Temperature/°C | Time/min | Temperature/°C | Time/min | Temperature/°C |
|---|---|---|---|---|---|
| 0 | 24.1 | 3.5 | 34.2 | 6.5 | 33.7 |
| 0.5 | 24.0 | 4.0 | 34.8 | 7.0 | 33.6 |
| 1.0 | 24.1 | 4.5 | 35.0 | 7.5 | 33.5 |
| 1.5 | 24.1 | 5.0 | 34.6 | 8.0 | 33.4 |
| 2.0 | 24.2 | 5.5 | 34.2 | 8.5 | 33.2 |
| 2.5 | 24.1 | 6.0 | 33.9 | 9.0 | 33.1 |
| 3.0 | – | | | | |

1 Plot a graph of temperature (vertically) against time (horizontally) using the results in Table 10.2.
2 Extrapolate the graph backwards from 9 minutes to 3 minutes, as shown in Figure 10.11, in order to estimate the maximum temperature. This assumes that all the zinc reacted at once and there was no loss of heat to the surroundings.
   a) What is the estimated maximum temperature at 3 minutes?
   b) What is the temperature rise, $\Delta T$, for the reaction?
3 Calculate the energy given out during the reaction using the equation:

   energy transferred = mass × specific heat capacity × temperature change

▶▶▶

Assume that:

- all the heat is transferred to the solution in the polystyrene cup
- the density of the solution is $1\,g\,cm^{-3}$
- the specific heat capacity of the solution is $4.2\,J\,g^{-1}\,K^{-1}$.

4 How many moles of each chemical reacted?
   a) $CuSO_4$
   b) Zn

5 What is the enthalpy change of the reaction, $\Delta H_{reaction}$, for the amounts of Zn and $CuSO_4$ in the equation?
   (State the value of $\Delta H_{reaction}$ in $kJ\,mol^{-1}$ with the correct sign.)

6 What are the main sources of error in the measurements and procedure for the experiment?

7 Look critically at the procedures in the experiment and suggest improvements to minimise errors and increase the reliability of the result.

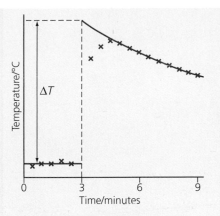

**Figure 10.11** Estimating the maximum temperature reached by the mixture when zinc reacts with copper sulfate solution.

## Reliability of thermochemical experiments

The results obtained from the experiments that have just been described are not at all reliable. The main reasons for this are likely to be:

- the lack of precision of the measuring equipment such as the thermometer.
- heat loss to the surroundings
- the slowness of the reaction

In the combustion experiment some of the heat will also be given to the container and probably there will be a deposit of soot resulting from the incomplete combustion of the fuel. The results obtained will clearly be inaccurate. This is a general problem with thermochemistry experiments that can only be partly addressed. It is often possible to improve matters by using improved insulation – for example if a polystyrene cup is used it could be surrounded by better insulation and the cup could include a lid – but nevertheless only more specialist apparatus such as the bomb calorimeter shown in Figure 10.12 can provide more reliable results.

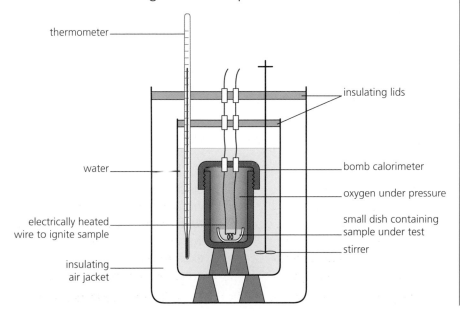

**Figure 10.12** Bomb calorimeter.

In questions 5, 6, 7 and 8 assume that the specific heat capacity, $c$, is $4.18\,J\,g^{-1}\,K^{-1}$ and that the density of water is $1.00\,g\,cm^{-3}$.

5 Burning butane, $C_4H_{10}$, from a Camping Gaz® container raised the temperature of $200\,g$ water from $18\,°C$ to $28\,°C$. The Gaz container was weighed before and after and the loss in mass was $0.29\,g$. Estimate the molar enthalpy change of combustion of butane.

6 When $2.0\,g$ of ammonium nitrate, $NH_4NO_3$, is dissolved in $50\,cm^3$ of water at $18\,°C$, the temperature falls to $15\,°C$.
   a) Calculate the energy absorbed when the ammonium nitrate dissolves in water.
   b) Calculate the enthalpy change per mole when ammonium nitrate dissolves in water

7 When $2.0\,g$ of sodium hydroxide is dissolved in $50\,cm^3$ of water at $18\,°C$ the temperature rises to $28.8\,°C$.
   Calculate the enthalpy change per mole when sodium hydroxide dissolves in water. Give your answer to **three** significant figures.

8 When $0.500\,g$ of magnesium is added to $100\,cm^3$ of $1.00\,mol\,dm^{-3}$ hydrochloric acid at $20.0\,°C$ the temperature rises to $37.5\,°C$.
   a) Write the equation for the reaction.
   b) Calculate the amounts in moles of the magnesium and hydrochloric acid used in the experiment.
   c) Calculate the enthalpy change of the reaction if $1\,mol$ of magnesium had been used.
   d) Suppose the experiment had been repeated using $100\,cm^3$ of $2.00\,mol\,dm^{-3}$ hydrochloric acid. What difference would this make to the temperature rise that would have been obtained?
   e) Suppose the experiment had been repeated using $1.00\,g$ of magnesium. What difference would this have made to:
      i) the temperature rise that would have been obtained
      ii) the enthalpy change per mole calculated in part (c)?

# Standard enthalpy changes

Enthalpy changes are measured experimentally at various temperatures and pressures. But if these enthalpy changes are to be compared it is important that they are measured under the same conditions. The conditions that are used for these thermochemical measurements are known as **standard conditions**.

Standard conditions for thermochemical measurements are:

- a temperature of $298\,K$ ($25\,°C$)

- a pressure of $100\,kPa$

- all substances in their most stable state at $298\,K$ and $100\,kPa$. This would be relevant if, for example, an element could exist in more than one crystalline form (called an allotrope). The allotrope which was the most stable at $298\,K$ and $100\,kPa$ would be selected. An example is carbon which can exist at $25\,°C$ either as diamond or graphite; but the more stable form is graphite.

The standard conditions for thermochemical measurements differ from those that can be used for reactions involving gases which refer to a temperature of $273\,K$. (See page 50.)

Although $298\,K$ is used as a standard, values are sometimes quoted at other temperatures in which case the temperature will be specifically indicated.

Standard pressure is sometimes quoted as $101.3\,kPa$ but OCR papers will use $100\,kPa$ as standard pressure.

To convert K to °C subtract 273.

The superscripted symbol, $\ominus$, is used with $\Delta H$, (i.e. $\Delta H^\ominus$) to denote that a change has been measured under or converted to standard conditions.

There are a number of thermochemical values that have been measured and tabulated by chemists and their precise definition needs to be known.

## Equations representing enthalpy change

It may be clearer to represent the enthalpy changes by means of an equation.

### Standard of enthalpy of formation

An equation relating to the standard enthalpy change of formation of a substance must reflect the standard definition. Therefore, it must:

- show the reactants as elements in their natural state

- produce 1 mol of the substance, even if this means that there are fractions in the equation

- include state symbols.

The equation for the standard enthalpy change of formation of ethane is:

$$2C(s) + 3H_2(g) \rightarrow C_2H_6(g)$$

The equation for the standard enthalpy change of formation of ethanol is:

$$2C(s) + 3H_2(g) + \tfrac{1}{2}O_2(g) \rightarrow CH_3CH_2OH(l)$$

### Standard enthalpy of combustion

An equation relating to the standard enthalpy change of combustion of a substance must reflect the standard definition. Therefore, it must:

- show 1 mol of the substance reacting completely with excess oxygen, even if this means having fractions in the equation

- include state symbols.

The products of combustion are usually carbon dioxide and water.

The equation for the standard enthalpy change of combustion of ethane is:

$$C_2H_6(g) + 3\tfrac{1}{2}O_2(g) \rightarrow 2CO_2(g) + 3H_2O(l)$$

The equation for the standard enthalpy change of combustion of ethanol is:

$$CH_3CH_2OH(l) + 3O_2(g) \rightarrow 2CO_2(g) + 3H_2O(l)$$

The equation for the standard enthalpy change of combustion of ethanamide ($CH_3CONH_2$) is more complex. However, the principles are the same – the equation must have 1 mol of the reactant and be balanced, even though this involves odd-looking fractions:

$$CH_3CONH_2(s) + 2\tfrac{3}{4}O_2(g) \rightarrow 2CO_2(g) + 2\tfrac{1}{2}H_2O(l) + \tfrac{1}{2}N_2(g)$$

### Standard enthalpy change of neutralisation

An equation relating to the standard enthalpy of neutralisation of a substance must reflect the standard definition. It must:

- show 1 mol of water being formed from the named base or acid, even if this means having fractions in the equation

- include state symbols.

Be careful to remember that if an acid is dibasic (diprotic) that 2 mol of water are produced if it is completely neutralised. So the enthalpy of neutralisation would be for $\frac{1}{2}$ a mole of the acid. For example:

$$\tfrac{1}{2}H_2SO_4(aq) + NaOH(aq) \rightarrow \tfrac{1}{2}Na_2SO_4(aq) + H_2O(l)$$

Usually the products are a salt and water.

The equation for the standard enthalpy change of neutralisation of dilute hydrochloric acid with aqueous sodium hydroxide is:

$$HCl(aq) + NaOH(aq) \rightarrow NaCl(aq) + H_2O(l)$$

An ionic equation could also be used to show the standard enthalpy change of neutralisation:

$$H^+(aq) + OH^-(aq) \rightarrow H_2O(l)$$

Strong acids and strong bases are fully ionised so that the enthalpy change of a reaction between them to form 1 mol of water will be the same regardless of which acid or base is chosen.

However, if the experiment is repeated using a weak acid such as ethanoic acid and a strong base such as sodium hydroxide, the enthalpy change measured has a lower value. The reaction is the same but the difference lies in the strength of the acid. At the start of the neutralisation the dissociation of ethanoic acid:

$$CH_3COOH(aq) \rightarrow CH_3COO^-(aq) + H^+(aq)$$

is not complete. As the $H^+$ is neutralised by the base the acid ionises further until it is completely neutralised but this requires a certain amount of energy that would otherwise have been released as heat.

The neutralisation reaction of a weak base such as aqueous ammonia and a strong acid has an even lower value for the enthalpy of neutralisation. This is because during the neutralisation the ammonia ionises:

$$NH_3(aq) + H_2O(aq) \rightarrow NH_4^+(aq) + OH^-(aq)$$

and this requires more energy than that required for the dissociation of ethanoic acid.

The neutralisation of the weak acid–weak base combination of ethanoic acid and ammonia results in an even lower value for the enthalpy of neutralisation.

Experiments of this type can give some indication of the relative strengths of acids and bases, even though the measurements are not particularly accurate.

# Hess' law

Hess' law states that, if a reaction can take place by more than one route, the enthalpy change for the reaction is the same irrespective of the route taken, provided that the initial and final conditions are the same.

Enthalpy changes for combustion reactions and for reactions that take place in solution can be measured experimentally. However, in reality, this is not the case for many chemical reactions. If the activation energy is particularly high or the reaction rate is very slow (these are usually related), it is unlikely that a reaction will take place readily. In these circumstances an enthalpy change cannot be measured directly.

Enthalpies that cannot be measured directly are calculated by using energy cycles based on enthalpies that can be measured. Since energy cannot be created or destroyed it follows that the enthalpy change for a reaction is independent of the route taken. This idea was first proposed by Germain Henri Hess and is known as Hess' law.

Hess' law can be illustrated using a simple enthalpy cycle:

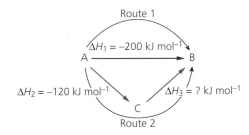

The enthalpy change for route 1 is equal to the enthalpy change for route 2:

$$\Delta H_1 = \Delta H_2 + \Delta H_3$$

Therefore, $\Delta H_3 = \Delta H_1 - \Delta H_1$

$$= -200 - (-120)$$

$$= -200 + 120$$

$$= -80 \, kJ \, mol^{-1}$$

## Enthalpy change of formation of carbon monoxide

The usefulness of Hess' law can be illustrated by the combustion of carbon. Carbon burns to form either $CO(g)$ or $CO_2(g)$. It is impossible to measure the enthalpy change of formation of $CO(g)$ directly because some $CO_2(g)$ would always be formed but it can be calculated using Hess' law.

The enthalpy changes for the following reactions can be measured experimentally:

$$C(s) + O_2(g) \rightarrow CO_2(g) \qquad \Delta_c H = -394 \, kJ \, mol^{-1}$$
$$CO(g) + \tfrac{1}{2}O_2(g) \rightarrow CO_2(g) \qquad \Delta_c H = -284 \, kJ \, mol^{-1}$$

This allows us to determine the enthalpy change for:

$$C(s) + \tfrac{1}{2}O_2(g) \rightarrow CO(g)$$

using Hess' law, as shown below.

Step 1 Start with the enthalpy change that has to be calculated. Call it $\Delta H_1$. Write an equation for the reaction. This is the top line of the cycle:

$$C(s) + \tfrac{1}{2}O_2 \xrightarrow{\quad \Delta H_1 \quad} CO(g)$$

Step 2 Construct an enthalpy cycle with two alternative routes:

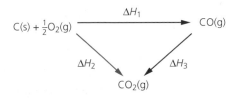

Step 3 Apply Hess' law to the cycle.

The simplest way to apply Hess' law is to look at the direction of the arrows. Route 1 has arrows that point in the clockwise direction; route 2 has an arrow that points anti-clockwise:

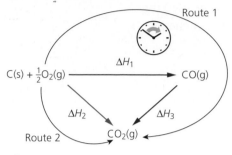

route 1 = route 2

$$\Delta H_1 + \Delta H_3 = \Delta H_2$$

Therefore, $\Delta H_1 = \Delta H_2 - \Delta H_3$

$$= -394 - (-284)$$

$$= -394 + 284$$

$$= -110\,\text{kJ}\,\text{mol}^{-1}$$

The enthalpy change of formation of carbon monoxide is $-110\,\text{kJ}\,\text{mol}^{-1}$.

# Construction of enthalpy cycles

There are two common enthalpy cycles that you may be asked to construct which require the use of Hess' law.

### Determining an enthalpy change of formation using enthalpies of combustion

In order to calculate the enthalpy change of formation using enthalpy changes of combustion, it is best to construct a cycle with the combustion products at the bottom. The arrows always then point downwards to the combustion products.

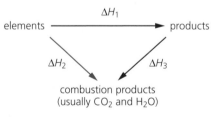

Applying Hess' law to the cycle:

$$\Delta H_1 + \Delta H_3 = \Delta H_2$$

Therefore, $\Delta H_1 = \Delta H_2 - \Delta H_3$

### Determining an enthalpy change of combustion using enthalpy changes of formation

For this cycle enthalpy changes of formation are used to calculate an enthalpy change of combustion. The elements are best shown at the bottom of the triangle; the arrows then point upwards.

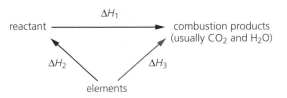

Applying Hess' law to the cycle:

$$\Delta H_1 + \Delta H_2 = \Delta H_3$$

Therefore, $\Delta H_1 = \Delta H_3 - \Delta H_2$

---

### Example 4

Use the enthalpy changes of combustion data in Table 10.3 to calculate the standard enthalpy change of formation of methane.

**Table 10.3** Enthalpy of combustion data.

| Substance | Formula | $\Delta H_c^{\ominus}$/kJ mol$^{-1}$ |
|---|---|---|
| Carbon | C(s) | −394 |
| Hydrogen | H$_2$(g) | −286 |
| Methane | CH$_4$(g) | −890 |

The enthalpy change of combustion of hydrogen will be to form water rather than steam as the standard conditions specify a temperature of 25 °C.

**Answer**

**Step 1** Write the equation for the required enthalpy change of formation:

$$C(s) + 2H_2(g) \rightarrow CH_4(g)$$

**Step 2** Construct the enthalpy cycle, writing the combustion products at the bottom. Both arrows point downwards:

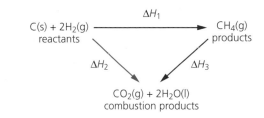

**Step 3** Apply Hess' law.

Using Hess' law, $\Delta H_1 + \Delta H_3 = \Delta H_2$

Therefore, $\Delta H_1 = \Delta H_2 - \Delta H_3$

$$\Delta H_2 = \Delta_c H(C) + 2 \times (\Delta_c H(H_2))$$

$$= -394 + 2(-286)$$

$$= -966 \text{ kJ mol}^{-1}$$

$$\Delta H_3 = \Delta_c H(CH_4) = -890 \text{ kJ mol}^{-1}$$

$$\Delta H_1 = -966 - (-890) = -76 \text{ kJ mol}^{-1}$$

Therefore, the standard enthalpy change of formation of methane, $\Delta_f H^{\ominus}(CH_4)$, is −76 kJ mol$^{-1}$.

## Example 5

Use the standard enthalpy change of formation data in Table 10.4 to calculate the standard enthalpy change of combustion of ethane.

**Table 10.4** Enthalpy of formation data.

| Substance | Formula | $\Delta_f H^\ominus$/kJ mol$^{-1}$ |
|---|---|---|
| Ethane | $C_2H_6(g)$ | −85 |
| Carbon dioxide | $CO_2(g)$ | −394 |
| Water | $H_2O(l)$ | −286 |

**Answer**

**Step 1** Write the equation for the required enthalpy change of combustion:

$$C_2H_6(g) + 3\tfrac{1}{2}O_2(g) \rightarrow 2CO_2(g) + 3H_2O(l)$$

**Step 2** Construct the enthalpy cycle, writing the combustion products at the bottom. Both arrows point upwards:

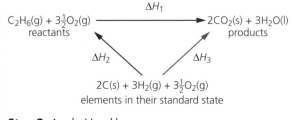

**Step 3** Apply Hess' law.

Using Hess' law, $\Delta H_1 + \Delta H_2 = \Delta H_3$

Therefore, $\Delta H_1 = \Delta H_3 - \Delta H_2$

$\Delta H_2$ for formation of $C_2H_6(g) = -85$ kJ mol$^{-1}$

$$\Delta H_3 = 2 \times \Delta_f H(CO_2) + 3 \times \Delta_f H(H_2O)$$

$$= 2(-394) + 3(-286)$$

$$= -1646 \text{kJ mol}^{-1}$$

(Be careful to multiply the enthalpies to correspond with the number of moles used in the equation for the combustion.)

$$\Delta H_1 = -1646 - (-85) = -1561 \text{kJ mol}^{-1}$$

Therefore, the standard enthalpy change of combustion of ethane, $\Delta_c H^\ominus$, is −1561 kJ mol$^{-1}$.

Hess' law can also be used to determine an enthalpy change for an inorganic reaction that cannot be measured in the laboratory. This is illustrated by the following worked example.

## Example 6

Reaction 1: When 2.60 g of calcium carbonate is added to 50.0 cm³ (an excess) of hydrochloric acid there is a temperature rise of 2.3 °C.

Reaction 2: When 1.50 g of calcium oxide is added to 50.0 cm³ (an excess) of hydrochloric acid there is a temperature rise of 11.1 °C.

Use this data to determine the enthalpy change of reaction for the decomposition of calcium carbonate into calcium oxide:

$$CaCO_3(s) \rightarrow CaO(s) + CO_2(s)$$

### Answer

The energy gained in reaction 1:

$$q = mc\Delta T$$

$m$ is the mass of hydrochloric acid = 50 g

$$q = 50.0 \times 4.18 \times 2.3 = 480.7 \, J \quad \text{or} \quad 0.4807 \, kJ$$

The molar mass of calcium carbonate is 100.1 g, so the amount in moles of calcium carbonate used was:

$$\frac{2.60}{100.1} = 0.0260$$

$$\Delta H_1 = \frac{0.4807}{0.0260} = -18.5 \, kJ \, mol^{-1}$$

The energy gained in reaction 2:

$$q = 50.0 \times 4.18 \times 11.1 = 2320 \, J \quad \text{or} \quad 2.320 \, kJ$$

The molar mass of calcium oxide is 56.1 g so the amount in moles of calcium oxide used was:

$$\frac{1.50}{56.1} = 0.0267 \, mol$$

$$\Delta H_2 = \frac{2.320}{0.0267} = -86.9 \, kJ \, mol^{-1}$$

The following cycle must now be used to determine the enthalpy change for the decomposition of calcium carbonate to calcium oxide.

$$\Delta H_1 = \Delta_r H + \Delta H_2$$

$$\Delta_r H = \Delta H_1 - \Delta H_2$$

$$\Delta_r H = -18.5 - (-86.9) = +68.4 \, kJ \, mol^{-1}$$

This answer is different from the theoretical value for this reaction as there would be substantial heat loss in an experiment of this kind.

### Tip

The mass ($m$) in reaction 1 can be 52.60 g which includes the mass of the calcium carbonate. Either 50.0 or 52.60 g is acceptable. Likewise in reaction 2 the mass can be 50.0 g or 51.5 g.

**9 a)** By writing a balanced equation, show that the standard enthalpy change of formation of carbon dioxide is the same as the standard enthalpy change of combustion of carbon (graphite).

**b)** Write equations for the standard enthalpy change of formation of:

**i)** aluminium oxide, $Al_2O_3$

**ii)** hydrogen chloride, HCl

**iii)** propane, $C_3H_8$.

**10** Use the data in the table to calculate the standard enthalpy change of formation for ethene.

| Substance | Formula | $\Delta_c H^\ominus$/kJ mol−1 |
|---|---|---|
| Carbon | C(s) | −394 |
| Hydrogen | $H_2$(g) | −286 |
| Ethene | $C_2H_4$(g) | −1411 |

**11** Use the data in the table to calculate the standard enthalpy change of combustion of butane.

| Substance | Formula | $\Delta_f H^\ominus$/kJ mol−1 |
|---|---|---|
| Butane | $C_4H_{10}$(g) | −127 |
| Carbon dioxide | $CO_2$(g) | −394 |
| Water | $H_2O$(l) | −286 |

**12** Use the standard enthalpy change of combustion data below to determine the standard enthalpy change for the following reaction.

$$C_2H_6(g) \rightarrow C_2H_4(g) + H_2(g)$$

| Substance | Formula | $\Delta_c H^\ominus$/kJ mol−1 |
|---|---|---|
| Ethane | $C_2H_6$(g) | −1561 |
| Ethene | $C_2H_4$(g) | −1411 |
| Hydrogen | $H_2$(g) | −286 |

What does this indicate about the likelihood of this reaction occurring under standard conditions?

# Practice questions

## Multiple choice questions 1–10

Use the reaction profile below to answer Questions 1 and 2.

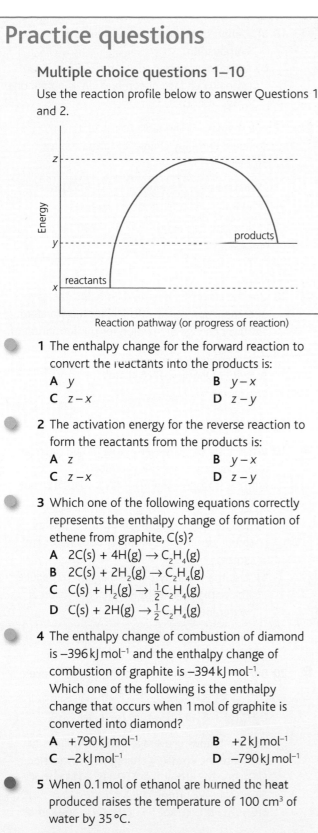

Reaction pathway (or progress of reaction)

**1** The enthalpy change for the forward reaction to convert the reactants into the products is:

    **A** $y$                **B** $y - x$

    **C** $z - x$           **D** $z - y$

**2** The activation energy for the reverse reaction to form the reactants from the products is:

    **A** $z$                **B** $y - x$

    **C** $z - x$           **D** $z - y$

**3** Which one of the following equations correctly represents the enthalpy change of formation of ethene from graphite, C(s)?

    **A** $2C(s) + 4H(g) \rightarrow C_2H_4(g)$

    **B** $2C(s) + 2H_2(g) \rightarrow C_2H_4(g)$

    **C** $C(s) + H_2(g) \rightarrow \frac{1}{2}C_2H_4(g)$

    **D** $C(s) + 2H(g) \rightarrow \frac{1}{2}C_2H_4(g)$

**4** The enthalpy change of combustion of diamond is $-396\,kJ\,mol^{-1}$ and the enthalpy change of combustion of graphite is $-394\,kJ\,mol^{-1}$. Which one of the following is the enthalpy change that occurs when 1 mol of graphite is converted into diamond?

    **A** $+790\,kJ\,mol^{-1}$     **B** $+2\,kJ\,mol^{-1}$

    **C** $-2\,kJ\,mol^{-1}$       **D** $-790\,kJ\,mol^{-1}$

**5** When 0.1 mol of ethanol are burned the heat produced raises the temperature of 100 cm³ of water by 35 °C.

The specific heat capacity of the water = $4.18\,Jg^{-1}K^{-1}$
Which one of the following gives the value (ignoring the sign) of the enthalpy of combustion of ethanol in $kJ\,mol^{-1}$?

    **A** $\dfrac{0.1 \times 4.18 \times 35}{1000}$

    **B** $0.1 \times 4.18 \times 35 \times 1000$

    **C** $\dfrac{100 \times 4.18 \times 35 \times 0.1}{1000}$

    **D** $\dfrac{100 \times 4.18 \times 35}{0.1 \times 1000}$

**6** 50 cm³ of 1.0 mol dm⁻³ sodium hydroxide is added to 50 cm³ of 1.0 mol dm⁻³ hydrochloric acid and the temperature rise as a result of the reaction is measured as $T_1$ °C.

The experiment is then repeated using 100 cm³ of 1.0 mol dm⁻³ sodium hydroxide and 100 cm³ of 1.0 mol dm⁻³ hydrochloric acid. The temperature rise is again measured this time as $T_2$ °C. Assuming any heat losses are the same in each experiment, when the values of $T_1$ °C and $T_2$ °C are compared it will be found that

    **A** $T_1 = T_2$         **B** $T_1 = 2 \times T_2$

    **C** $T_2 = 2 \times T_1$     **D** $T_2 = 4 \times T_1$

**7** The following are enthalpy changes for the reactions shown.

$$C(s) + O_2(g) \rightarrow CO_2(g) \qquad \Delta_r H = -394\,kJ\,mol^{-1}$$

$$2CO(g) + O_2(g) \rightarrow 2CO_2(g) \qquad \Delta_r H = -566\,kJ\,mol^{-1}$$

Which one the following gives the enthalpy of formation of carbon monoxide in $kJ\,mol^{-1}$?

    **A** $+566 - (\frac{1}{2} \times 394) = +369$

    **B** $+394 - (\frac{1}{2} \times 566) = +111$

    **C** $-394 + (\frac{1}{2} \times 566) = -111$

    **D** $-566 + (\frac{1}{2} \times 394) = -369$

**8** Hydrogen and chlorine react to form hydrogen chloride:

$$H_2(g) + Cl_2(g) \rightarrow 2HCl(g)$$

The bond enthalpy of $H_2$ is $+436\,kJ\,mol^{-1}$ and the bond enthalpy of $Cl_2$ is $243\,kJ\,mol^{-1}$. If the enthalpy change of the reaction is $-185\,kJ\,mol^{-1}$, which one of the following is the bond enthalpy of HCl?

    **A** $+864\,kJ\,mol^{-1}$

    **B** $+432\,kJ\,mol^{-1}$

    **C** $-432\,kJ\,mol^{-1}$

    **D** $-864\,kJ\,mol^{-1}$

Use the key below to answer Questions 9 and 10.

| A | B | C | D |
|---|---|---|---|
| 1, 2 & 3 correct | 1, 2 correct | 2, 3 correct | 1 only correct |

**9** Which of the following processes will be exothermic?
1 $CH_3(g) + H(g) \rightarrow CH_4(g)$
2 $2Ca(s) + O_2(g) \rightarrow 2CaO$
3 $H_2SO_3(aq) + Ca(OH)_2(aq) \rightarrow CaSO_3(aq) + 2H_2O(l)$

**10** Which of the following is always true for an endothermic reaction?
1 The total enthalpy of the products is greater than the reactants.
2 Heat will be absorbed from the surroundings when the reaction takes place.
3 The activation energy for the conversion of the products into the reactants is less than the activation energy for the conversion of the reactants into the products.

**11** Explain, with the aid of enthalpy profile diagrams, what is meant by an exothermic reaction and an endothermic reaction. On each diagram, label both $\Delta H$ and $E_a$. (6)

**12 a)** Explain, with the aid of an equation, what is meant by bond enthalpy.
**b)** Some bond enthalpies are given in the table.

| Bond | Bond enthalpy/kJ mol⁻¹ |
|---|---|
| H–H | 436 |
| Br–Br | 193 |
| H–Br | 366 |

Calculate the enthalpy change for the following reaction:
$H_2(g) + Br_2(g) \rightarrow 2HBr(g)$ (6)

**13** When 2.0 cm³ of concentrated sulfuric acid is added to 48.0 cm³ of water, the temperature of the mixture rises by 6.7 °C. Calculate the amount of heat evolved. Assume that the specific heat capacity of the mixture is 4.18 J g⁻¹ K⁻¹.

**14** State the standard condition for:
**a)** temperature　**b)** pressure. (2)

**15 a)** Define standard enthalpy change of formation.
**b)** Write an equation to illustrate the standard enthalpy change of formation of propanal, $CH_3CH_2CHO(l)$. (3)

**16 a)** Define standard enthalpy change of combustion.
**b)** Write an equation to illustrate the standard enthalpy change of combustion of propanone, $CH_3COCH_3(l)$. (3)

**17** Use the data in the table to calculate the standard enthalpy change of formation for propane.

| Substance | Formula | $\Delta_c H^{\ominus}$/kJ mol⁻¹ |
|---|---|---|
| Carbon | C(s) | −394 |
| Hydrogen | $H_2(g)$ | −286 |
| Propane | $C_3H_8(g)$ | −2219 |

(4)

**18** Use the values for standard enthalpy changes of combustion below to calculate the standard enthalpy change of reaction for the formation of methanol,

$\Delta_c H^{\ominus}[CH_3OH(l)] = -726\ kJ\ mol^{-1}$

$\Delta_c H^{\ominus}[C_{graphite}] = -394\ kJ\ mol^{-1}$

$\Delta_c H^{\ominus}[H_2(g)] = -286\ kJ\ mol^{-1}$ (3)

**19** When a spark ignites a mixture of hydrogen gas and oxygen gas, water is produced:
$H_2(g) + \frac{1}{2}O_2(g) \rightarrow H_2O(l)$ 　$\Delta H = -285.8\ kJ\ mol^{-1}$
$H_2(g) + \frac{1}{2}O_2(g) \rightarrow H_2O(g)$ 　$\Delta H = -241.8\ kJ\ mol^{-1}$
**a)** On the same diagram, draw the enthalpy profile for the formation of both water and steam.
**b)** Use the enthalpy-profile diagram to deduce the enthalpy change for the conversion of water into steam:

$H_2O(l) \rightarrow H_2O(g)$ (3)

**20** Butane, $C_4H_{10}$, burns readily on a camp cooker. The equation for the reaction is:

$C_4H_{10}(g) + 6\frac{1}{2}O_2(g) \rightarrow 4CO_2(g) + 5H_2O(g)$

Use the following average bond enthalpies to calculate the enthalpy change of the reaction.
$E(C–C)$　347 kJ mol⁻¹　　$E(C–H)$　413 kJ mol⁻¹
$E(O=O)$　498 kJ mol⁻¹
$E(C=O)$　805 kJ mol⁻¹　　$E(H–O)$　464 kJ mol⁻¹
(3)

**21** A student suggested that ethane might react with bromine in two different ways in bright sunlight.
Reaction 1: $C_2H_6(g) + Br_2(l) \rightarrow C_2H_5Br(l) + HBr(g)$
Reaction 2: $C_2H_6(g) + Br_2(l) \rightarrow 2CH_3Br(g)$

a) Use the following average bond enthalpies to calculate the enthalpy changes for the two possible reactions.

$E(C–C)$    $347\,kJ\,mol^{-1}$    $E(Br–Br)$    $193\,kJ\,mol^{-1}$
$E(C–H)$    $413\,kJ\,mol^{-1}$    $E(H–Br)$    $366\,kJ\,mol^{-1}$
$E(C–Br)$    $290\,kJ\,mol^{-1}$

b) Use your calculations to explain which of the reactions is more likely to occur.

c) Suggest two reasons why your calculated enthalpy changes may not agree with the accurately determined experimental values.

(10)

**22** When $25.0\,cm^3$ of nitric acid of concentration $1.00\,mol\,dm^{-3}$ is mixed with $50\,cm^3$ of $0.50\,mol\,dm^{-3}$ sodium hydroxide solution, the temperature increases by $4.6\,°C$. Calculate the standard enthalpy change of neutralisation of nitric acid. Assume that the specific heat capacity, $c$, of the mixture is $4.18\,J\,g^{-1}\,K^{-1}$ and that the density of both the nitric acid and sodium hydroxide solutions is $1.00\,g\,cm^{-3}$. (6)

**23** Use the data in the table to calculate the standard enthalpy change of combustion for ethanol.

| Substance | Formula | $\Delta_f H^{\ominus}/kJ\,mol^{-1}$ |
|---|---|---|
| Ethanol | $C_2H_5OH(g)$ | −277 |
| Carbon dioxide | $CO_2(g)$ | −394 |
| Water | $H_2O(l)$ | −286 |

(4)

**24** Use the data in the table to calculate the standard enthalpy change for the reduction of iron(III) oxide by carbon monoxide to make iron.

| Substance | Formula | $\Delta_f H^{\ominus}/kJ\,mol^{-1}$ |
|---|---|---|
| Iron(III) oxide | $Fe_2O_3(s)$ | −824 |
| Carbon monoxide | $CO(g)$ | −110 |
| Carbon dioxide | $CO_2(g)$ | −394 |

(5)

**25** Tin is manufactured by heating tinstone, $SnO_2$, at high temperatures with coke (carbon).
There are two possible reactions for the process.

Reaction 1: $SnO_2(s) + C(s) \rightarrow Sn(s) + CO_2(g)$

Reaction 2: $SnO_2(s) + 2C(s) \rightarrow Sn(s) + 2CO(g)$

a) Calculate the standard enthalpy change for each of the possible reactions using the data below.

$\Delta_f H^{\ominus}[SnO_2(s)] = -581\,kJ\,mol^{-1}$

$\Delta_f H^{\ominus}[CO_2(g)] = -394\,kJ\,mol^{-1}$

$\Delta_f H^{\ominus}[CO(g)] = -110\,kJ\,mol^{-1}$

b) Use your calculations to explain which of the reactions, on the basis of the enthalpy changes, would be most economic for industry. (9)

**26** A student is given an unknown alcohol that has a standard enthalpy change of combustion of $-726\,kJ\,mol^{-1}$.

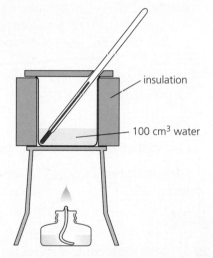

insulation

$100\,cm^3$ water

Using the apparatus shown in the diagram, the student obtains the following results:
- mass of water = $100\,g$
- initial temperature = $18.69\,°C$
- final temperature = $25.66\,°C$
- mass of spirit burner + alcohol at the start = $117.416\,g$
- mass of spirit burner + alcohol at the end = $117.288\,g$

The specific heat capacity, $c$, of the apparatus is $4.18\,J\,g^{-1}\,K^{-1}$. Calculate the relative molecular mass of the alcohol and, hence, identify it. (6)

**27** In an experiment, $50\,cm^3$ of $2.0\,mol\,dm^{-3}$ hydrochloric acid is mixed with $100\,cm^3$ of $2.0\,mol\,dm^{-3}$ of sodium hydroxide. Both solutions were initially at $21\,°C$ but, after mixing, the temperature rose to $30.0\,°C$.
The specific heat capacity, $c$, of the apparatus is $4.18\,J\,g^{-1}\,K^{-1}$.

a) Calculate $\Delta_r H$ for the reaction:

$NaOH(aq) + HCl(aq) \rightarrow NaCl(aq) + H_2O(l)$

**b)** If the experiment was repeated using 100 cm³ of 1.0 mol dm⁻³ of sodium hydroxide, what would be the temperature rise obtained when it was mixed with the 50 cm³ of 2.0 mol dm⁻³ hydrochloric acid ?
Explain your answer.

**c)** If the experiment was repeated using 50 cm³ of 2.0 mol dm⁻³ of sodium hydroxide, what would be the temperature rise obtained when it was mixed with the 50 cm³ of 2.0 mol dm⁻³ hydrochloric acid ?
Explain your answer.                                    (12)

**28** Reaction 1: When 8.11 g of anhydrous magnesium sulfate is added to 100.0 cm³ of water the temperature of the water goes up by 7.5 °C.
Reaction 2: When 8.23 g of magnesium sulfate crystals ($MgSO_4.7H_2O$) are added to 100 cm³ of water the temperature of the water falls by 1.0 °C.
Use this data to determine the enthalpy change for the reaction:

$$MgSO_4 + 7H_2O \rightarrow MgSO_4.7H_2O$$

Give your answer to **3** significant figures.

The specific heat capacity, $c$, of the apparatus is 4.2 J g⁻¹ K⁻¹.                    (8)

## Challenge

**29 a)** Write equations to illustrate the standard enthalpy changes of formation of $N_2O_5(g)$ and $H_2O(l)$.

**b)** Use the following data to calculate the standard enthalpy change of formation of $N_2O_5(g)$:
- $N_2O_5(g)$ reacts with water to produce nitric acid. When 1 mol of $N_2O_5(g)$ reacts with water $\Delta H = -74$ kJ mol⁻¹
- $\Delta_f H^\ominus(H_2O(l)) = -286$ kJ mol⁻¹
- $\Delta_f H^\ominus(HNO_3(l)) = -174$ kJ mol⁻¹     (5)

**30** The *Apollo 11* lunar module was the first manned spacecraft to land on the moon, in July 1969. It was propelled by Aerozine 50, a 50:50 mixture, by mass, of hydrazine ($N_2H_4$) and 1,1-dimethylhydrazine (($CH_3)_2$ $NNH_2$), with dinitrogen tetroxide ($N_2O_4$) as an oxidising agent.

The products of the reaction between Aerozine 50 and dinitrogen tetroxide are nitrogen, water and carbon dioxide.

**a)** Write an equation for the reaction between hydrazine and dinitrogen tetroxide.

**b)** Write an equation for the reaction between 1,1-dimethylhydrazine and dinitrogen tetroxide.

**c)** If 2200 kg of Aerozine 50 is required for the lunar module to take off, what mass of oxidising agent is needed?

**d)** Assuming complete oxidation of the fuel, use the following standard enthalpies of formation to calculate the enthalpy released at take-off.
$\Delta_f H^\ominus(N_2H_4) = 50.6$ kJ mol⁻¹
$\Delta_f H^\ominus((CH_3)_2NNH_2) = 49.3$ kJ mol⁻¹
$\Delta_f H^\ominus(H_2O(g)) = -241.8$ kJ mol⁻¹
$\Delta_f H^\ominus(N_2O_4) = -19.5$ kJ mol⁻¹
$\Delta_f H^\ominus(CO_2) = -393.7$ kJ mol⁻¹

**e)** A swimming pool contains 350 000 dm³ of water. Suppose the enthalpy generated at take-off was used to heat this water, which had an initial temperature of 20 °C. What would be the final temperature reached? Ignore heat losses to the surroundings and assume that the specific heat capacity of the water is 4.2 J g⁻¹ K⁻¹ and that its density is 1 g cm⁻³.                    (22)

# Rates and equilibria

### Test yourself on prior knowledge

a) Explain what is meant if a reaction is described as endothermic.
b) What is meant by the activation energy of a reaction?
c) Draw a fully labelled diagram to illustrate the reaction profile of an endothermic reaction.

## Rates of reaction

Reactions vary enormously in the rate at which they occur. Some are so fast that the reaction mixture explodes while others proceed at such a slow rate that they may take years to complete. In between these extremes there are some reactions whose rate can reasonably be measured in the laboratory. Understanding what causes these different rates and how they might be controlled is important to chemists, particularly on an industrial scale where the rate at which a product can be obtained is of economic importance. It may make the difference between successfully marketing a chemical or deciding that it is not viable to produce it at all.

### Measuring the rate of a reaction

There are a variety of ways in which the rate of reaction can be measured in the laboratory. Typically, it might be possible to measure a change in mass, volume or concentration or possibly a change in acidity as a reaction proceeds (a pH meter could do this) or maybe a change in colour.

It is particularly helpful if one of the products is a gas as the rate at which it is produced can be readily measured using a syringe.

Take for example the reaction between calcium carbonate and hydrochloric acid:

$$CaCO_3(s) + 2HCl(aq) \rightarrow CaCl_2(aq) + CO_2(g) + H_2O(l)$$

In an experiment, the volume of carbon dioxide being produced could be measured as the reaction proceeds in a syringe, as shown in Figure 11.1.

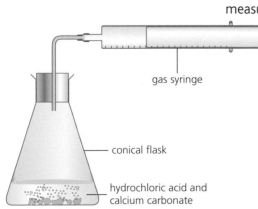

gas syringe

conical flask

hydrochloric acid and calcium carbonate

**Figure 11.1** Collecting and measuring the volume of carbon dioxide produced during the course of the reaction between calcium carbonate and hydrochloric acid.

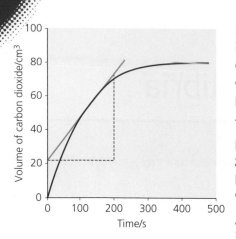

**Figure 11.2** Graph of the volume of carbon dioxide collected during the course of the reaction between calcium carbonate and hydrochloric acid.

It would initially be produced quite quickly but, as the reaction proceeded, it would gradually slow down as the reactants were used up. Eventually, once one or both of the reactants had been used up the production of carbon dioxide would cease. A graph of the volume of carbon dioxide produced against time would appear as shown in Figure 11.2.

The rate is not constant but can be given a numerical value at a particular time by drawing a tangent to the curve and measuring its gradient. For example, in Figure 11.2 the rate at which the reaction is proceeding at 110 s is given by the gradient of the tangent at that point on the curve. In this case its value is $52\,cm^3/200\,s = 0.26\,cm^3\,s^{-1}$.

A rate of particular importance is the initial rate of the reaction. This is because the concentrations of the reactants are precisely known at the start of the reaction whereas it is likely to be difficult to measure the concentrations while the reaction is progressing. The initial rate is obtained by taking the gradient of the concentration/time graph as the reaction starts (i.e. time = zero).

> **Tip**
>
> The experiment could also be followed by measuring the loss in mass of the reactants as the carbon dioxide is produced.

> **Test yourself**
>
> **1 a)** Draw a sketch of a graph of concentration against time for a reaction in which the concentration of a reactant was measured as the reaction proceeded.
>
> **b)** How would you determine the rate just as the reaction started?

## Activity

### Investigation of the effect of concentration on the rate of a reaction

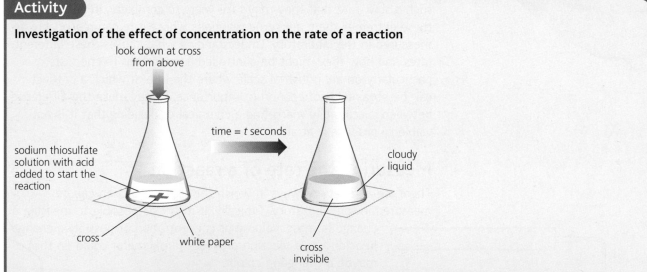

**Figure 11.3** Investigating the effect of the concentration of thiosulfate ions on the rate of reaction in acid solution. The hydrogen ion concentration is the same in each experiment.

Figure 11.3 illustrates an investigation of the effect of concentration on the rate at which thiosulfate ions in solution react with hydrogen ions to form a precipitate of sulfur.

$$S_2O_3{}^{2-}(aq) + 2H^+(aq) \rightarrow S(s) + SO_2(aq) + H_2O(l)$$

The observer records the time taken for the sulfur precipitate to obscure the cross on the paper under the flask. The rate of the reaction is proportional to $1/t$.

▶▶▶

The results of the investigation are shown in Table 11.1.

**Table 11.1** Results of the investigation in Figure 11.3.

| Experiment | Concentration of thiosulfate ions/mol dm$^{-3}$ | Time, $t$, for the cross to be obscured/s | Rate of reaction, $1/t/s^{-1}$ |
|---|---|---|---|
| 1 | 0.15 | 43 | 0.023 |
| 2 | 0.12 | 55 | – |
| 3 | 0.09 | 66 | 0.015 |
| 4 | 0.06 | 105 | 0.0095 |
| 5 | 0.03 | 243 | 0.0041 |

**1** How would you prepare 50 cm$^3$ of a solution of sodium thiosulfate solution with a concentration of 0.12 mol dm$^{-3}$ from a solution with a concentration of 0.15 mol dm$^{-3}$?

**2** Calculate the value for the rate of reaction when the concentration of thiosulfate ions is 0.12 mol dm$^{-3}$.

**3** Plot a graph to show how the rate of reaction varies with the concentration of thiosulfate ions.

**4** What is the relationship between reaction rate and concentration of thiosulfate for this reaction according to the graph?

## Factors affecting the rate of a reaction

Experimental observations show that the rate of reaction is influenced by temperature, concentration, the use of a catalyst and, for gaseous reactions, pressure.

The collision theory of reactivity helps to explain these observations. A reaction cannot take place unless a collision occurs between the reacting particles although not all collisions lead to a successful reaction. For reaction to occur:

- the colliding particles must be correctly orientated

- the energy of a collision between reacting particles must exceed the activation energy, $E_a$, for that particular reaction.

The need for colliding particles to be correctly orientated is illustrated by the reaction between iodomethane, $CH_3I$, and hydroxide ions, $OH^-$ (Figure 11.4). In iodomethane, the C–I bond is polar. The carbon has a δ+ charge; the iodine has a δ− charge. If a negatively charged hydroxide ion approaches iodomethane towards the $I^{δ-}$ there is mutual repulsion and reaction does not occur. If the approach is towards the $C^{δ+}$ and the energy is high enough, the collision leads to a reaction.

> **Key term**
>
> The **activation energy** is the minimum energy required for a reaction to occur.

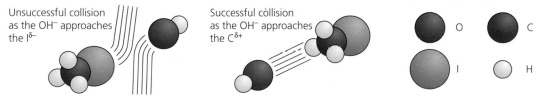

Unsuccessful collision as the $OH^-$ approaches the $I^{δ-}$

Successful collision as the $OH^-$ approaches the $C^{δ+}$

O   C   I   H

**Figure 11.4** Effect of orientation on reaction outcome.

The frequency of collisions can be increased by:

- increasing the concentration
- increasing the surface area of a solid
- increasing the temperature
- increasing the pressure in a gaseous reaction.

Increasing the temperature is the most significant factor because this increases the energy of the collision.

## Effect of concentration

Increasing the concentration increases the chance of a collision between the reacting particles. The more collisions there are, the faster the reaction.

For a gaseous reaction, increasing the pressure has the same effect as increasing the concentration because a higher pressure compresses a mixture of gases and increases their concentration. When gases react, they react faster at high pressure because there is an increased chance of a collision.

## Effect of surface area

If a reactant is a solid, then its surface area will affect the rate at which it is able to react. Clearly breaking a solid into smaller particles means that more of its surface is exposed to the reagent it is reacting with and the reaction will proceed faster.

## Effect of temperature, the Boltzmann distribution

Raising the temperature increases the energy of the particles – they move faster and collide more often. However, this increased frequency of collisions does not explain fully the effect that raising the temperature has on the rate of reaction. When a collision occurs, the particles exchange (gain or lose) energy.

The Austrian scientist, Ludwig Boltzmann, showed mathematically that the energies of molecules at a constant temperature are distributed as shown in Figure 11.5. This type of graph is called a **Boltzmann distribution**.

**Tip**

You won't be asked to explain why there is a distribution of energy but you should be able to sketch the Boltzmann distribution.

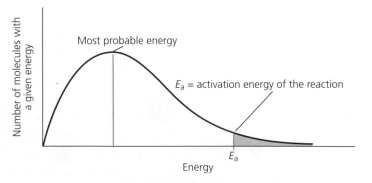

**Figure 11.5** Boltzmann distribution of energies.

This was an important contribution to our understanding of how reactions occur. Instead of all reacting particles having a fixed energy at a fixed temperature, Boltzmann showed that that the energies followed the pattern shown by the distribution curve. In particular:

- The distribution curve goes through the origin, which shows that there are no particles with zero energy.

- At high energy, the distribution approaches, but never touches, the horizontal axis. This shows that there is no theoretical maximum energy.

- The area under the curve represents the *total* number of particles.

The shaded area in Figure 11.5 represents the number of particles with energy greater than or equal to the activation energy, $E \geq E_a$. This shaded area, therefore, indicates the number of particles with sufficient energy to react.

A reaction can occur only if the colliding particles have an energy greater than, or equal to, the activation energy, $E_a$. Raising the temperature increases the energy of the particles and, therefore, changes the number of particles with energy greater than or equal to the activation energy. This can be shown by comparing a typical distribution of energies at an initial temperature, $T_1$, with the distribution of energies when the temperature is raised to $T_2$ (Figure 11.6). The number of particles remains the same so the area under both curves is the same. However, at the higher temperature ($T_2$), the distribution flattens and shifts to the right. This means that there are more particles with higher energy and a greater proportion of particles have an energy that exceeds the activation energy.

Lowering the temperature has the opposite effect.

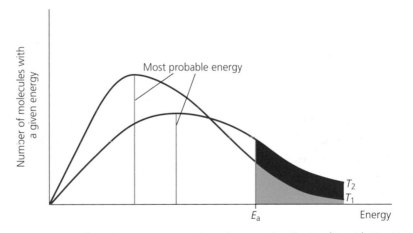

**Figure 11.6** Effect of temperature on the Boltzmann distribution ($T_2 > T_1$). The blue area indicates the proportions of particles with energy greater than the activation energy at temperature $T_1$. The red and blue areas indicate the proportion of particles with energy greater than the activation energy at temperature $T_2$.

## Effect of the use of a catalyst

Catalysts alter the rate of reactions without themselves being used up by the overall reaction. Although catalysts are sometimes used to slow reactions down their purpose is usually to increase the rate of a reaction. They are divided into two types: **homogeneous** and **heterogeneous**. Both work by providing an alternative route, or mechanism, for the reaction that has a different activation energy. This is illustrated in Figure 11.7 using an enthalpy profile and in Figure 11.8 with a Boltzmann distribution. On these diagrams, $E_a$ is the activation energy of the uncatalysed reaction; $E_{cat}$ is the activation energy of the catalysed reaction.

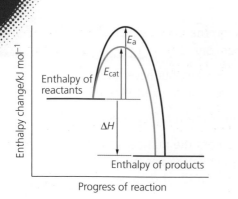

**Figure 11.7** Enthalpy-profile diagram showing the activation energy.

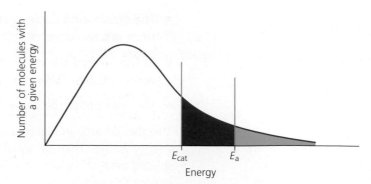

**Figure 11.8** Boltzmann distribution showing the effect of a catalyst on The blue area indicates the proportions of particles with energy greater than the activation energy without the catalyst. The red and blue areas indicate the proportion of particles with energy greater than the activation energy. effect of a catalyst on activation energy.

When a catalyst is used the enthalpy change and the distribution of energies remain the same but the activation energy is lowered so that more particles now have more energy than the new lower activation energy, $E_{cat}$. Hence the reaction goes faster.

### Homogeneous catalysts

The most common example is the formation of an ester, $R_1COOR_2$, by reacting an alcohol with a carboxylic acid using sulfuric acid as the catalyst. Esterification will be covered fully in the A level Book 2.

Another example involves the oxidation of iodide ions using persulfate ions (peroxosulfate ions).

Iodide ions, $I^-$(aq) are easily oxidised and the persulfate ion, $S_2O_8^{2-}$(aq) is a powerful oxidising agent but surprisingly the reaction between them in water is slow. The equation for the reaction is:

$$S_2O_8^{2-} + 2I^- \rightarrow 2SO_4^{2-} + I_2$$

For the reaction to occur the two ions must collide, however, the rate of collision is slow because both ions are negative and repel each other.

The reaction can be catalysed by adding aqueous solutions of either $Fe^{2+}$(aq) ions or $Fe^{3+}$(aq) ions.

The transition metal ions catalyse the reaction by providing an alternate two-step mechanism:

### Tip

You will not be expected to know the details of this reaction but it serves as an illustration of how a homogeneous catalyst might work.

| catalyst | $Fe^{2+}$(aq) | $Fe^{3+}$(aq) |
|---|---|---|
| Step 1 | $S_2O_8^{2-} + 2Fe^{2+} \rightarrow 2SO_4^{2-} + 2Fe^{3+}$ | $2Fe^{3+} + 2I^- \rightarrow 2Fe^{2+} + I_2$ |
| Step 2 | $2Fe^{3+} + 2I^- \rightarrow 2Fe^{2+} + I_2$ | $S_2O_8^{2-} + 2Fe^{2+} \rightarrow 2SO_4^{2-} + 2Fe^{3+}$ |
| | $Fe^{2+}$ is reformed and acts as a catalyst | $Fe^{3+}$ is reformed and acts as a catalyst |
| Net reaction | $S_2O_8^{2-} + 2I^- \rightarrow 2SO_4^{2-} + I_2$ | $S_2O_8^{2-} + 2I^- \rightarrow 2SO_4^{2-} + I_2$ |

The $Fe^{2+}$ and the $Fe^{3+}$ ions provide an alternative mechanism that involves an intermediate step of lower activation energy. The activation energy is lowered because the initial step in the mechanism now involves ions of opposite charge. The iron ions from the catalyst take part in the reaction, but are released at the end of the reaction, so the final amount of the catalyst is the same as when the reagents were mixed.

Another example of homogeneous catalysis occurs in the loss of ozone from the upper atmosphere (stratosphere). This is a gas-phase reaction in which the catalyst is the chlorine radical (Chapter 14, page 237).

### Heterogeneous catalysts

A **heterogeneous catalyst** is in a different phase from the reactants. The most common type of heterogeneous catalysis involves reactions of gases in the presence of a solid catalyst.

The mode of action of a heterogeneous catalyst is different from that of a homogeneous catalyst. A heterogeneous catalyst works by adsorbing the gases onto its solid surface. This adsorption weakens the bonds within the reactant molecules, which lowers the activation energy for the reaction. Bonds are broken and new bonds are formed. The product molecules are then desorbed from the surface of the solid catalyst.

**Tip**

Make sure you don't confuse adsorb with absorb.

Transition metals are often used as heterogeneous catalysts. Iron is the catalyst in the Haber process for the manufacture of ammonia (page 176). The iron is either finely divided (and therefore has a large surface area) or in a porous form containing a small amount of metal oxide promoters:

$$N_2(g) + 3H_2(g) \xrightarrow{\text{Fe catalyst}} 2NH_3(g)$$

The production of poly(ethene) and other polymers from alkenes requires the use of a Ziegler–Natta catalyst. This is a mixture of titanium(IV) chloride and an organic compound of aluminium, $Al_2(CH_3)_6$. The mode of action of such catalysts is complicated and not altogether understood. Another important heterogeneous catalyst is the alloy of platinum, palladium and rhodium used as catalytic converters in vehicles.

## Economic importance of catalysts

Catalysts have great importance economically and industry spends considerable time trying to develop efficient and cost-effective versions. The reduction in the energy requirements of reactions means that they operate at lower temperatures and with a consequent saving of energy. This not only results in lower costs but may also reduce the demand for fossil fuels whose combustion is a normal source of the energy supply. Burning of fossil fuels also creates pollution and the consequent reduction in the emission of carbon dioxide in particular is an important consideration.

### Test yourself

2 Define the term 'activation energy' ($E_a$).
3 If the temperature is changed, the distribution of molecular energies changes.
   a) Copy the distribution curve on the right and add two new curves:
      i)  at temperature $T_2$, which is higher than $T_1$
      ii) at temperature $T_3$, which is lower than $T_1$.
   b) What do the three curves at $T_1$, $T_2$ and $T_3$ have in common?
   c) Explain why increasing the temperature increases the rate of reaction.

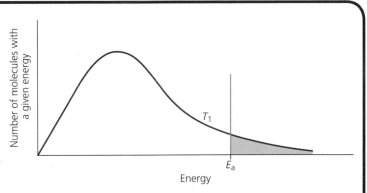

4 a) Explain what is meant by the term 'catalyst'.
  b) Use a Boltzmann distribution curve to explain how a catalyst works.
  c) Explain what is meant by a heterogeneous catalyst. Give an example.
  d) Explain what is meant by homogeneous catalyst. Give an example.

5 Describe how, and explain why, increasing the pressure on a gaseous reaction affects the rate of reaction.

6 The graph illustrates the effect of changing conditions on the reaction of zinc metal with sulfuric acid. The red line shows the volume of hydrogen plotted against time using an excess of zinc and $50\,cm^3$ of $2.0\,mol\,dm^{-3}$ sulfuric acid at $20\,°C$.
  a) Write a balanced equation for the reaction.
  b) Identify which line of the graph shows the effect of carrying out the same reaction under the same conditions with the following changes:
    i) adding a few drops of copper(II) sulfate solution to act as a catalyst
    ii) increasing the temperature to $30\,°C$
    iii) using the same mass of zinc but in larger pieces
    iv) using $50\,cm^3$ of $1.0\,mol\,dm^{-3}$ sulfuric acid.

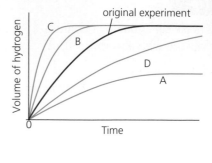

# Chemical equilibrium

## Reversible reactions

There are many everyday examples of reactions or processes that are reversible, the best known being the changes of the physical states of water. When the temperature of water falls below $0\,°C$, it freezes and ice forms. When the temperature rises above $0\,°C$, the ice melts and water forms again. This process can be represented as:

$$H_2O(s) \rightleftharpoons H_2O(l)$$

The symbol '$\rightleftharpoons$' indicates that a reaction is reversible.

Another reversible reaction is esterification.

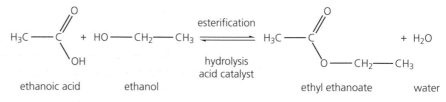

In the presence of an acid catalyst, ethanoic acid reacts with ethanol to produce the ester ethyl ethanoate and water but ethyl ethanoate is hydrolysed by water, also in the presence of an acid catalyst, to produce ethanoic acid and ethanol again.

**Tip**

In a reversible reaction as shown, the reaction from left to right is called the **forward reaction**; the reaction from right to left is the **reverse reaction**.

Refluxing allows the reactants to mix together at the boiling point of the mixture without letting any vapour escape. This is achieved by placing a condenser on top of the flask containing the reactants (see page 229).

## Dynamic equilibrium

If ethanoic acid and ethanol are refluxed in the presence of an acid catalyst, the forward reaction is initially fast because the concentrations of both reagents are high.

However, as they react, the concentration of each reagent decreases, which lowers the rate of the forward reaction.

The reverse reaction is initially slow because the amount of ethyl ethanoate and water present is small. However, the concentrations of ethyl ethanoate and water build up, increasing the rate of the reverse reaction.

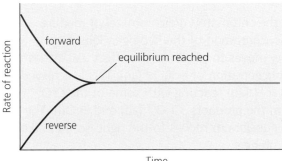
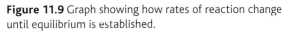

**Figure 11.9** Graph showing how rates of reaction change until equilibrium is established.

In summary, the forward reaction starts rapidly but slows down; the reverse reaction starts slowly and speeds up. A point is eventually reached when the rate of the forward reaction equals the rate of the reverse reaction (Figure 11.9). When this happens, the system is in dynamic equilibrium.

- The system is in **equilibrium** because the amount of each chemical remains constant.

- The equilibrium is **dynamic** because the reactants and the products are both constantly interacting.

Another example of a dynamic equilibrium occurs during the formation of stalactites and stalagmites (Figure 11.10).

**Key term**

A dynamic equilibrium is reached when the rate of the forward reaction equals the rate of the reverse reaction. The concentrations of the reagents and products remain constant; the reactants and the products react continuously.

**Figure 11.10** Stalactites and stalagmites form in limestone caves because the reaction of calcium carbonate with carbon dioxide and water is reversible.
$CaCo_{3(s)} + H_2O_{(l)} + CO_2 (g) \rightleftharpoons Ca(HCO_3)_{2(aq)}$

Dynamic equilibrium can only be achieved in a closed system (i.e. no reagent is allowed to escape).

A visual representation of a dynamic equilibrium is a 'Mexican wave', often seen at major sporting events. The wave moves around the stadium (dynamic), but as one person stands to wave, another sits down (equilibrium). The number of people waving and the number of people sitting is constant but the individuals are constantly changing.

# le Chatelier's principle

**Key term**

le Chatelier's principle states that if a closed system at equilibrium is subject to a change, the system will move to minimise the effect of that change.

**Tip**

By 'system' le Chatelier meant the equilibrium mixture containing both the reactants and products.

The French chemist Henri le Chatelier studied dynamic equilibria and suggested a general qualitative rule that could be used to predict the movement of the position of the equilibrium. This is known as le Chatelier's principle.

The factors that can be readily changed are concentration, pressure and temperature and le Chatelier's principle tells us that:

- If the concentration of a component is increased, the system will move to decrease the concentration of that component.

- If the pressure is increased, the system will move to decrease the pressure.

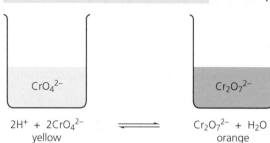

$$2H^+ + 2CrO_4^{2-} \rightleftharpoons Cr_2O_7^{2-} + H_2O$$
yellow                    orange

**Figure 11.11** Equilibrium between the chromate ion, $CrO_4^{2-}$, and the dichromate ion, $Cr_2O_7^{2-}$.

• If the temperature is increased, the system will move to decrease the temperature.

# Effect of changing concentration on the equilibrium position

Le Chatelier's principle applies to all reactions at equilibrium. However, there are only a few reactions where the principle can be observed in action. One such mixture is the equilibrium between the yellow chromate ion, $CrO_4^{2-}$, and the orange dichromate ion, $Cr_2O_7^{2-}$ (Figure 11.11). Each ion is coloured, so it is possible to observe the movement of the position of equilibrium.

Adding acid to the chromate(VI)/dichromate(VI) mixture increases the concentration of the hydrogen ions, $H^+(aq)$. The system now moves to minimise the effect, i.e. it tries to decrease the concentration of acid, $H^+(aq)$. This is achieved by the additional $H^+(aq)$ reacting with some of the $CrO_4^{2-}$ (aq) ions to form the products, $Cr_2O_7^{2-}(aq)$ and $H_2O(l)$. The position of the equilibrium moves to the right, so the mixture becomes more orange.

This can be reversed by adding a small amount of base, $OH^-$ (aq). The base reacts with the $H^+(aq)$, which decreases the concentration of the $H^+(aq)$ on the left-hand side of the equilibrium. The system moves to minimise the effect, i.e. it tries to increase the concentration of the acid, $H^+(aq)$. This is achieved by the $H_2O(l)$ reacting with some of the $Cr_2O_7^{2-}(aq)$ ions to form $CrO_4^{2-}(aq)$ ions and $H^+(aq)$ ions. The equilibrium position moves to the left and the orange solution becomes yellow.

Both these changes to the equilibrium mixture are, of course, a consequence of a change in the rate of reaction on either side of the equilibrium. Adding $H^+(aq)$ increases the rate of reaction between $CrO_4^{2-}$ (aq) and $H^+(aq)$ so that there will be an increase in the formation of $Cr_2O_7^{2-}$ (aq) and $H_2O(l)$. The equilibrium then proceeds to re-establish itself but the end result will nevertheless be an increase in the $Cr_2O_7^{2-}(aq)$ present.

# Effect of changing pressure on the equilibrium position

The pressure of a mixture of gases depends on the number of gas molecules present in a given volume. The greater the number of gas molecules, the greater is the pressure in an equilibrium mixture. If the pressure is increased in a system at equilibrium, the equilibrium position alters to resist the increased pressure by reducing the total number of gas molecules. Using le Chatelier's principle this is achieved by the equilibrium position moving to the side of the reaction mixture that has fewer gaseous molecules.

If the pressure is increased on a system such as $2SO_2(g) + O_2(g) \rightleftharpoons 2SO_3(g)$, the equilibrium position moves to the right. This is because there are three molecules of gas ($2SO_2(g) + 1O_2(g)$) on the left-hand side of the equilibrium but only two molecules of gas ($2SO_3(g)$) on the right-hand side. The result of this shift in equilibrium position is to reduce the total number of molecules in the mixture and hence the increase in pressure is resisted.

If the pressure is increased on a system such as $2HI(g) \rightleftharpoons H_2(g) + I_2(g)$, the position of the equilibrium does not move because there are equal numbers of gaseous molecules on each side of the equilibrium.

## Effect of changing temperature on the equilibrium position

Temperature influences the rate of reaction and also plays an important role in determining the equilibrium position. The effect of temperature can only be predicted if the sign of $\Delta H$ for value of the reaction is known.

Consider again the equilibrium between sulfur dioxide, oxygen and sulfur trioxide.

The forward reaction is exothermic:

$$2SO_2(g) + O_2(g) \rightleftharpoons 2SO_3(g) \qquad \Delta H = -197 \, kJ \, mol^{-1}$$

and the reverse reaction is endothermic.

If the temperature of the reaction mixture is increased, le Chatelier's principle predicts that the mixture will respond by attempting to lower the temperature again. This will be achieved by moving the equilibrium to the left. It can perhaps be best understood by rewriting the forward reaction to emphasise that heat is given to the surroundings when the reaction occurs.

$$2SO_2(g) + O_2(g) \rightleftharpoons 2SO_3(g) + \text{heat}$$

So if heat is supplied by increasing the temperature, the reaction responds by trying to remove some of the heat and therefore more $SO_2(g)$ and $O_2(g)$ are formed.

### Example 1

At 500°C, hydrogen iodide forms an equilibrium with hydrogen and gaseous iodine.

$$2HI(g) \rightleftharpoons H_2(g) + I_2(g) \qquad \Delta H = +51 \, kJ \, mol^{-1}$$

Predict the effect on the rate at which the equilibrium is formed and the position of the equilibrium if

a) the temperature is increased to 700°C
b) the equilibrium is cooled to room temperature.

**Answer**

a) Increasing the temperature will increase the rate at which the equilibrium will be established.
The equilibrium can be written as:

$$\text{heat} + 2HI(g) \rightleftharpoons H_2(g) + I_2(g)$$

So increasing the temperature will also mean more hydrogen and iodine will form

b) As the temperature is decreased, the equilibrium will be established more slowly and the equilibrium will initially shift so that more hydrogen iodide is formed.
However, iodine is a solid at room temperature so once this starts forming, iodine is being removed from the equilibrium and all the hydrogen iodide will then decompose into hydrogen and iodine.

## Effect of using a catalyst on the equilibrium position

A catalyst is a substance that speeds up the rate of reaction by providing an alternative route, or mechanism, that has a lower activation energy. A catalyst does not alter the amount of product.

In a system at equilibrium, a catalyst speeds up the forward and the reverse reactions equally. Therefore, a catalyst has no effect on the *position* of the equilibrium.

However, catalysts play an important part in reversible reactions because they reduce the time taken to reach equilibrium. In the presence of a catalyst, the same amount of product is produced more quickly.

## The use of compromise conditions – the Haber process

Large amounts of nitrogenous compounds, particularly fertilisers, are needed by humans. Atmospheric nitrogen is in plentiful supply but it cannot be used directly. It has to be converted (fixed) into a usable compound. The Haber process 'fixes' nitrogen, converting it into ammonia:

$$N_2(g) + 3H_2(g) \rightleftharpoons 2NH_3(g) \qquad \Delta H = -93\,kJ\,mol^{-1}$$

Le Chatelier's principle allows the optimum conditions for this industrial process to be determined.

The enthalpy change is $-93\,kJ\,mol^{-1}$, so the forward reaction is exothermic. Therefore, if the temperature is increased, the equilibrium moves to the left and less ammonia is produced. It follows that a low temperature is optimum for the formation of ammonia.

If the pressure is increased on the system, the equilibrium position moves to the right to reduce the number of molecules. Therefore, high pressure is optimum for the formation of ammonia.

The effect of changing temperature and pressure on the yield of ammonia is shown in Table 11.2.

**Table 11.2** Percentage yield of ammonia under different conditions.

| Temperature/K | 25 atm | 50 atm | 100 atm | 200 atm |
|---|---|---|---|---|
| 373 | 92 | 94 | 96 | 98 |
| 573 | 28 | 40 | 53 | 67 |
| 773 | 3 | 6 | 11 | 18 |

The lowest temperature (373 K = 100 °C) gives the highest percentage yield. However, at low temperature, the reaction rate is slow. A compromise has to be reached between yield and rate of reaction.

The highest pressure (200 atm) gives the highest percentage yield and also increases the reaction rate. However, at high pressure the operating costs increase. A compromise has to be reached between yield/rate and costs.

Tip

Atm. is an abbreviation for atmosphere. 1 atm is approximately equal to 101 kPa (or 100 kPa).

## Manufacturing conditions

The manufacture of ammonia in a modern plant is highly efficient. The operating conditions vary but are typically:

● a temperature of around 700 K (427 °C)

● a pressure of around 200 atm.

These conditions are a compromise.

The rate of reaction is also increased by using a catalyst of either finely divided iron, or porous iron incorporating metal oxide promoters.

Additionally, if the equilibrium mixture is cooled to about −233 K (−40 °C), the ammonia gas liquefies but both nitrogen and hydrogen remain as gases (see Table 11.3). Therefore, ammonia gas is lost from the equilibrium mixture. The system moves to minimise the effect of this loss so more nitrogen and hydrogen react to replace the ammonia gas that was liquefied and removed from the system, and any unreacted nitrogen and hydrogen are recycled and reacted again.

Most manufacturing processes involving gases require compromise conditions to determine the best operating conditions to achieve maximum efficiency for the formation of the desired product.

**Table 11.3** Boiling points of nitrogen, hydrogen and ammonia.

| Gas | Boiling point |
|---|---|
| Nitrogen | 77 K (−196 °C) |
| Hydrogen | 20 K (−253 °C) |
| Ammonia | 240 K (−33 °C) |

### Test yourself

7 a) Explain what is meant by the term 'reversible reaction'.
  b) Explain what is meant by the term 'dynamic equilibrium'.
  c) State four external variables that could be changed to affect an equilibrium.
8 Use le Chatelier's principle to deduce what happens to the following equilibrium when it is subjected to the changes below:
$$CH_3OH(g) \rightleftharpoons CO(g) + 2H_2(g) \qquad \Delta H = -129\,kJ\,mol^{-1}$$
  a) the temperature is increased
  b) the pressure is decreased
  c) a catalyst is added.

# The equilibrium constant, $K_c$

For any reversible reaction that has reached dynamic equilibrium it is possible to relate the concentrations of the reactants and products present using an important constant known as the **equilibrium constant**. The equilibrium constant varies with temperature but its value is the same for any other change in condition.

In general terms for the reaction:

$$aA + bB \rightleftharpoons cC + dD$$

the equilibrium constant $K_c$ is given by:

$$K_c = \frac{[C]^c[D]^d}{[A]^a[B]^b}$$

In the expression for $K_c$ the square brackets '[ ]' indicate that the concentrations of the reactants and products are expressed in units of $mol\,dm^{-3}$. The concentrations of the chemicals on the right-hand side of the equation appear on the top line of the expression. The concentrations of reactants on the left appear on the bottom line. Each concentration term is raised to the power of the number in front of its formula in the balanced equation.

For the equilibrium:

$$3H_2(g) + N_2(g) \rightleftharpoons 2NH_3(g)$$

$$K_c = \frac{[NH_3(g)]^2}{[H_2(g)]^3[N_2(g)]}$$

If there is a solid reactant or product in the equilibrium then it is not included in the expression for $K_c$.

For the equilibrium (which forms at high temperatures):

$$H_2O(g) + C(s) \rightleftharpoons H_2(g) + CO(g)$$

$$K_c = \frac{[H_2(g)][CO(g)]}{[H_2O(g)]}$$

The value of $K_c$ is important because it gives an indication of the balance of reactants and products present in the equilibrium mixture. If $K_c$ has a large value it will mean that there are more of the products present than the reactants, if $K_c$ is small then it is the reactants that must be present in larger quantity. The values of $K_c$ do, in fact, vary enormously from large values down to values that are very small.

For example in the equilibrium:

$$CH_3COOH(l) + C_2H_5OH(l) \rightleftharpoons CH_3COOC_2H_5(l) + H_2O(l)$$

$$K_c = \frac{[CH_3COOC_2H_5(l)][H_2O(l)]}{[CH_3COOH(l)][C_2H_5OH(l)]}$$

The value of $K_c$ is approximately 4 at room temperature so there is quite an equal balance between the reactants and products. $K_c$ will not have any units because the concentrations of each component will cancel out in the equilibrium expression.

For the equilibrium between sulfur dioxide, oxygen and sulfur trioxide:

$$2SO_2(g) + O_2(g) \rightleftharpoons 2SO_3(g)$$

$$K_c = \frac{[SO_3(g)]^2}{[SO_2(g)]^2[O_2(g)]}$$

The equilibrium constant $K_c$ for this reaction has a size of approximately $1.0 \times 10^{12}$ at 230°C, so in this case the equilibrium mixture has vastly more of the product, $SO_3$, than the reactants. In this case $K_c$ will have units of 1/mol dm$^{-3}$ which should be written as dm$^{+3}$ mol$^{-1}$ because the concentrations do not completely cancel in the equilibrium expression.

## The effect on $K_c$ of a change in temperature

The effect of a change in temperature on the value of $K_c$ is best explained using an example.

The equilibrium:

$$2SO_2(g) + O_2(g) \rightleftharpoons 2SO_3(g) \qquad \Delta H = -197\,kJ\,mol^{-1}$$

is exothermic, and has already been explained on page 175, an increase in temperature causes more $SO_2$ and $O_2$ to be formed within the equilibrium mixture. It must therefore follow that the value of the equilibrium constant will decrease as the temperature rises.

## Calculating an equilibrium constant

If the equilibrium concentrations of each component of an equilibrium mixture is known the equilibrium constant can be calculated by substituting these values into the expression for the equilibrium constant.

### Example 2

At 450°C, hydrogen and gaseous iodine form an equilibrium mixture with hydrogen iodide.

$$H_2(g) + I_2(g) \rightleftharpoons 2HI(g)$$

When an equilibrium mixture is analysed it is found to contain $0.015\,mol\,dm^{-3}$ of HI, $0.0012\,mol\,dm^{-3}$ of $I_2$ and $0.0038\,mol\,dm^{-3}$ of $H_2$.

Calculate the value of the equilibrium constant to 2 significant figures.

**Answer**
The equilibrium constant for the formation of HI is:

$$K_c = \frac{[HI(g)]^2}{[H_2(g)][I_2(g)]}$$

Therefore $K_c = (0.015)^2/(0.0012)(0.0038) = 49$

### Test yourself

9 a) Write the equilibrium constant for the equilibrium:
$$2NO_2(g) \rightleftharpoons N_2O_4(g)$$
   b) At a certain temperature the equilibrium constant for this reaction has a value of 0.0025.
   (i) What does this indicate about the position of the equilibrium?
   (ii) If the concentration of $NO_2$ is $1.0\,mol\,dm^{-3}$, calculate the concentration of the $N_2O_4$.
10 When the equilibrium $2NOCl(g) \rightleftharpoons 2NO(g) + Cl_2(g)$ is analysed it is found that the equilibrium concentrations of the gases are $NOCl = 3.42\,mol\,dm^{-3}$, $NO = 0.32\,mol\,dm^{-3}$, $Cl_2 = 0.16\,mol\,dm^{-3}$. Use these figures to calculate $K_c$ for the equilibrium.

# Practice questions

## Multiple choice questions 1–10

**1** The equilibrium concentration of the products of the reaction

$$H_2O(g) + C(s) \rightleftharpoons H_2(g) + CO(g) \quad \Delta H = +180.5 \, kJ$$

can be increased both by:

A raising the temperature and lowering the pressure

B raising the temperature and increasing the pressure

C lowering the temperature and increasing the pressure

D lowering the temperature and pressure.

**2** In which one of the following reactions would a decrease in pressure most increase the equilibrium yield of the products?

A $NH_4Cl(g) \rightleftharpoons NH_3(g) + HCl(g)$

B $Fe_3O_4(s) + 4H_2(g) \rightleftharpoons 3Fe(s) + 4H_2O(g)$

C $2HBr(g) \rightleftharpoons H_2(g) + Br_2(g)$

D $2NO_2(g) \rightleftharpoons N_2O_4(g)$

**3** Which one of the following gives the correct expression for the equilibrium constant for the reaction shown by the equation below?

$$4CuO(s) \rightleftharpoons 2Cu_2O(s) + O_2(g)$$

A $\dfrac{[CuO]^4}{[Cu_2O]^2[O_2]}$

B $[O_2]$

C $\dfrac{[Cu_2O]^2 \, [O_2]}{[CuO]^4}$

D $\dfrac{[Cu_2O]^2}{[CuO]^4}$

**4**

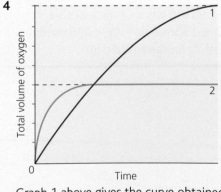

Graph 1 above gives the curve obtained when 40 cm³ of 2.0 mol dm⁻³ hydrogen peroxide is decomposed.

Graph 2 is obtained in another experiment to decompose hydrogen peroxide. Assuming all other conditions of the experiment are kept constant, which one of the following would result in curve 2 being obtained.

A 80 cm³ of 1.0 mol dm⁻³ hydrogen peroxide is decomposed

B 20 cm³ of 2.0 mol dm⁻³ hydrogen peroxide is decomposed

C 20 cm³ of 4.0 mol dm⁻³ hydrogen peroxide is decomposed

D 10 cm³ of 4.0 mol dm⁻³ hydrogen peroxide is decomposed

**5** A student carries out an experiment to measure the loss in mass of a container containing **excess** marble (calcium carbonate) chips when reacted with 50 cm³ of 1.0 mol dm⁻³ hydrochloric acid. The results are shown below.

| Time/min | 1 | 2 | 3 | 4 | 5 | 6 | 7 |
|---|---|---|---|---|---|---|---|
| Loss in mass/g | 0.45 | 0.80 | 1.02 | 1.09 | 1.10 | 1.10 | 1.10 |

The student repeats the experiment again using **excess** marble chips and hydrochloric acid and obtains the following results.

| Time/min | 1 | 2 | 3 | 4 | 5 | 6 | 7 |
|---|---|---|---|---|---|---|---|
| Loss in mass/g | 0.36 | 0.64 | 0.82 | 0.95 | 1.04 | 1.10 | 1.10 |

Which one of the following gives the volume and concentration of the hydrochloric acid that the student used when the experiment was repeated?

A 25 cm³ of 2.0 mol dm⁻³

B 50 cm³ of 0.5 mol dm⁻³

C 100 cm³ of 0.5 mol dm⁻³

D 100 cm³ of 1.0 mol dm⁻³

**6** Which one of the following would react to give off 0.01 mol of hydrogen most quickly when added to excess magnesium ribbon?

A 25 cm³ of 1.0 mol dm⁻³ $H_2SO_4$

B 25 cm³ of 1.0 mol dm⁻³ HCl

C 20 cm³ of 2.0 mol dm⁻³ HCl

D 15 cm³ of 2.0 mol dm⁻³ $H_2SO_4$

**7** Two equilibria are shown below:

first equilibrium: $2SO_2(g) + O_2(g) \rightleftharpoons 2SO_3(g)$

second equilibrium: $2NH_3(g) \rightleftharpoons 3H_2(g) + N_2(g)$

Each of these equilibria is formed by starting with the reactants on the left-hand side of the equilibria.

The rate at which each equilibrium is formed will be increased by:

A using decreased pressure in the first equilibrium and increased pressure in the second equilibrium

B using increased pressure in the first equilibrium and decreased pressure in the second equilibrium

C using decreased pressure in both cases

D using increased pressure in both cases.

Use the key below to answer Questions 8, 9 and 10.

| A | B | C | D |
|---|---|---|---|
| 1, 2 & 3 correct | 1, 2 correct | 2, 3 correct | 1 only correct |

**8**

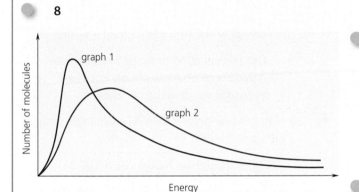

Graph 1 above represents the Boltzmann distribution at a given temperature.

Graph 2 would be obtained:

1 by raising the temperature
2 by adding a catalyst
3 by reducing the number of molecules present.

**9** Equilibrium is reached in a chemical reaction when:

1 the activation energy of the reaction changes until it becomes the same for both reactants and products

2 the products react together at the same rate as they are formed

3 the concentrations of the reactants and products are no longer changing.

**10** Which of the following affect the rate at which an equilibrium is established?

1 an increase in pressure
2 the addition of a catalyst
3 the size of the equilibrium constant

**11** When 4 g of zinc foil are reacted with 50 cm³ of 1.0 mol dm⁻³ sulfuric acid, 100 cm³ of hydrogen are collected in a syringe in 50 s. In this reaction the sulfuric acid is in excess.

Explain whether the reaction would be faster or slower if the experiment was repeated using:

a) 4 g of zinc foil and 50 cm³ of 2.0 mol dm⁻³ sulfuric acid

b) 2 g of zinc foil are reacted with 50 cm³ of 1.0 mol dm⁻³ sulfuric acid

c) 4 g of zinc foil are reacted with 100 cm³ of 1.0 mol dm⁻³ sulfuric acid.  (3)

**12** Heating limestone, $CaCO_3$, in a closed furnace produces an equilibrium mixture of calcium carbonate with calcium oxide, CaO, and carbon dioxide gas. Heating the solid in an open furnace decomposes the solid completely into the oxide. How do you account for this difference?  (2)

**13** NO(g) and $Cl_2$(g) form an equilibrium with NOCl(g):

$$NO(g) + \tfrac{1}{2}Cl_2(g) \rightleftharpoons NOCl(g) \quad \Delta H = -77.1\,kJ\,mol^{-1}$$

Explain the effect on the equilibrium of:

a) decreasing the temperature
b) decreasing the pressure
c) adding a catalyst.  (7)

**14** Methane and steam form an equilibrium mixture with carbon dioxide and hydrogen.

$$CH_4(g) + 2H_2O(g) \rightleftharpoons CO_2(g) + 4H_2(g)$$
$$\Delta H = +165\,kJ\,mol^{-1}$$

Explain the effect on the rate at which equilibrium forms and the position of the equilibrium of:

a) increasing the temperature
b) decreasing the pressure
c) adding a catalyst.  (8)

**15** Write an expression for the equilibrium constant for each of the following:

a) $H_2(g) + CO_2(g) \rightleftharpoons H_2O(g) + CO(g)$
b) $H_2(g) + I_2(g) \rightleftharpoons 2HI(g)$
c) $CH_4(g) + 2H_2O(g) \rightleftharpoons CO_2(g) + 4H_2(g)$  (4)

**16 a)** The value of $K_c$ for the equilibrium:

$$H_2(g) + CO_2(g) \rightleftharpoons H_2O(g) + CO(g)$$

is $1.0 \times 10^{-5}$ at 298 K.

What does this indicate about the balance of the equilibrium at 298 K?  (2)

**b)** What will be the value of the equilibrium constant for:

$$H_2O(g) + CO(g) \rightleftharpoons H_2(g) + CO_2(g)$$  (1)

**17** An experiment is carried out to measure how the rate of reaction changes when 6.54 g of zinc chips reacts with 25.0 cm³ of 1.00 mol dm⁻³ hydrochloric acid in a conical flask.

$$Zn(s) + 2HCl(aq) \rightarrow ZnCl_2(aq) + H_2(g)$$

A graph is plotted of the volume of hydrogen obtained every 30 s. This is shown below.

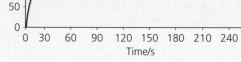

**a) i)** Calculate the amounts in mol of the zinc and hydrochloric acid used in this experiment.

**ii)** Which of the reactants is in excess?

**b)** On a copy of the graph, sketch and label the lines you would expect to get if the experiment was repeated using:
   **i)** 10.00 g of zinc and 25 cm³ of 1.00 mol dm⁻³ hydrochloric acid – label this line A
   **ii)** 6.54 g of zinc and 20 cm³ of 1.00 mol dm⁻³ hydrochloric acid – label this line B
   **iii)** the original reactants but with the 6.54 g of zinc broken into smaller pieces – label this line C.
   Explain each of your answers above. (10)

**18** Under what conditions are the following in equilibrium?
**a)** water and ice
**b)** water and steam
**c)** copper(II) sulfate crystals and copper(II) sulfate solution (3)

**19** Carbon dioxide is dissolved in water under pressure to make sparkling mineral water. In a bottle of sparkling mineral water, there is an equilibrium between carbon dioxide dissolved in the drink, $CO_2(aq)$, and carbon dioxide in the gas above the drink, $CO_2(g)$.
**a)** Write an equation to represent the equilibrium between carbon dioxide gas and carbon dioxide in solution.
**b)** Use this example to explain the term 'dynamic equilibrium'.
**c)** Explain why lots of bubbles of gas form when a bottle of sparkling mineral water is opened.
**d)** Less than 1% of the dissolved carbon dioxide reacts with water to form hydrogencarbonate ions:

$$CO_2(g) + H_2O(l) \rightleftharpoons HCO_3^-(aq) + H^+(aq)$$

Use this equation to explain why carbon dioxide is much more soluble in sodium hydroxide solution than in water. (7)

**20** Nitrogen and oxygen form an equilibrium with nitrogen(II) oxide.

$$\tfrac{1}{2}N_2(g) + \tfrac{1}{2}O_2(g) \rightleftharpoons NO(g) \quad \Delta H = +90.3 \, kJ \, mol^{-1}$$

**a)** Explain the effect on the equilibrium of:
   **i)** decreasing the temperature
   **ii)** increasing the pressure.
**b)** Suggest why this equilibrium does not readily form at room temperature. (7)

**21** At 600 °C an equilibrium is established between nitrogen(IV) oxide and a mixture of nitrogen(II) oxide and oxygen. When the temperature is raised to 700 °C it is found that there is less nitrogen(IV) oxide in the equilibrium mixture.
**a)** Write the equation for the equilibrium showing the decomposition of nitrogen(IV) oxide into nitrogen(II) oxide and oxygen
**b)** Explain whether the decomposition is endothermic or exothermic. (3)

**22** When the indicator methyl orange is dissolved in water, the following dynamic equilibrium is set up:

a) Hydrochloric acid is added to the equilibrium mixture. State, and explain, the colour change that occurs.

b) Then, aqueous potassium hydroxide, KOH(aq), is added dropwise to the solution in (a) until no further colour change occurs. Deduce the colour changes that occur. Explain your answer.

c) Aqueous sodium carbonate, $Na_2CO_3$(aq), is added to the original equilibrium mixture. State what you would see. Explain your answer. (11)

**23 a)** State le Chatelier's principle.

b) The gases $NO_2$ and $N_2O_4$ form an equilibrium mixture.

$$2NO_2(g) \rightleftharpoons N_2O_4(g) \qquad \Delta H = -57.2\,kJ\,mol^{-1}$$

$NO_2$ is brown in colour but $N_2O_4$ is colourless. An equilibrium mixture of these gases is placed in a syringe. Explain what colour change you would expect to observe if:

i) the pressure on the mixture of gases is increased

ii) the syringe is cooled. (10)

**24** When steam is passed over heated carbon the following equilibrium is established:

$$C(s) + H_2O(g) \rightleftharpoons CO(g) + H_2(g)$$

a) If the pressure is increased, explain what effect this would have on this equilibrium.

b) The enthalpy of formations of $H_2O(g)$ and $CO(g)$ are $-241.8\,kJ\,mol^{-1}$ and $-110.5\,kJ\,mol^{-1}$ respectively.

i) Calculate the enthalpy change for the forward reaction to make $CO(g)$ and $H_2(g)$.

ii) Explain what effect increasing the temperature would have on the equilibrium.

iii) What would happen if the temperature was decreased below 100 °C? (10)

**25** The values of $K_c$ for the equilibrium $N_2O_4(g) \rightleftharpoons 2NO_2(g)$ are $1.44\,mol\,dm^{-3}$ at 400 K and $41\,mol\,dm^{-3}$ at 500 K.

a) What is the effect of increasing the temperature on the position of equilibrium?

b) What is the sign of $\Delta H$ for the forward reaction? (3)

## Challenge

**26** In the UK, ethanol is manufactured by the reaction of ethene with steam in the presence of a catalyst. The reaction is reversible and exothermic, with an enthalpy change of reaction of $-46\,kJ\,mol^{-1}$. The catalyst is phosphoric acid. It is supported on the surface of an inert solid (silica). The proportion of steam in the reaction has to be controlled to prevent the catalyst taking up water so that it is diluted and runs off the surface of the solid.
The process is carried out at 300 °C at 60–70 times atmospheric pressure. The water:ethene ratio is around 0.6 : 1.

a) Which alternative method of making ethanol is mainly used in other parts of the world?

b) How can ethene for this process be made?

c) i) Write an equation for the reaction of ethene with steam and show the enthalpy change.

ii) What type of reaction is this?

d) Suggest a reason for supporting the phosphoric acid on the surface of silica.

e) State three conditions that favour the formation of ethene at equilibrium according to le Chatelier's principle.

f) Suggest reasons why the working conditions are not those suggested by your answer to part (a).

g) About 5% of the ethene is converted to ethanol each time the reaction mixture passes through the catalyst bed. Suggest how a yield of 95% conversion is achieved. (8)

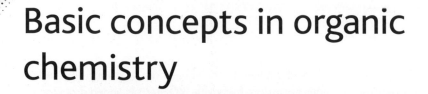

# Chapter 12

# Basic concepts in organic chemistry

## Functional groups

Carbon is found in millions of different compounds including the molecules of most living systems. Carbohydrates, fats and proteins all contain carbon and hydrogen and the study of compounds that contain the C–H bond is referred to as organic chemistry. There are three main reasons why carbon can form so many different compounds.

- Carbon, unlike most other elements, has the ability to form chains, branched chains and rings of varying sizes.

- The C–C and the C–H bonds present in most organic compounds have strong bonds with very low polarity and are therefore not very reactive.

- When carbon atoms form chains or rings linked by single covalent bonds no more than two of the bonds of each carbon atom are used. The remaining two bonds can bond to other atoms such as hydrogen, oxygen, nitrogen and the halogens.

This gives rise to millions of organic compounds.

Ethane, $CH_3CH_3$, and ethanol, $CH_3CH_2OH$ have very different properties despite having very similar structures.

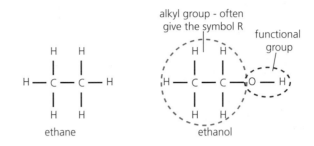

- Ethane is a gas, ethanol is a liquid.

- Ethane is unreactive, ethanol is readily oxidised.

Clearly the O–H group in ethanol has a big effect on the properties (Figure 12.1).

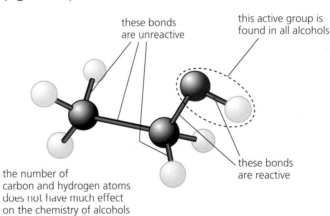

**Figure 12.1** The structure of ethanol labelled to show the reactive functional group and the unreactive hydrocarbon skeleton.

The O–H group in ethanol is an example of a functional group which is responsible for most of the reactions of ethanol and for other alcohols. If the O–H group is the only functional group present it reacts in more or less the same way irrespective of the hydrocarbon chain and it is not uncommon to represent all alcohols by the formula R–OH.

This makes the study of organic chemistry easier as all molecules that contain the same functional group have similar properties and react in a similar way. Molecules with the same functional group form a homologous series.

Initially four homologous series are studied; alkanes, alkenes, alcohols and haloalkanes. However, the reactions of these four series produce other functional groups which you will have to recognise. These include aldehydes, ketones, carboxylic acids and amines (Table 12.1).

### Key term

A **functional group** is either a structural feature (e.g. a carbon-to-carbon double bond, C=C), a group of atoms (e.g. a hydroxyl group, O–H) or a single atom (e.g. Cl). It is the functional group that determines much of the chemistry of a compound.

### Key term

A **homologous series** is a group of compounds:

- that have the same general formula
- that contain the same functional group
- in which each member of the homologous series differs from the next by $CH_2$.

**Table 12.1** Functional groups.

| Name | Functional group | General formula* | Prefix or suffix | First member | Second member† |
|---|---|---|---|---|---|
| Alkane | None – contains C–C and C–H single bonds | $C_nH_{2n+2}$ | -ane | $CH_4$ Methane | $C_2H_6$ Ethane |
| Alkene | 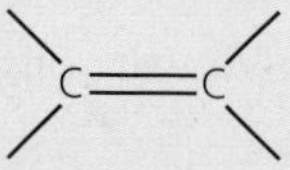 $\diagdown C=C \diagup$ | $C_nH_{2n}$ $n \neq 1$ | -ene | $C_2H_4$ Ethene | $C_3H_6$ Propene |
| Alcohol | 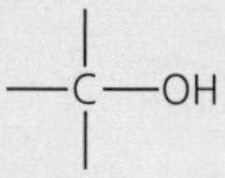 —C—OH | R–OH | -ol Hydroxy- | $CH_3OH$ Methanol | $C_2H_5OH$ Ethanol |
| Haloalkane | 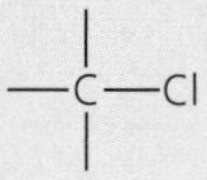 —C—Cl | R–X | Chloro- Bromo- Iodo- | $CH_3Cl$ Chloromethane | $C_2H_5Cl$ Chloroethane |
| Aldehyde | 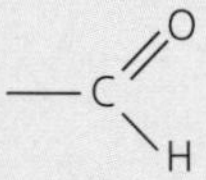 —C(=O)H | 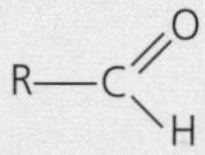 R—C(=O)H | -al | HCHO Methanal | $CH_3CHO$ Ethanal |
| Ketone | 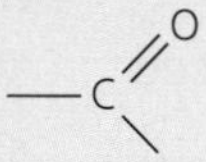 —C(=O)— | 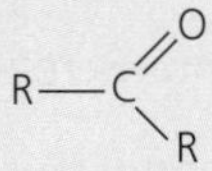 R—C(=O)—R | -one | $CH_3COCH_3$ Propanone | $C_2H_5COCH_3$ Butanone |
| Carboxylic acid | 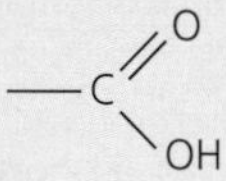 —C(=O)OH | R–COOH | -oic acid | HCOOH Methanoic acid | $CH_3COOH$ Ethanoic acid |
| Amine | 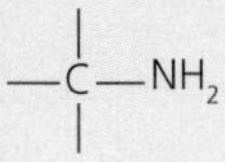 —C—$NH_2$ | R–$NH_2$ | -amine or amino ...-ane | $CH_3NH_2$ Methylamine or aminomethane | $C_2H_5NH_2$ Ethylamine or aminoethane |

*R = alkyl group = $C_nH_{2n+1}$.　　†Add $CH_2$ to first member.

---

**Test yourself**

1 Why do all alcohols have similar properties?
2 Identify the functional groups in each of the following compounds and the homologous series to which they belong:
   a) $CH_3CH_2CHO$
   b) $CH_3CH_2CH_2OH$
   c) $CH_3CH_2Cl$
   d) $CH_3CH=CHCH_3$
   e) $CH_3COCH_3$
   f) $CH_3COOH$

## Naming organic compounds

With such a wide variety of compounds it is necessary to have an organised way of naming them. Nomenclature should follow the International Union of Pure and Applied Chemistry (IUPAC) rules. The IUPAC naming rules are based around the systematic names given to the alkanes and the prefix or suffix given to each functional group. The prefixes and suffixes for the functional groups are given in Table 12.1.

**Table 12.2** Formulae and names of the first ten alkanes.

| Formula | Name |
| --- | --- |
| $CH_4$ | Methane |
| $C_2H_6$ | Ethane |
| $C_3H_8$ | Propane |
| $C_4H_{10}$ | Butane |
| $C_5H_{12}$ | Pentane |
| $C_6H_{14}$ | Hexane |
| $C_7H_{16}$ | Heptane |
| $C_8H_{18}$ | Octane |
| $C_9H_{20}$ | Nonane |
| $C_{10}H_{22}$ | Decane |

**Tip**

The names of the first four alkanes simply have to be learnt. From the fifth alkane onwards all straight-chain alkanes get their name by using the Greek numerical prefix followed by -ane.

First, you need to learn the names of the alkanes (Table 12.2).

The name of the alkane is derived from the name of the longest continuous carbon chain. The alkyl, R, groups derived from the name of the alkanes are then used to name any side chains or branches (Table 12.3).

**Table 12.3** Naming organic compounds.

| Number of carbon atoms in the longest straight chain | Stem name of longest straight chain | Number of carbon atoms in a side chain or branch | Name of alkyl, R side chain or branch |
| --- | --- | --- | --- |
| 1 | Meth- | 1 | Methyl |
| 2 | Eth- | 2 | Ethyl |
| 3 | Prop- | 3 | Propyl |
| 4 | But- | 4 | Butyl |
| 5 | Pent- | 5 | Pentyl |
| 6 | Hex- | 6 | Hexyl |
| 7 | Hept- | 7 | Heptyl |
| 8 | Oct- | 8 | Octyl |
| 9 | Non- | 9 | Nonyl |
| 10 | Dec- | 10 | Decyl |

**Example 1**

Give the full name of the compound:

$$CH_3-CH_2-CH_2-\overset{\overset{\textstyle CH_3}{|}}{CH}-CH_3$$

**Answer**

The longest carbon chain contains five atoms, so the name includes **pent-**.

This is an alkane, so it ends with **-ane**.

There is a branch containing one carbon atom, hence the name contains **methyl**.

It is possible to number the carbon chain from the left or the right.

Numbering from the left, the methyl group is on the fourth carbon atom: 4-methyl:

$$\overset{1}{C}H_3-\overset{2}{C}H_2-\overset{3}{C}H_2-\overset{\overset{\textstyle CH_3}{|}}{\overset{4}{C}H}-\overset{5}{C}H_3$$

Numbering from the right, the methyl group is on the second carbon atom: 2-methyl.

$$\overset{5}{C}H_3-\overset{4}{C}H_2-\overset{3}{C}H_2-\overset{\overset{\textstyle CH_3}{|}}{\overset{2}{C}H}-\overset{1}{C}H_3$$

The convention is to number from the chain end that results in the lowest number. In this case, numbering is from the right.

The full name of the compound is **2-methylpentane**.

## Example 2

Give the full name of the compound:

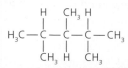

$$\overset{6}{C}H_3 — \overset{5}{C}H_2 — \overset{4}{C}H — \overset{3}{C}H_2 — \overset{2}{C}H — \overset{1}{C}H_3$$

with Br on carbon 4 and OH on carbon 2.

**Answer**

The longest carbon chain contains six atoms, so the name includes **hex-**.

The alcohol group is on the second carbon, hence **-2-ol**.

The bromine atom is on the fourth carbon, hence **4-bromo-**.

This is an alcohol, so the name ends in **-ol**. It is also a bromoalkane, so the name starts with **bromo-**.

The full name is **4-bromohexan-2-ol**.

## Example 3

Give the full name of

$$H_3C — \overset{H}{\underset{CH_3}{C}} — \overset{CH_3}{\underset{H}{C}} — \overset{H}{\underset{CH_3}{C}} — CH_3$$

**Answer**

It is an alkane so the name ends in –*ane*.

The longest carbon chain contains five atoms, so the name includes *pent-*.

There are three side chains each with one carbon, so the name contains *trimethyl-*.

The methyl side chains are on atoms 2, 3 and 4, so the name contains *2,3,4 trimethyl-*.

The full name is **2,3,4 trimethylpentane**.

## Test yourself

3 Name each of the following:

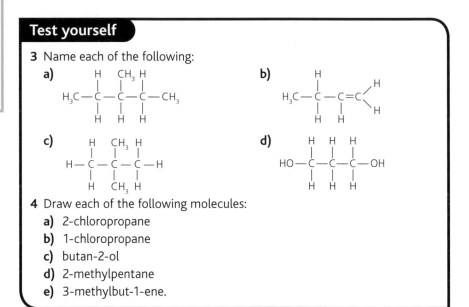

a)

b)

c)

d)

4 Draw each of the following molecules:
   a) 2-chloropropane
   b) 1-chloropropane
   c) butan-2-ol
   d) 2-methylpentane
   e) 3-methylbut-1-ene.

## Key terms

An **empirical formula** gives the simplest ratio of the elements in a compound. These can be calculated from the amounts of each element.

A **molecular formula** represents the actual number of atoms of each element in the molecule. It does not provide any detail of the arrangement of the atoms. For example, the molecular formula of ethanol is $C_2H_6O$.

# Types of formula

Different types of formula are used in organic chemistry. You must be familiar with both empirical formulae and molecular formulae and be able to draw structural formulae, displayed formulae and skeletal formulae.

## Calculation of empirical formulae

Analysis of an organic compound may only provide information about the ratio of the numbers of atoms present. Therefore, all that can be established is its empirical formula. This might be the same as the molecular formula (as is the case with methane), but usually it is not possible to be certain.

The mass in grams of each element in a sample can be determined by experiment. To find the empirical formula, the mass in grams of each component must be converted into an amount in moles.

Suppose a sample of a compound contains 1.5 g of carbon and 0.5 g of hydrogen.

$$\text{amount in moles} = \frac{\text{mass in grams}}{\text{molar mass}}$$

Therefore, the amount in moles of carbon atoms must be:

$$\frac{1.5}{12.0} = 0.125$$

The amount in moles of hydrogen atoms must be:

$$\frac{0.5}{1.0} = 0.50$$

Therefore, the ratio of number of moles of carbon to number of moles of hydrogen is 0.125 : 0.5. This compound contains four times as many moles of hydrogen atoms as it does carbon atoms. Therefore, each molecule contains four times as many hydrogen atoms as it does carbon atoms.

The empirical formula of the compound is $CH_4$.

---

**Tip**

In elemental analysis, amounts will always be given in terms of atoms. Hydrogen refers to hydrogen atoms (H – mass = 1) and *not* hydrogen molecules ($H_2$ – mass = 2).

---

**Example 3**

A 2.8 g sample of a hydrocarbon is found to contain 2.4 g of carbon and 0.4 g of hydrogen(H). What is its empirical formula?

**Answer**

$$\text{amount in moles of carbon atoms} = \frac{2.4}{12.0} = 0.2$$

$$\text{amount in moles of hydrogen atoms} = \frac{0.4}{1.0} = 0.4$$

The ratio of carbon atoms : hydrogen atoms is 0.2 : 0.4.

For every 1 carbon atom there are 2 hydrogen atoms.

The empirical formula is **$CH_2$**.

The empirical formula, $CH_2$, cannot be the molecular formula of the compound because a carbon atom has to have four bonds.

---

**Tip**

Do not be put off by the use of percentages instead of mass in grams.

---

**Example 4**

A sample of a compound is found to contain 48.65% of carbon, 8.11% of hydrogen and 43.24% of oxygen. What is its empirical formula?

**Answer**

It follows that 100 g of the compound will contain 48.65 g of carbon, 8.11 g of hydrogen and 43.24 g of oxygen.

| | C | H | O |
|---|---|---|---|
| $\dfrac{\text{Amount of atoms}}{\text{mol}}$ | $\dfrac{48.65}{12.0} = 4.05$ | $\dfrac{8.11}{1.0} = 8.11$ | $\dfrac{43.24}{16.0} = 2.70$ |
| Simplest ratio | $\dfrac{4.05}{2.70} = 1.5$ | $\dfrac{8.11}{2.70} = 3$ | $\dfrac{2.70}{2.70} = 1$ |

The simplest ratio gives the empirical formula as $C_{1.5}H_3O$. It is not possible to have 1.5 C atoms so the formula is doubled to give the empirical formula as **$C_3H_6O_2$**.

---

**Tip**

Do not round during a calculation. The 1.5 carbon atoms **cannot** be rounded up to 2.

---

## Calculation of molecular formula

The steps usually involved in the calculation of a molecular formula are as follows:

- Calculate the empirical formula from the percentage composition by mass.

- Use the empirical formula unit and the relative molecular mass to deduce the molecular formula.

### Example 5

Compound A has relative molecular mass 62. Its composition by mass is carbon 38.7%, hydrogen 9.7%, oxygen 51.6%. Calculate the empirical formula and the molecular formula of compound A.

**Answer**

|  | C | H | O |
|---|---|---|---|
| Percentage | 38.7 | 9.7 | 51.6 |
| Divide by relative atomic mass | $\frac{38.7}{12.0} = 3.23$ | $\frac{9.7}{1.0} = 9.7$ | $\frac{51.6}{16.0} = 3.23$ |
| Divide by the smallest | $\frac{3.23}{3.23} = 1$ | $\frac{9.7}{3.23} = 3$ | $\frac{3.23}{3.23} = 1$ |

The simplest ratio of C:H:O is 1:3:1. The empirical formula is **$CH_3O$**.

empirical formula mass = 12 + 3 + 16 = 31

The number of empirical units needed to make up the molecular mass is found by dividing the relative molecular mass by the empirical mass:

Therefore, the molecular formula is made up of two empirical units of $CH_3O$.

The molecular formula is **$C_2H_6O_2$**.

### Test yourself

5 A compound contained only carbon, hydrogen and oxygen. When analysed it consisted of 38.7% carbon and 9.68% hydrogen. Its relative molecular mass is 62.0.
   a) What percentage of oxygen did it contain?
   b) What is the empirical formula?
   c) What is the molecular formula?

6 A hydrocarbon consists of 82.8% carbon and its relative molecular mass is 58.
   a) Calculate its empirical formula.
   b) Calculate its molecular formula.

7 A sample of a hydrocarbon was burnt completely in oxygen. 1.69 g of carbon dioxide and 0.346 g of water were produced.
   a) What is the percentage of carbon in carbon dioxide?
   b) What is the mass of carbon in 1.69 g of carbon dioxide?
   c) What is the percentage of hydrogen in water?
   d) What is the mass of hydrogen in 0.346 g of water?
   e) Use your answers to (b) and (d) to calculate the empirical formula of the hydrocarbon.

### Tip

Do not round during the calculation.

# Representing chemical formulae

A **structural formula** uses the least amount of detail needed for an unambiguous structure. There are two possibilities for the structure of a compound of molecular formula $C_4H_{10}$: butane and methylpropane. The formula $C_4H_{10}$ is, therefore, ambiguous.

- The structural formula of butane is $CH_3CH_2CH_2CH_3$, which can be simplified to $CH_3(CH_2)_2CH_3$.

- The structural formula of methylpropane is $CH_3CH(CH_3)CH_3$, which can be simplified to $(CH_3)_3CH$.

The structural formula, therefore, makes it clearer how the atoms are bonded.

A **displayed formula** gives yet more information. It shows the relative positions of atoms and the number of bonds between them. Displayed formulae sometimes indicate the bond angles around each carbon atom. For methane and ethane the full displayed formulae shows the 3D structure:

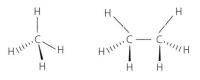

Displayed formulae are usually simplified:

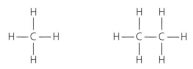

The molecular formula of ethanol is $C_2H_6O$ and the structural formula is $CH_3CH_2OH$. The displayed formula is:

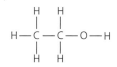

A **skeletal formula** is used to show a simplified organic structure by not writing the carbon and the hydrogen atoms in the alkyl chains, leaving just a skeleton which shows the bonds between each carbon and the associated functional groups. The skeletal formulae for butane and 2-methylpropane are shown below:

butane      methylpropane

No symbols are shown because butane and 2-methylpropane are both alkanes. The alcohol group (−OH), present in both butan-1-ol and 2-methylpropan-1-ol, is shown as part of the skeletal formula and the OH is attached to the carbon–carbon bonds that make up the skeleton as shown below:

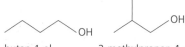

butan-1-ol      2-methylpropan-1-ol

Some cyclic alkanes and benzene are represented as shown below:

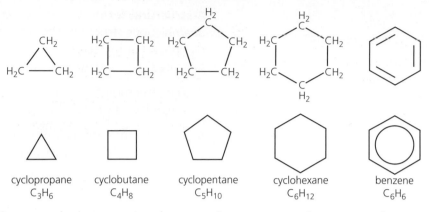

| cyclopropane $C_3H_6$ | cyclobutane $C_4H_8$ | cyclopentane $C_5H_{10}$ | cyclohexane $C_6H_{12}$ | benzene $C_6H_6$ |

Compounds that contain a benzene ring are **aromatic** compounds whilst compounds that contain a hydrocarbon ring which is not benzene are **alicyclic** compounds.

---

### Test yourself

8  Calculate the molecular formula of each of the following compounds:
   a) contains 80% C and 20% H and has $M_r = 60$
   b) contains 92.3% C and 7.7% H and has $M_r = 78$
   c) contains 2.4 g C, 0.6 g H and 1.6 g O and has $M_r = 46$
   d) contains 62.1% C, 10.3% H and 27.6% O and has $M_r = 58$
   e) contains 0.36 g C, 0.06 g H and 0.48 g O and has $M_r = 180$.
9  Draw the displayed formula and the structural formula for each of the following molecules:
   but-1-ene, 2-bromopropane, propan-2-ol and propan-1-ol.
10  Draw the skeletal formula of but-1-ene, 2-bromopropane, propan-2-ol and propan-1-ol.

---

# Isomerism

## Structural isomerism

One reason why carbon forms so many compounds is because it is sometimes possible to join the atoms together in different ways. $C_4H_{10}$ is the first alkane in which the atoms can be joined to form two different structures shown below:

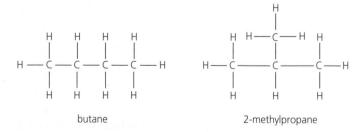

| butane | 2-methylpropane |

As the number of carbon atoms increase it is possible to find more and more different ways of joining them together. It has been calculated that for the molecular formula $C_{14}H_{30}$ there are 1858 different ways of joining the atoms together, hence there are 1858 different compounds with the formula $C_{14}H_{30}$. These different forms are called isomers. They can occur with all functional groups.

### Tip

You must be able to draw *and* name structural isomers.

There are three types of structural isomer: chain, positional and functional group.

## Chain isomerism

Chain isomers have differences in chain length. Examples include:

● butane and methylpropane, as shown above

● but-1-ene and methylpropene.

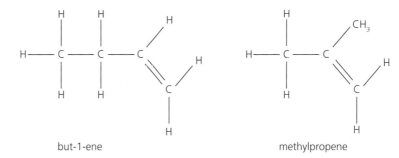

but-1-ene          methylpropene

## Positional isomerism

Positional isomers differ by the position of their functional group or groups. Examples include:

● 1-bromobutane and 2-bromobutane

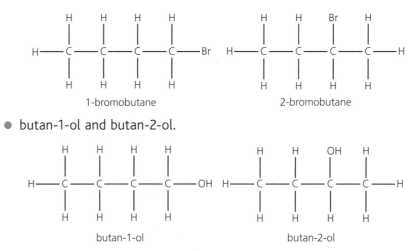

1-bromobutane          2-bromobutane

● butan-1-ol and butan-2-ol.

butan-1-ol          butan-2-ol

## Functional-group isomerism

Functional-group isomers have different functional groups. An example is propanal (an aldehyde) and propanone (a ketone):

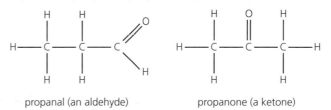

propanal (an aldehyde)          propanone (a ketone)

## Stereoisomerism

Structural isomerism is common within organic chemistry; however, stereoisomerism is more limited and requires certain key structural features. Stereoisomerism in alkenes will be dealt with fully in Chapter 13.

## Test yourself

11 Draw and name isomers of:
   **a)** $C_5H_{12}$
   **b)** $C_6H_{14}$.
12 Draw skeletal formulae and name the structural isomers of:
   **a)** $C_5H_{10}$ that are alkenes
   **b)** $C_5H_{10}$ that are cycloalkanes.
13 Draw skeletal formulae and name the alcohols that are isomers of:
   **a)** $C_4H_9OH$
   **b)** $C_5H_{11}OH$.

# Bonding in organic compounds

Organic compounds are essentially covalent. A covalent bond is a shared pair of electrons between two atoms. Each atom provides one electron of the shared pair.

All organic compounds contain carbon. The electron configuration of carbon is $1s^2 2s^2 2p^2$. At first sight, it appears that only the two 2p-electrons are available to form covalent bonds. However, when carbon forms bonds with other atoms, one of the 2s-electrons is promoted to the third 2p-orbital, as shown in Figure 12.2.

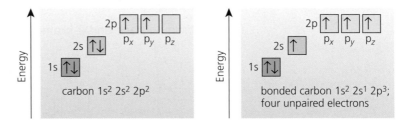

**Figure 12.2** Electron configuration for unbonded and bonded carbon.

Carbon now has four unpaired electrons available for sharing and, therefore, can form four covalent bonds which are referred to as σ-bonds (sigma bonds), where the two shared electrons are between the centres of the two nuclei.

Alkanes consist of carbon and hydrogen only (e.g. methane, $CH_4$). The bonds are all σ-bonds. The four electron pairs involved mean that the bonds around the carbon atom point to the corners of a regular tetrahedron (pages 100 and 102). The bond angle is 109° 28′, which is about 109.5°.

Carbon has an electronegativity of 2.5 (see pages 105–106); hydrogen has an electronegativity of 2.1. Therefore, the C–H bond is slightly polar, with the electrons drawn towards the carbon atom:

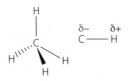

This applies to all four bonds. Therefore, in a molecule of methane the electrons in each C–H bond are drawn towards the central carbon atom:

Since the $CH_4$ molecule is symmetrical, the dipoles cancel each other out. Methane is non-polar, as are all alkanes.

Functional groups such as those in haloalkanes and alcohols disturb the alkane symmetry. This, together with the electronegativities of the halogen and oxygen, makes the molecules polar. In polar molecules such as chloromethane and methanol the dipoles can be determined using Pauling electronegativity values (Table 12.2; see also page 105).

**Table 12.2** Pauling electronegativity values of elements commonly found in organic compounds.

| Element | C | H | O | Cl | Br | N |
|---|---|---|---|---|---|---|
| Electronegativity | 2.5 | 2.1 | 3.5 | 3.0 | 2.8 | 3.0 |

In chloromethane the differences in electronegativity, and the lack of symmetry, results in the shared pair of electrons in each σ-bond moving along the σ-bonds as shown below:

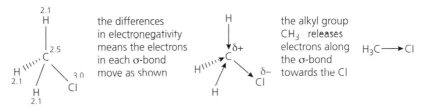

This movement of the shared pair of electrons in a σ-bond (covalent bond) is known as the inductive effect. Alkyl groups like $CH_3-$, $CH_2CH_3-$, $CH_2CH_2CH_3-$ (methyl, ethyl, propyl) are described as electron releasing and are said to have a positive inductive effect. By contrast the Cl in $CH_3Cl$ withdraws electrons from the C–Cl bond and has a negative inductive effect.

In methanol the differences in electronegativity, and the lack of symmetry, results in the shared pair of electrons in each σ-bond moving along the σ-bonds as shown below:

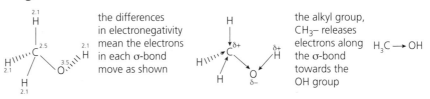

Non-polar molecules such as alkanes are unreactive compared with polar molecules such as chloromethane and methanol. The strength and the polarity of the bonds affect the chemical properties of a molecule.

## Important terms

In order to understand the chemistry of compounds containing functional groups, you need to know the following terms:

● **Reagents**   These are the chemicals involved in a reaction.

● **Conditions**   These normally describe the temperature, pressure, solvent and the use of a catalyst.

**Tip**

For a reaction of any functional group you are expected to know the reagents and conditions. You also need and to be able to write a balanced equation and, often, the mechanism.

- **Aliphatic**   a compound containing carbon and hydrogen in straight chains, branched chains or non-aromatic rings.
- **Aromatic**   a compound containing a benzene ring.

# Calculations in organic chemistry

Most chemistry exams contain questions that require a calculation. Organic chemistry calculations can include the determination of:

- empirical and molecular formulae – which were explained earlier on pages 188–189
- percentage yield
- atom economy.

## Percentage yield

Very few reactions result in the complete conversion of reactants to products. Many preparations of organic substances are inefficient and yield only a small amount of product. Sometimes, losses occur as a result of experimental error. In these cases, the percentage yield of the reaction is recorded:

$$\% \text{ yield} = \frac{\text{amount (in moles) of product obtained} \times 100}{\text{theoretical amount (in moles) of product}}$$

Percentage yield can also be expressed as:

$$\% \text{ yield} = \frac{\text{mass of product obtained}}{\text{theoretical mass of product}} \times 100$$

### Example 6

Ethanol, $C_2H_5OH$, can be oxidised to form ethanoic acid, $CH_3COOH$. In an experiment, 2.3 g of ethanol is oxidised to produce 2.4 g of ethanoic acid. Calculate the percentage yield.

**Answer**

In this example, the symbol [O] is used to represent the oxidising agent.

$$C_2H_5OH + 2[O] \rightarrow CH_3COOH + H_2O$$

The equation shows that 1 mol of ethanol produces 1 mol of ethanoic acid.

**Step 1**   Calculate the amount in moles of ethanol used:

$$\text{amount in moles of ethanol used} = n = \frac{\text{mass of ethanol}}{\text{molar mass of ethanol}}$$

$$= \frac{2.3}{46} = 0.050 \text{ mol}$$

The mole ratio ethanol:ethanoic acid is 1:1, hence the maximum amount of ethanoic acid that could be made is 0.050 mol.

**Step 2**   Calculate the amount in moles of ethanoic acid produced:

$$\text{amount in moles of ethanoic acid produced} = n = \frac{\text{mass of ethanoic acid}}{\text{molar mass of ethanoic acid}}$$

$$= \frac{2.4}{60} = 0.040 \text{ mol}$$

**Step 3**   Calculate the percentage yield:

$$\text{percentage yield} = \frac{\text{actual yield}}{\text{maximum yield}} \times 100 = \frac{0.040}{0.050} \times 100 = 80\%$$

In Example 6 there is one organic starting material and one product, and it is assumed that there is enough oxidising agent for the reaction to go to completion. If there are two reagents, it may first be necessary to establish whether the reagents are in the exact proportions for reaction or whether one of them is in excess.

### Example 7

In a reaction between methane and chlorine, the percentage yield of tetrachloromethane, $CCl_4$, is 72%. Calculate the mass of tetrachloromethane that could be obtained by reacting 10.0 g of methane with excess chlorine.

**Answer**

The equation of the reaction is:

$$CH_4 + 4Cl_2 \rightarrow CCl_4 + 4HCl$$

**Step 1** Calculate the amount in moles of methane present:

$$\text{amount in moles of methane} = \frac{10}{16} = 0.625 \text{ mol}$$

Chlorine is in excess. Therefore, the equation indicates that If the reaction were 100% efficient, 0.625 mol of tetrachloromethane would be produced.

**Step 2** Calculate the amount in moles of tetrachloromethane produced:

percentage yield = 72%

amount in moles of tetrachloromethane

$$\text{produced} = \frac{72}{100} \times 0.625 = 0.45 \text{ mol}$$

**Step 3** Calculate the mass of tetrachloromethane produced:

mass of 1 mol of $CCl_4$ = 154 g

mass of 0.45 mol = 0.45 × 154 = 69.3 g

## Atom economy

The percentage yield of a reaction is one measure of the efficiency of the process. However, a reaction with a good yield of product can still be inefficient and wasteful if there are other products that have to be discarded.

Consider the substitution reaction of sodium hydroxide and 1-bromobutane. This produces butan-1-ol and sodium bromide. The equation is:

$$C_4H_9Br + NaOH \rightarrow C_4H_9OH + NaBr$$

A percentage yield of butanol of 70% may be a good yield, but the process might be considered less than satisfactory if there were no use for the other product, sodium bromide. There is, therefore, another measure that can be applied to reactions to take account of wastage. This is called the **atom economy**.

$$\text{atom economy} = \frac{\text{molecular mass of the desired product}}{\text{sum of the molecular masses of all products}} \times 100$$

In the butanol, $C_4H_9OH$, example given above, the atom economy would be calculated as follows:

$M_r$ of $C_4H_9OH$ is $(4 \times 12.0) + (9 \times 1.0) + 16.0 + 1.0 = 74.0$

$M_r$ of NaBr is $23.0 + 79.9 = 102.9$

$$\text{atom economy} = \frac{74.0}{74.0 + 102.9} \times 100 = 41.8\%$$

This result indicates that the reaction is not ideal.

A reaction can be inefficient but have a high atom economy. For example, the yield of the ester, ethyl ethanoate, from the reaction between ethanol and ethanoic acid is usually only about 40%.

The equation is as follows:

$$CH_3COOH + C_2H_5OH \rightarrow CH_3COOC_2H_5 + H_2O$$

However, the atom economy is good:

$M_r$ of ethyl ethanoate = 88.0

$M_r$ of water = 18.0

$$\text{atom economy} = \frac{88.0}{88.0 + 18.0} \times 100 = 83.0\%$$

Addition reactions of alkenes have a 100% atom economy because there is only one product, for example:

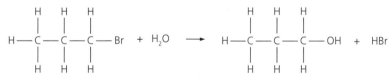

However, the percentage yield may be less than 100%.

Using industrial procedures with a high atom economy would produce less waste. Atom economy is a relatively new concept, but its importance is likely to grow as society becomes more concerned about the need to conserve resources and avoid unwanted by-products.

## Tip

It is important to recognise and understand the command words in a question. In Question 14 the command words are 'calculate' and 'deduce'. As you might expect 'calculate' will involve using your calculator but 'deduce' can be done without the use of a calculator. In this case the amount in moles of propan-1-ol can be deduced by using the mole ratio in the balanced equation.

## Test yourself

14 Propan-1-ol was prepared by the reaction shown below:

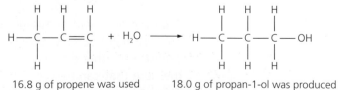

2.46 g of 1-bromopropane was used      0.96 g of propan-1-ol was produced

a) Calculate the amount in moles of 1-bromopropane used.
b) Deduce the amount in moles of propan-1-ol that could be produced.
c) Calculate the amount in moles of propan-1-ol actually produced.
d) Calculate the percentage yield.
e) Calculate the atom economy.

Propan-1-ol can also be produced by the reaction between propene and steam shown below:

16.8 g of propene was used      18.0 g of propan-1-ol was produced

f) Calculate the percentage yield and the atom economy of this reaction.

15 2.3 g of ethanol, $C_2H_5OH$ reacts with ethanoic acid to produce 2.2 g of ethyl ethanoate as shown below.

$$C_2H_5OH + CH_3COOH \rightarrow CH_3COOC_2H_5 + H_2O$$

Calculate the percentage yield and the atom economy.

# Practice questions

## Multiple choice questions 1–10

**1** Which one of the following is a ketone?

  **A** $C_6H_5CH_2CHO$

  **B** $C_6H_5COCH_3$

  **C** $CH_3CH_2COOH$

  **D** $CH_3CH_2OCH_2CH_3$

**2** Which one of the following is the molecular formula of compound X, shown below?

compound X

  **A** $C_9H_{18}$

  **B** $C_9H_{20}$

  **C** $C_8H_{16}$

  **D** $C_8H_{18}$

**3** Which one of the following is the number of possible isomers of $C_6H_{14}$?

  **A** 3

  **B** 4

  **C** 5

  **D** 6

**4** Which one of the following is the name of compound X, shown below?

compound X

  **A** 5-ethyl-3-methylhex-3-ene

  **B** 2-ethyl-4-methylhex-3-ene

  **C** 3,5-dimethylhept-3-ene

  **D** 3,5-dimethylhept-4-ene

**5** Which one of the following compounds contains a double bond?

  **A** $C_5H_{11}Br$

  **B** $C_4H_9Br$

  **C** $C_3H_5Br$

  **D** $C_2H_5Br$

**6** Ethanol, $C_2H_5OH$, can be prepared by reacting ethene, $C_2H_4$, with steam. Under certain condition 28.0 g of ethene reacted with steam and 28.0 g of ethanol was obtained. The approximate percentage yield for this reaction was:

  **A** 100%

  **B** 80%

  **C** 60%

  **D** 40%

**7** There are nine isomers with molecular formula $C_7H_{16}$. How many of the isomers have 'butane' in their systematic name?

  **A** 1

  **B** 2

  **C** 3

  **D** 4

**8** Which one of the following is the number of possible structural isomers of $C_4H_8$?

  **A** 3

  **B** 4

  **C** 5

  **D** 6

Use the key below to answer Questions 9 and 10.

| A | B | C | D |
| --- | --- | --- | --- |
| 1, 2 & 3 correct | 1, 2 correct | 2, 3 correct | 1 only correct |

**9** Chemicals within the same homologous series:

  1  have the same the same functional group

  2  have the same general formula

  3  differ from the next in the series by $CH_2$.

**10** Which of the following hydrocarbons contains 85.7% carbon by mass?

  1  $C_3H_6$

  2  $C_5H_{10}$

  3  $C_7H_{14}$

**11** Identify the functional group in each of the following molecules:

  **a)** dodecane

  **b)** pentene

  **c)** propane

  **d)** propan-2-ol

  **e)** 2-bromobutane

  **f)** cyclobutane. (6)

**12** For each of the following, give the formula of the next member in the homologous series:

  **a)** $C_2H_5OH$

  **b)** $C_3H_6$

  **c)** $CH_3I$

  **d)** $C_7H_{16}$. (4)

**13** Draw the displayed formula and the structural formula for the following molecules:

  **a)** 2 chloropropane

  **b)** 1-chloropropane

  **c)** butan-2-ol

  **d)** 2-methylpentane

  **e)** 3-methylbut-1-ene. (5)

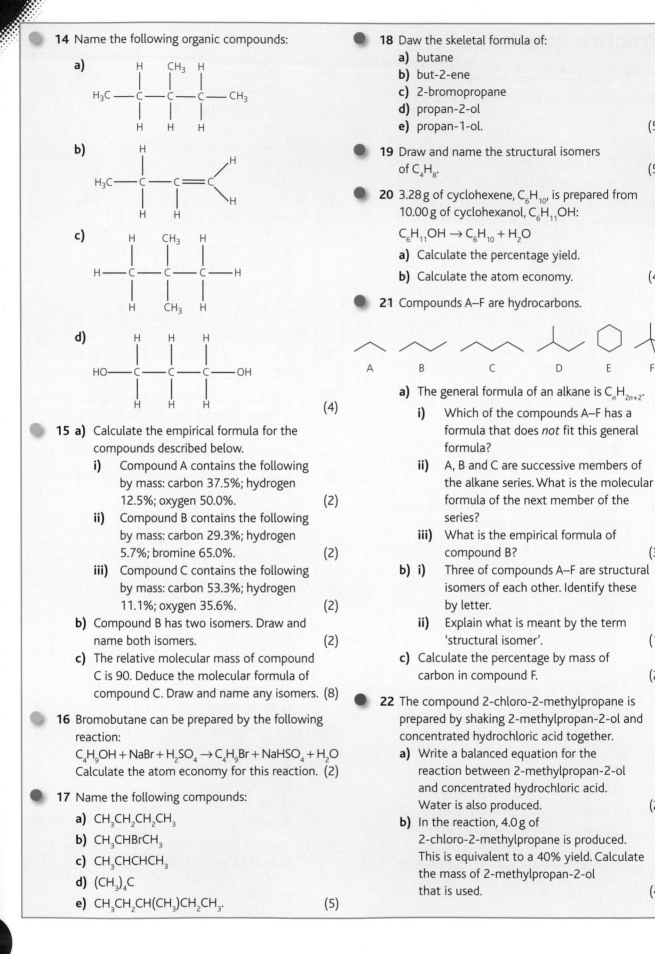

**14** Name the following organic compounds:

a)

b)

c)

d)

(4)

**15 a)** Calculate the empirical formula for the compounds described below.

i) Compound A contains the following by mass: carbon 37.5%; hydrogen 12.5%; oxygen 50.0%. (2)

ii) Compound B contains the following by mass: carbon 29.3%; hydrogen 5.7%; bromine 65.0%. (2)

iii) Compound C contains the following by mass: carbon 53.3%; hydrogen 11.1%; oxygen 35.6%. (2)

b) Compound B has two isomers. Draw and name both isomers. (2)

c) The relative molecular mass of compound C is 90. Deduce the molecular formula of compound C. Draw and name any isomers. (8)

**16** Bromobutane can be prepared by the following reaction:

$$C_4H_9OH + NaBr + H_2SO_4 \rightarrow C_4H_9Br + NaHSO_4 + H_2O$$

Calculate the atom economy for this reaction. (2)

**17** Name the following compounds:

a) $CH_3CH_2CH_2CH_3$

b) $CH_3CHBrCH_3$

c) $CH_3CHCHCH_3$

d) $(CH_3)_4C$

e) $CH_3CH_2CH(CH_3)CH_2CH_3$. (5)

**18** Daw the skeletal formula of:

a) butane

b) but-2-ene

c) 2-bromopropane

d) propan-2-ol

e) propan-1-ol. (5)

**19** Draw and name the structural isomers of $C_4H_8$. (5)

**20** 3.28 g of cyclohexene, $C_6H_{10}$, is prepared from 10.00 g of cyclohexanol, $C_6H_{11}OH$:

$$C_6H_{11}OH \rightarrow C_6H_{10} + H_2O$$

a) Calculate the percentage yield.

b) Calculate the atom economy. (4)

**21** Compounds A–F are hydrocarbons.

a) The general formula of an alkane is $C_nH_{2n+2}$.

i) Which of the compounds A–F has a formula that does *not* fit this general formula?

ii) A, B and C are successive members of the alkane series. What is the molecular formula of the next member of the series?

iii) What is the empirical formula of compound B? (3)

b) i) Three of compounds A–F are structural isomers of each other. Identify these by letter.

ii) Explain what is meant by the term 'structural isomer'. (1)

c) Calculate the percentage by mass of carbon in compound F. (2)

**22** The compound 2-chloro-2-methylpropane is prepared by shaking 2-methylpropan-2-ol and concentrated hydrochloric acid together.

a) Write a balanced equation for the reaction between 2-methylpropan-2-ol and concentrated hydrochloric acid. Water is also produced. (2)

b) In the reaction, 4.0 g of 2-chloro-2-methylpropane is produced. This is equivalent to a 40% yield. Calculate the mass of 2-methylpropan-2-ol that is used. (4)

**23** Compound X is a hydrocarbon with molar mass 68 g mol⁻¹. It contains 88.24% carbon, deduce its molecular formula. (Hint: this looks easy, but be careful.) Compound X is a cyclic alkene. Draw and name all possible isomers of compound X. (10)

## Challenge

**24** Compound X is a bromoalkene. It has the following percentage composition by mass: carbon 12.9%; hydrogen 1.1%; bromine 86.0%.

**a) i)** Calculate the empirical formula of compound X. Show your working. (2)

**ii)** The relative molecular mass of compound X is 278.7. Show that the molecular formula is $C_3H_3Br_3$. (1)

**b)** Compound X is one of six possible structural isomers of $C_3H_3Br_3$ that are bromoalkenes. Two of these isomers are shown below as isomer 1 and isomer 2.

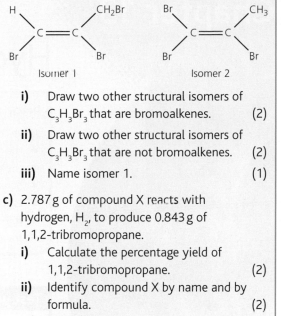

Isomer 1                    Isomer 2

**i)** Draw two other structural isomers of $C_3H_3Br_3$ that are bromoalkenes. (2)

**ii)** Draw two other structural isomers of $C_3H_3Br_3$ that are not bromoalkenes. (2)

**iii)** Name isomer 1. (1)

**c)** 2.787 g of compound X reacts with hydrogen, $H_2$, to produce 0.843 g of 1,1,2-tribromopropane.

**i)** Calculate the percentage yield of 1,1,2-tribromopropane. (2)

**ii)** Identify compound X by name and by formula. (2)

**iii)** What is the atom economy of this reaction? (1)

# Chapter 13

# Hydrocarbons

## Prior knowledge

*In this chapter it is assumed that you are familiar with:*

- the structure and names of the first five straight-chain alkanes
- the combustion products of hydrocarbons
- balancing equations
- empirical and molecular formula calculations
- the existence of fossil fuels
- fractional distillation and cracking.

## Test yourself on prior knowledge

1  Draw and name the first five straight-chain alkanes.
2  Write equations for the complete combustion of:
   a)  $C_3H_8$
   b)  $C_8H_{18}$
   c)  $C_{24}H_{50}$.
3  Explain what is meant by a fossil fuel.
4  Explain how the different fractions of crude oil are separated.
5  Suggest why the combustion of fossil fuels might be bad for the environment.
6  Write an equation to illustrate the cracking of $C_{10}H_{22}$ into $C_8H_{18}$.

## Hydrocarbons as fuel

The Earth has large, though depleting, reserves of hydrocarbons. These compounds are particularly important as fuels. Hydrocarbons are compounds that contain only hydrogen and carbon. The carbon atoms in hydrocarbons can be linked by single, double or triple bonds and can be in a chain or a ring.

A fuel is a source of energy. A useful fuel should:

- have a high energy output

- be either abundant or manufactured easily

- be easy to transport and store

- be easy to ignite

- cause minimum environmental damage.

Alkanes are called saturated hydrocarbons.

# Alkanes

## Physical properties

### Melting point and boiling point

The variation in boiling point depends on the degree of intermolecular bonding (page 108). Alkanes are covalent compounds with C–H bonds that have a very low polarity and the only intermolecular bonding is a consequence of induced dipole–dipole interactions. The strength of the induced dipole–dipole interactions depends on two key factors: the number of electrons in the molecules and the shape of the molecules. There are two important trends in the variation of boiling point:

- As the relative molecular mass of the alkanes gets larger, so does the boiling point. This is due to the increasing chain length and therefore an increase in the number of electrons present. Hence there will be more induced dipole–dipole interactions. The boiling points of the first six alkanes are shown in Table 13.1 and the trend is shown graphically in Figure 13.1.

Table 13.1 Boiling points of the alkanes.

| Alkane | Molecular formula | Boiling point/K | Boiling point/°C | State at r.t.p. |
| --- | --- | --- | --- | --- |
| Methane | $CH_4$ | 109 | −164 | Gas |
| Ethane | $C_2H_6$ | 185 | −88 | Gas |
| Propane | $C_3H_8$ | 231 | −42 | Gas |
| Butane | $C_4H_{10}$ | 273 | 0 | Gas |
| Pentane | $C_5H_{12}$ | 309 | 36 | Liquid |
| Hexane | $C_6H_{14}$ | 342 | 69 | Liquid |

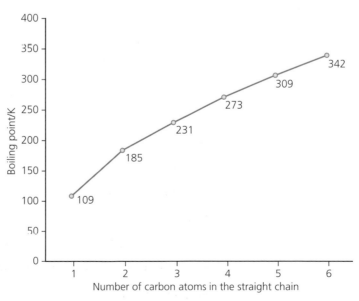

Figure 13.1 Trend in boiling point of the first six alkanes.

- For isomers with the same relative molecular mass, the boiling point decreases as the degree of branching increases. This is illustrated in Table 13.2 for the three isomers of $C_5H_{12}$, which all have relative molecular mass 72.

**Table 13.2** Relationship between boiling point and degree of branching.

| Name | Structure | Boiling point/K | Boiling point/°C | State at r.t.p. |
|------|-----------|-----------------|------------------|-----------------|
| Pentane | | 309 | 36 | Liquid |
| Methylbutane | | 301 | 28 | Liquid/gas |
| Dimethylpropane | | 283 | 10 | Gas |

As shown in Figure 13.2, molecules of straight-chain alkanes, such as pentane, pack together more closely than molecules of branched-chain alkanes, such as methylbutane and dimethylpropane. This increased amount of surface contact creates more intermolecular forces (induced dipole–dipole interactions). More energy is needed to overcome these intermolecular forces, so the boiling point of pentane is the highest of the three isomers.

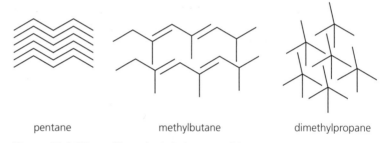

pentane                methylbutane                dimethylpropane

**Figure 13.2** Effect of branched chains on packing.

Branched-chain alkanes, such as methylbutane and dimethylpropane, cannot pack together as tightly. This reduces the intermolecular forces and, hence lowers the boiling point and melting point.

## Solubility

Alkanes have very low polarity, which means that they do not mix with polar molecules. Therefore, alkanes are insoluble in polar solvents, such as water. Liquid alkanes are less dense than water and when mixed together the alkanes will form an immiscible upper layer.

Alkanes dissolve readily in each other because the molecules link through induced dipole–dipole interactions. Crude oil is a mixture, or solution, of over 100 different alkanes.

## Chemical properties

Alkanes are unreactive because:

● the C–H bonds have very low polarity

● the single bonds linking together the carbon and hydrogen atoms are relatively strong.

However, they do undergo two important reactions – combustion and reactions with halogens.

## Combustion of alkanes

Combustion of alkanes in excess oxygen produces carbon dioxide and water:

$$CH_4 + 2O_2 \rightarrow CO_2 + 2H_2O$$

$$C_2H_6 + 3\tfrac{1}{2}O_2 \rightarrow 2CO_2 + 3H_2O$$

$$C_{11}H_{24} + 17O_2 \rightarrow 11CO_2 + 12H_2O$$

Combustion of alkanes in a limited supply of oxygen produces carbon monoxide (or carbon) and water:

$$CH_4 + 1\tfrac{1}{2}O_2 \rightarrow CO + 2H_2O \qquad CH_4 + O_2 \rightarrow C + 2H_2O$$

$$C_2H_6 + 2\tfrac{1}{2}O_2 \rightarrow 2CO + 3H_2O \qquad C_2H_6 + 1\tfrac{1}{2}O_2 \rightarrow 2C + 3H_2O$$

$$C_{11}H_{24} + 11\tfrac{1}{2}O_2 \rightarrow 11CO + 12H_2O \qquad C_{11}H_{24} + 6O_2 \rightarrow 11C + 12H_2O$$

Carbon monoxide is poisonous and restricts the effectiveness of the blood to transport oxygen in the body. If the supply of oxygen is even more limited carbon (soot) and water are the major products. Therefore, when alkanes are burnt there must be good ventilation, and car engines must be effectively tuned to ensure that the volume of carbon monoxide in the exhaust is within prescribed limits. Clearly the engine of the lorry in Figure 13.3 needs to be re-tuned.

# Radical substitution

A radical is a particle with an unpaired electron. The unpaired electron in a radical can be shown as a dot, e.g. Cl·, ·CH₃. Radicals are highly reactive and can attack non-polar compounds such as alkanes in a substitution reaction.

The mechanism of such reactions (i.e. the steps involved) is important and illustrates some general features of radical chemistry.

### Substitution by chlorine or bromine to form halogenoalkanes
The reaction between methane and chlorine is used as an example.

**Reagent:**          chlorine

**Conditions:**          ultraviolet radiation

**Equation:**          $CH_4 + Cl_2 \rightarrow CH_3Cl + HCl$

**Mechanism:**          radical substitution

There are three distinct steps that make up this mechanism:

● initiation – starts the reaction by forming a radical

● propagation – maintains the reaction by forming one of the products and a new radical

● termination – ends the reaction by removing the radicals and forming molecules.

**Figure 13.3** Serious faults with the engine of this lorry means that it is emitting dangerous and illegal levels of carbon (soot) and toxic carbon monoxide from its exhaust system.

---

**Key terms**

A **radical** is a particle that has at least one unpaired electron.
A **substitution** reaction is one in which an atom (or group) is replaced by another atom (or group).

---

**Tip**

In an alkyl radical such as a methyl radical. the 'dot' should be on the carbon.

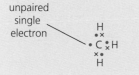

unpaired single electron

---

## Key term

**Homolysis** or **homolytic fission** occurs when a covalent bond is broken so that the atoms joined by the covalent bond each take one of the shared electrons.

## Tip

The first propagation step **always** produces a hydrogen halide (HBr or HCl) and an alkyl radical; the second propagation step **always** produces the organic product and re-generates the halogen radical.

In the **initiation** step, radicals are generated. The ultraviolet radiation provides sufficient energy to break the Cl–Cl bond homolytically to produce radicals. Homolysis or homolytic fission occurs when a covalent bond breaks and the atoms joined by the bond separate, each atom taking one of the shared pair of electrons:

$$Cl\text{–}Cl \rightarrow Cl\cdot + Cl\cdot$$

$$Cl\text{–}Cl \xrightarrow{\text{UV}} 2Cl\cdot$$

**Propagation** involves more than one step, each of which maintains the radical concentration. Usually a $Cl\cdot$ or a $Br\cdot$ radical causes the release of an alkyl radical, or vice versa. The propagation steps are difficult to control and can lead to a complex mixture of products.

For mono-substitution there are two essential propagation steps:

**Propagation step 1:** $\quad CH_4 + Cl\cdot \rightarrow HCl + \cdot CH_3$

**Propagation step 2:** $\quad \cdot CH_3 + Cl_2 \rightarrow CH_3Cl + Cl\cdot$

An important feature of the propagation steps is that the second reaction generates another $Cl\cdot$. This means that, in principle, the reaction might never stop and many molecules of methane might be destroyed from a single initiation but the chain can be stopped by a termination reaction.

**Termination** involves any two radicals reacting together to form a covalent bond:

**Possible termination steps:** $\quad CH_3\cdot + CH_3\cdot \rightarrow C_2H_6$

$$CH_3\cdot + Cl\cdot \rightarrow CH_3Cl$$

$$Cl\cdot + Cl\cdot \rightarrow Cl_2$$

Radicals are extremely reactive. They react with almost any molecule, atom or radical with which they collide. When $CH_4(g)$ and $Cl_2(g)$ react a mixture is formed and will include $CH_2Cl_2$, $CHCl_3$ and $CCl_4$ as well as $CH_3Cl$. Radical substitution reactions almost always result in a mixture of products which are difficult to separate. The sequence below shows how $CH_2Cl_2$ might be formed.

A chlorine radical could collide with the product molecule, $CH_3Cl$, in which case the following sequence would occur:

$$CH_3Cl + Cl\cdot \rightarrow HCl + \cdot CH_2Cl$$

$$\cdot CH_2Cl + Cl_2 \rightarrow CH_2Cl_2 + Cl\cdot$$

**Test yourself**

1 a) Draw a dot-and-cross diagram to represent a molecule of:
   i) ethane
   ii) propane.
  b) Give the bond angles in:
   i) ethane
   ii) propane.

2 a) What is meant by electronegativity?
  b) Explain why alkanes are unreactive.

3 Draw and name the isomers of $C_6H_{14}$. State which isomer has the highest boiling point and suggest which two isomers are likely to have low boiling points. Explain your answers.

4 Why is a series of organic compounds, such as the alkanes, comparable with a group of elements in the periodic table?

5 a) Write an equation for the complete combustion of propane in Calor gas.
  b) What are the products when propane burns in a poor supply of oxygen?

6 Why is it dangerous to allow a car engine to run in a garage with the door closed?

7 Write equations for all four possible substitution reactions when chlorine reacts with methane and name the products.

8 a) Explain why a mixture of bromine in hexane remains orange in the dark but fades and becomes colourless in sunlight.
  b) Write an equation for the reaction in part (a).
  c) Why can acid fumes be detected above the solution once the colour has faded?

9 a) Write an equation for the reaction between hexane and bromine in the presence of UV light.
  b) Describe the mechanism for the reaction between hexane and bromine in the presence of UV light.
  c) Explain why the product from the reaction between hexane and bromine contains many different organic compounds.

# Alkenes

**Key term**

Unsaturated molecules contain one or more double (or triple) bonds.

**Tip**

You will recall that cracking breaks long chain alkanes into a shorter chain alkane and an alkene; $C_{10}H_{22} \rightarrow C_8H_{18} + C_2H_4$

Alkenes are unsaturated hydrocarbons and contain carbon–carbon double bonds.

**Alkenes** are unsaturated molecules that contain at least one C=C double bond. They are invaluable to chemists because they can be used to make a wide range of useful products such as alcohols and polymers. Alkenes can be produced from the cracking of heavier fractions of crude oil. They also occur naturally in plant and animal tissue such as oil, fat and pigments.

## Physical properties

### Boiling point

As with alkanes, the boiling points of the alkenes increase with increasing chain length. However, the boiling point of an alkene is lower than that of the corresponding alkane (Table 13.4).

**Table 13.4** Comparison of the boiling points of alkanes and alkenes.

| Alkane | Boiling point/K | Alkene | Boiling point/K |
|---|---|---|---|
| Ethane | 185 | Ethene | 169 |
| Propane | 231 | Propene | 226 |
| Butane | 273 | But-1-ene | 267 |
| Pentane | 309 | Pent-1-ene | 303 |

The alkenes have lower boiling points because their molecules cannot pack together as efficiently as those of alkanes. Hence there are fewer intermolecular forces (induced dipole–dipole) to overcome.

## Solubility

Alkenes, like alkanes, are insoluble in water.

## Bonding in alkenes

An alkene molecule contains a C=C double bond. The double bond consists of a σ-bond and a π-bond.

- A **σ-bond** is a single covalent bond made up of two shared electrons with the electron density concentrated between the two nuclei.

- A **π-bond** is formed by the 'sideways' overlap of two p-orbitals on adjacent carbon atoms, each providing one electron.

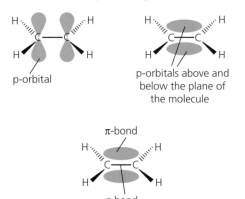

On average the energy required to break a C–C single bond is 347 kJ mol⁻¹ whereas the energy needed to break a C=C bond is 612 kJ mol⁻¹. This indicates that the π-bond is considerably weaker than the σ-bond.

The C–H bond angle at each side of the C=C double bond is approximately 120° (usually in the region 116–124°), which results in a trigonal planar arrangement of bonds round each carbon atom:

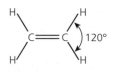

In other alkenes, for example propene and but-1-ene, the equivalent bond angles on either side of the C=C double bond are approximately 120°, but the bond angles in the alkyl groups are 109.5°, as expected:

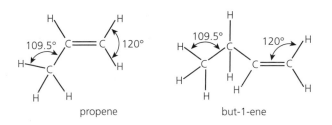

propene          but-1-ene

## Stereoisomerism in alkenes

The C=C double bond in alkenes prevents freedom of rotation, which under certain circumstances can lead to the existence of stereoisomers known as *E/Z* isomers.

The **two** key features of ***E/Z* isomerism** are:

● the presence of a C=C double bond

● each carbon atom in the C=C double bond is bonded to two different atoms or groups.

Substituents on the molecules shown below cannot twist around the double bond and move from one side to the other without breaking the π bond. This means that the molecules are in fact, different.

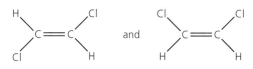

But-1-ene and but-2-ene both have a C=C double bond that prevents free rotation. However, one carbon atom in the C=C double bond in but-1-ene is bonded to two hydrogen atoms, so but-1-ene does not exhibit *E/Z* isomerism. In but-2-ene, both carbon atoms in the double bond are bonded to different atoms/groups:

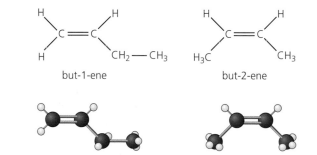

but-1-ene          but-2-ene

But-2-ene possesses both essential key features, and hence has a *Z*- and an *E*- isomer:

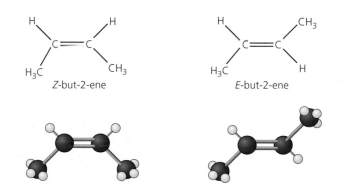

*Z*-but-2-ene          *E*-but-2-ene

**Key term**

Stereoisomers have the same molecular and structural formulae but a different three-dimensional spatial arrangement.

**Tip**

The prefixes '*Z*' and '*E*' come from the German words '*zusammen*', meaning together (i.e. on the same side of the molecule), and '*entgegen*', meaning opposite.

The skeletal formula of the two stereoisomers of but-2-ene clearly shows the difference in shape.

Although the difference in geometry changes the shape of the molecule, it has little or no effect on the general chemistry of the two isomers. However, Z-but-2-ene has a slightly higher boiling point than E-but-2-ene.

## Rules for assigning *E/Z* isomerism

Not all alkenes exist as *E/Z* isomers. It is essential that each C atom in the C=C double bond is bonded to two different atoms or groups. Each C atom in the C=C double bond must be asymmetric. Each of the alkenes below meets this criterion.

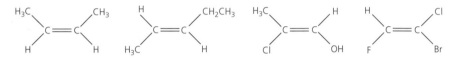

Each C in the double bond is bonded to two different atoms or groups. So it is possible to form at least one stereoisomer.

### Cahn, Ingold and Prelog (CIP) rules

There are a set of rules known as the Cahn, Ingold and Prelog (CIP) rules that enable you to decide whether you should name each isomer as an *E*-isomer or a *Z*-isomer.

According to these rules, the atomic number of each atom bonded to the C in the C=C double bond determines whether or not it is an *E*-isomer or a *Z*-isomer.

1   If the two attached atoms with the highest atomic numbers are on the diagonally opposite sides of the double bonds it is an *E*-isomer.

2   If the two attached atoms with the highest atomic numbers are **not** diagonally opposite each other across the double bonds it is a *Z*-isomer.

3   If two attached atoms have the same atomic number, then the adjacent atoms with the highest atomic number are taken into account. With alkyl groups it therefore follows that $CH_3CH_2CH_2 > CH_3CH_2 > CH_3$.

In a compound such as $CH_3(Cl)C=CHOH$, the C on the left-hand side of the C=C double bond is bonded to a Cl (atomic number 17) and a C (atomic number 6) whilst the C on the right-hand side is bonded to a H (atomic number 1) and a O (atomic number 8). So we can apply rules 1 and 2:

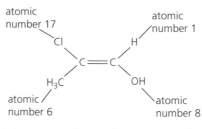

The two with the highest atomic numbers are on diagonally opposite sides of the C=C double bond so this is an *E*-isomer.

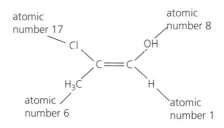

The two with the highest atomic numbers are on the same side of the C=C double bond so this is a *Z*-isomer.

In the compounds below, the C atom on the right of the C=C bond is bonded to two carbon atoms, each with atomic number 6. We therefore need to apply rule 3.

It is still possible to have *E/Z* isomers by considering the adjacent atoms: $CH_3CH_2$ has priority over $CH_3$ so it follows that the isomer on the left is the *Z*-stereoisomer and the isomer on the right is the *E*-isomer.

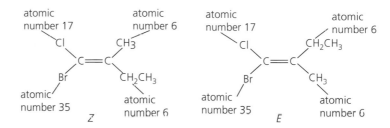

The molecules shown on page 210 can each be assigned as either *E* or *Z*, as shown below.

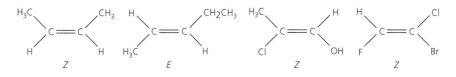

## Cis–trans isomerism

*Cis–trans* isomerism is a special case of *E/Z* isomerism in which the substituent groups attached to each carbon atom of the C=C group are the same.

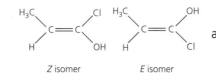

are *E/Z* isomers but **not** *cis–trans* isomers

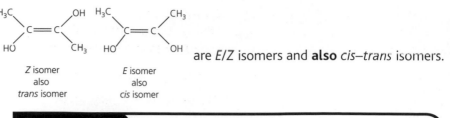

Z isomer
also
trans isomer

E isomer
also
cis isomer

are *E/Z* isomers and **also** *cis–trans* isomers.

> **Test yourself**
>
> 10 a) Draw and name the structural isomers of alkenes that have the
>    formula:
>    i) $C_4H_8$
>    ii) $C_5H_{10}$
>    iii) $C_6H_{12}$.
>    b) Explain what is meant by the term structural isomer.
> 11 a) Explain what is meant by the term stereoisomer.
>    b) Use your answers to Question 10 above and decide which, if any,
>       can form *E/Z* isomers.
>    c) Draw and name each *E/Z* isomer.
>    d) 2-methylhept-3-ene exists as *E/Z* isomers.
>       Draw and label the skeletal form of each isomer.

## Chemical properties

The C=C double bond is unsaturated, which means that the main reactions of alkenes are addition reactions.

**Key term**

An addition reaction is a reaction in which two molecules combine together to form a single product.

The reactivity of alkenes can be explained in terms of the relatively low bond enthalpy of the π-bond. In this type of reaction, the π-bond breaks and an atom, or group, adds to each of the two carbon atoms. The general reaction can be summarised as:

$$CH_2{=}CH_2 + X{-}Y \rightarrow CH_2XCH_2Y$$

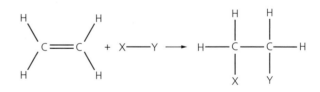

**Tip**

For most organic reactions, you need to know the balanced equation, the reagents, the conditions and what, if anything is observed.

## Reactions of ethene

### Hydrogenation
Ethene reacts with hydrogen to form ethane, according to the following equation:

$$CH_2CH_2 + H_2 \rightarrow CH_3CH_3$$

**Reagent:**       hydrogen

**Conditions:**    nickel catalyst at 150 °C

**Observations:**  no visible change

Hydrogenation of polyunsaturated compounds derived from plant oils is used in the production of margarine.

## Bromination

Ethene reacts with bromine ($Br_{2(aq)}$) to produce 1,2-dibromoethane, according to the equation:

$$CH_2CH_2 + Br_2 \rightarrow CH_2BrCH_2Br$$

**Reagent:** bromine

**Conditions:** mix at room temperature

**Observations:** decolorisation of bromine

This reaction is used to test for the presence of a C=C double bond.

**Tip**

Remember that the bromine is **decolorised**. In an exam, you will be penalised if you write that it 'goes clear' or is 'discoloured'.

## Reaction of ethene with hydrogen bromide

Ethene reacts with hydrogen bromide to produce bromoethane, according to the following equation:

$$CH_2CH_2 + HBr \rightarrow CH_3CH_2Br$$

**Reagent:** hydrogen bromide

**Conditions:** mix gases at room temperature

**Observations:** no visible change

## Hydration of ethene

A **hydration** reaction involves the addition of water and results in a new organic compound. Ethene is hydrated according to the following equation:

$$CH_2CH_2 + H_2O \rightarrow CH_3CH_2OH$$

**Reagent:** water (steam)

**Conditions:** an acid catalyst, such as phosphoric(v) acid, $H_3PO_4$, at 300 °C and high pressure (6 MPa)

**Observations:** no visible change

All alkenes undergo similar reactions under similar conditions. However, unsymmetrical alkenes, such as propene, produce two isomers when reacted with hydrogen chloride or water:

**Tip**

It is easy to confuse hydration with hydrolysis, which also involves reacting with water.

**Hydration** – water reacts with an organic molecule to produce a *single* product.
**Hydrolysis** – water reacts with an organic molecule and *two* new compounds are produced.

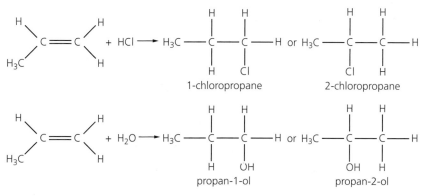

**Key term**

An **electrophile** is an electron-pair acceptor that forms a covalent bond.

## Mechanism of electrophilic addition

The mechanism of an electrophilic addition reaction involves an electrophile.

The movement of electron pairs is tracked by the use of curly arrows. The curly arrow always points from an area that is electron-rich to an area that is electron-deficient.

a curly arrow ⟳ shows the movement of an electron pair

The reaction is initiated by the presence of a dipole (charge separation along a bond) on the molecule that reacts with the alkene. This dipole could be:

● induced by the presence of the alkene, as is the case with bromine

● permanent, as in hydrogen bromide.

When describing mechanisms, it is essential that you show:

● relevant dipoles

● lone pairs

● curly arrows.

The key features of this mechanism are as follows:

● When the Br–Br molecule approaches the ethene, a temporary induced dipole is formed, resulting in polarisation of the Br–Br bond:

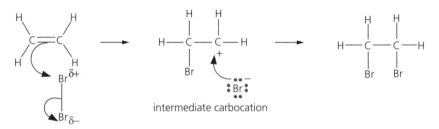

intermediate carbocation

● The initial curly arrow starts at the $\pi$-bond (within the C=C double bond) and points to the $Br^{\delta+}$.

● The second curly arrow shows the movement of the bonded pair of electrons in the Br–Br bond towards the $Br^{\delta-}$, resulting in heterolytic fission of the Br–Br bond.

● An intermediate carbonium ion (also called a carbocation) is formed together with a $Br^-$ ion, which contains the pair of electrons from the Br–Br bond.

● The third curly arrow from the lone pair of electrons on the $Br^-$ to the positively charged carbonium ion shows the formation of 1,2-dibromoethane.

When bromine reacts with an alkene, the Br–Br bond undergoes **heterolytic** fission; when it reacts with an alkane it undergoes **homolytic** fission. Compare the two ways in which the Br–Br bond is broken:

Br ⟳ Br ⟶ $Br^+$ + $Br^-$          Br ⟳ Br ⟶ Br• + Br•
heterolytic fission                    homolytic fission

Hydrogen bromide (and hydrogen chloride) are polar molecules and have a permanent dipole. The hydrogen atom has a $\delta+$ charge and acts as the electrophile.

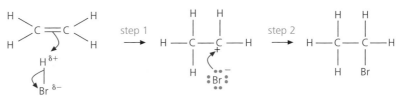

In the first step of the reaction between hydrogen bromide and ethene, an HBr molecule approaches an ethene molecule. The δ+ hydrogen end of the H–Br is attracted to the electron-dense double bond. As the H–Br molecule gets even closer, heterolytic fission of the π-bond occurs. The electrons in the π-bond form a covalent bond to the hydrogen atom and, at the same time, heterolytic fission of the H–Br bond also occurs. Electrons in the H–Br bond are taken over by the bromine atom producing a Br⁻ ion. The other product of step 1 is the highly reactive carbocation, or carbonium ion, $CH_3CH_2^+$. This reacts immediately with the Br⁻ ion in the second step of the reaction to form $CH_3CH_2Br$.

## The reaction of asymmetric alkenes with HBr and with $H_2O$

Alkenes such as propene react with both HBr and with $H_2O$ in exactly the same way as ethene reacts with both HBr and $H_2O$, but with propene in each reaction two possible products are formed.

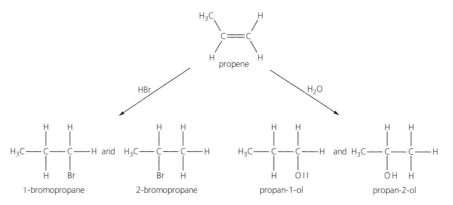

One would expect that the yield of each would be approximately 50%, but this is not the case. In the reaction with HBr about 90% of 2-bromopropane is formed and only approximately 10% of 1-bromopropane. The reaction with $H_2O$ is similar and the major product is always propan-2-ol and not propan-1-ol.

This was first notice in the late nineteenth century by the Russian chemist Vladimir Vasilevich Markownikoff (sometimes spelt Markovnikov), who studied thousands of alkene reactions and noted that whenever a compound with general formula HX reacted with an alkene the H atom in HX, always became attached to the carbon atom which had the larger number of H atoms. This rule became known as Markownikoff's rule.

This can be explained by considering the mechanism carefully. There are two alternatives, as follows:

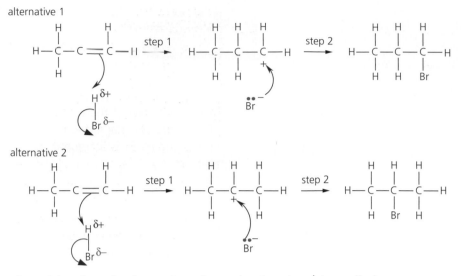

Step 1 involves the formation of a carbonium ion (also called a carbocation). This is the slow step in the mechanism.

Step 2 involves the reaction between the bromide ion, Br⁻ and the carbonium ion. This is the fast step in the mechanism.

Markownikoff's rule predicts that 2-bromopropane will be the major product indicating that the preferred mechanism is 'Alternative 2' in which a secondary carbonium ion is formed. Carbonium ions are unstable and can be stabilised by the inductive effect of adjacent alkyl groups (see page 186). The alkyl groups release electrons along the σ-bonds as shown below.

**Tip**

The classification of primary, secondary and tertiary is explained fully on page 226

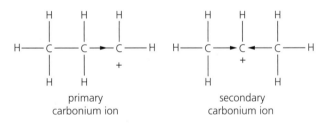

The primary carbonium ion is stabilised by one inductive effect whereas the secondary carbonium ion is stabilised by two inductive effects. As you might expect, a tertiary carbonium ion is even more stable and more readily formed.

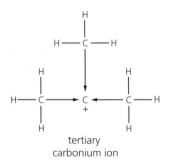

# Addition polymerisation of alkenes

Alkenes can undergo addition reactions in which one alkene molecule joins to another until a long molecular chain is built up. The individual alkene molecule is referred to as a **monomer**; the long-chain molecule is called a **polymer.**

Polymerisation can be initiated (started) in a number of ways. Often, the initiator is incorporated at the start of the long molecular chain. However, if the initiator is disregarded, the empirical formulae of the monomer and the polymer are the same.

## Poly(ethene)

The simplest monomer is ethene, which can be polymerised to produce poly(ethene). In the formula for poly(ethene), a large number of monomers, *n*, join together – this number may be as high as 10 000, such that a long-chain polymer is formed.

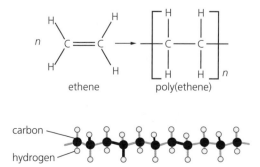

ethene          poly(ethene)

Poly(ethene), or polythene as it is commonly called, was developed in the 1930s. Today, it is widely used in everyday applications. It is tough, durable and unreactive. Poly(ethene) is non-polar and the molecules contain only strong covalent bonds. It is an excellent insulator and, because it is unaffected by the weather, is used for insulating electricity cables.

Changing the monomer gives polymers with different properties for different applications (Table 13.5).

The properties of a polymer can also be altered by the addition of a **plasticiser**. This is usually a liquid with a high boiling point (e.g. an ester see page 223). The plasticiser molecules are located between the polymer molecules, which reduces the intermolecular forces and makes the polymer more flexible (less rigid).

**Table 13.5** The monomer, polymer structure and uses of some important polymers.

| Monomer | Repeat unit of polymer | Properties | Uses |
|---|---|---|---|
| Propene | Poly(propene) | Tough<br>Very resistant to water, acids and alkalis<br>Easily moulded<br>Easily coloured | Food boxes<br>Ropes<br>Carpets<br>Toys |
| Chloroethane (vinylchloride) | Poly(chloroethane) (polyvinyl chloride/PVC) | Tough<br>Rigid<br>Flexible (when plasticiser added)<br>Very resistant to water, acids and alkalis | Clingfilm<br>Guttering<br>Window frames<br>Raincoats<br>Shower curtains<br>Insulation for cables |
| Tetrafluoroethene | Poly(tetrafluoroethene)/PTFE Also known as Teflon® | Tough<br>Very resistant to water, acids and alkalis<br>Low friction solid/slippery | Coating for non-stick pans/ skis<br>Gaskets<br>Plumbing tape<br>Low friction surfaces for bridges to move as metals expand and contract |
| Phenylethene (also known as styrene) | Poly(phenylethene) or polystyrene | Excellent insulator<br>Resistant to water, acids and alkalis. | Styrofoam® cups<br>Fast food containers<br>Refrigerator insulation<br>Packaging |
| Ethenol (also known as vinyl alcohol) | Poly(ethenol) or polyvinyl alcohol | Water soluble | Hospital laundry bags which can be handled without touching the infected contents |

**Figure 13.4** Clingfilm is just a thin film of PVC (polyvinylchloride). Its correct name is poly(chloroethene).

**Figure 13.5** This little boy is wearing a PVC waterproof jacket.

## Identification of the monomer from which a polymer is produced

It is possible to deduce the repeat unit of an addition polymer and, therefore, to identify the monomer from which the polymer is produced. For example:

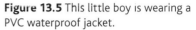

Therefore, the monomer is propene:

The equation for the polymerisation can be written as

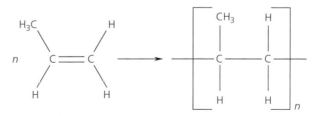

17 There are four alkenes that are isomers of $C_4H_8$.
   a) Draw and name the four alkenes.
   b) Draw the repeat units of each of the polymers that could be formed from each alkenes in (a).
   c) Name each of the polymers formed.
18 Poly(styrene) is made from the following monomer.

   a) What is the common name for the monomer?
   b) The systematic name of the monomer is phenylethene. Write a systematic name for the polymer.
   c) Draw a section of the polymer composed of three monomer units.
19 A section of an addition polymer is shown below. Identify the monomer from which it was made.

# Waste polymers and alternatives

The widespread use of these polymers has created a major disposal problem. The bonds in addition polymers are strong, covalent and non-polar, making most of the polymers resistant to chemical attack. In addition, since they are not broken down by bacteria, they are referred to as being **non-biodegradable**.

Plastic waste has in the past been buried in landfill sites, where it remains unchanged for decades. Some local authorities now collect waste plastic separately from other refuse.

Plastic waste is much more difficult to recycle than waste metals. There are many different types of plastic which have to be first separated and sorted before they can be recycled.

An alternative to dumping is incineration. Polymers are made from hydrocarbons and are, therefore, potentially good fuels. When burned, they release useful energy which could be used to generate electricity. In the UK, only about 10% of plastic is incinerated, but in countries such as Japan and Denmark up to 70% is burned. Some plastics, such as PVC, produce poisonous gases (e.g. HCl), so the incinerators have to be fitted with gas-scrubbers to prevent these toxins from being released into the environment.

Another possibility is recycling the polymers and using them as feedstock for the production of new polymers. Different types of polymer have to be separated from each other because mixtures yield inferior plastic products.

Oil-derived polymers also deplete finite resources and considerable effort has been put into developing biodegradable polymers that have useful properties. The aim is to create a polymer that:

- contains an active functional group that can be attacked by bacteria,
- is sustainable and doesn't use finite resources.

Other biodegradable compounds are based on condensation polymers, which are studied in the second year of the A Level course.

---

### Activity

**A brighter future for plastics**

Biodegradable plastics are now being synthesised and produced commercially as replacements for non-biodegradable plastics. One of the most important of these new biodegradable plastics is polylactic acid (PLA), which is produced from starch.

Starch is a natural polymer produced during photosynthesis. Cereal plants, such as maize and wheat, and tubers, such as potatoes, contain starch in large proportions. Starch can be processed directly into a bioplastic, but because this is soluble in water, articles made from starch swell and deform when exposed to moisture. The problem can be overcome by modifying the starch into a different polymer.

First, the starch is harvested from maize, wheat or potatoes. Micro-organisms then transform it into the monomer lactic acid. Finally, the lactic acid is chemically treated so that the molecules of lactic acid link into long chains or polymers, which bond together forming PLA.

Polylactic acid can be formed into containers and packaging for food and consumer goods as diverse as plant pots and disposable nappies. The biodegradable and compostable nature of PLA means that it will break down into harmless natural products under normal outdoor conditions. This gives it big political and environmental advantages over conventional plastic packaging, which uses an estimated 500 000 barrels of oil every day throughout the world. Its biodegradable qualities could also take the pressure off the world's mounting landfills, in which plastics take up an estimated 25% by volume. However, as PLA is significantly more expensive than conventional oil-based plastics, it has failed to gain widespread use, although it is becoming a more viable alternative as the price of crude oil continues to rise.

1 What do you understand by the following terms used in the introduction?
  a) Bioplastic
  b) Compostable
2 What limits the use of starch itself as a plastic?
3 The structural formula of lactic acid is:

  a) Write down the molecular formula of lactic acid.
  b) Give the names of the functional groups in lactic acid.
4 Starch reacts with water when micro-organisms transform it into lactic acid. Assuming that the formula of starch can be written as $(C_6H_{10}O_5)n$, write an equation for the conversion of starch to lactic acid.
5 The $-OH$ group and the $-COOH$ group can react as shown below:

$$R-OH + HOOC-R^1 \rightarrow \begin{matrix} R-O-C-R^1 \\ \| \\ O \end{matrix} + H_2O$$

Use this to explain how lactic acid molecules can link up to form long chains or polymers.
6 State three advantages of PLA over oil-based plastics.
7 a) What is the major disadvantage of PLA compared with oil-based plastics?
  b) Why is this disadvantage becoming less important?

221

# Practice questions

## Multiple choice questions 1–10

**1** Put the following alkanes in order of increasing boiling points. Start with the lowest boiling point first.

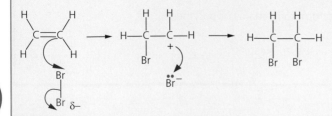

   1        2        3     4

  **A**  1, 2, 3, 4

  **B**  4, 3, 2, 1

  **C**  4, 3, 1, 2

  **D**  2, 1, 3, 4

**2** The complete combustion of methane is shown below:

$$CH_4(g) + 2O_2(g) \rightarrow CO_2(g) + 2H_2O(l)$$

At room temperature and pressure, which one of the following is the minimum volume of oxygen required to burn 16 g of methane?

  **A**  $2\,dm^3$

  **B**  $64\,dm^3$

  **C**  $24\,dm^3$

  **D**  $48\,dm^3$

**3** When fossil fuels such as petrol are burnt, carbon dioxide and water are produced. Assume that a petrol contains only octane, $C_8H_{18}$:

$$C_8H_{18}(g) + 12\tfrac{1}{2}O_2(g) \rightarrow 8CO_2(g) + 9H_2O(l)$$

Calculate the volume of carbon dioxide produced when 1 litre ($1\,dm^3$) of petrol is burnt.

(Assume the mass of 1 litre of $C_8H_{18}$ is 684 g and all volumes are measured room temperature and pressure where 1 mol of gas has a volume of $24\,000\,cm^3$.)

  **A**  $8000\,cm^3$

  **B**  $48\,000\,cm^3$

  **C**  $576\,000\,cm^3$

  **D**  $1\,152\,000\,cm^3$

**4** Students were asked to describe the electrophilic addition mechanism when ethene reacts with bromine using curly arrows and showing all relevant dipoles. One student's response is shown below.

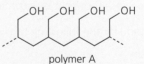

Identify the number of errors in the student's response.

  **A**  0

  **B**  1

  **C**  2

  **D**  3

**5** 2,2-dimethylbutane reacts with bromine, in the presence of UV light, to form a mono-substituted compound with molecular formula $C_6H_{13}Br$. Which one of the following is the number of possible isomers that could have been formed?

  **A**  4 isomers

  **B**  3 isomers

  **C**  2 isomers

  **D**  1 isomer

**6** When pent-1,4-diene, $CH_2{=}CHCH_2CH{=}CH_2$, is reacted completely with steam at high temperature and high pressure in the presence of a suitable acid catalyst, the number of different alcohols that could be formed is:

  **A**  1

  **B**  2

  **C**  3

  **D**  4

**7** Polymer A, shown below,

could be made from:

  **A**  $CH_3CH_2OH$.

  **B**  $CH_3CH_2CH_2OH$

  **C**  $CH_3CHCHOH$

  **D**  $CH_2CHCH_2OH$

Use the key below to answer Questions 8 to 10.

| A | B | C | D |
|---|---|---|---|
| 1, 2 & 3 correct | 1, 2 correct | 2, 3 correct | 1 only correct |

**8** Ethane reacts with $Cl_2$ in the presence of UV light. The initiation step in the mechanism is $Cl_2 \rightarrow 2Cl\cdot$. Which of the following correctly represent a possible propagation or termination step in the mechanism?

  1  $2CH_3CH_2\cdot \rightarrow CH_3CH_2CH_2CH_3$

  2  $CH_3CH_2\cdot + Cl\cdot \rightarrow CH_3CH_2Cl$

  3  $Cl\cdot + CH_3CH_3 \rightarrow CH_3CH_2Cl + H\cdot$

**9** Which of the following statements about the reactions of 2-methylpropene are correct?
  **1** It reacts with hydrogen to form a saturated hydrocarbon with empirical formula $C_2H_5$.
  **2** When it reacts with steam in the presence of an acid catalyst one of the products obtained is a tertiary alcohol.
  **3** It has a higher boiling point than but-1-ene.

**10** Which of the following statements is correct?
  **1** When HBr reacts with but-1-ene the major product is 2-bromobutane.
  **2** But-1,2-diene, $CH_2=CHCH=CH_2$ reacts with steam to form $HOCH_2CH(OH)CH(OH)CH_2OH$.
  **3** Cyclobutane, $C_4H_8$, reacts with $H_2(g)$ in the presence of a Ni catalyst.

**11** Name each of the following:

  **a)**

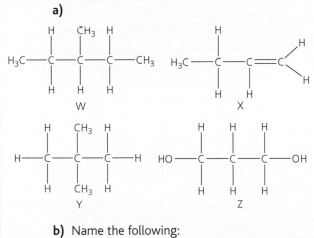

  W    X

  Y    Z

  **b)** Name the following:
    $CH_3CHBrCH_3$, $CH_3CHCHCH_3$, $(CH_3)_4C$ and $H_3CCH_2CH(CH_3)CH_2CH_3$. (2)

**12** Explain each of the following terms:
  **a)** radical
  **b)** homolytic fission
  **c)** initiation
  **d)** propagation
  **e)** termination
  **f)** substitution. (6)

**13 a)** Write the balanced equation and the mechanism for the reaction between:
    **i)** propane and bromine
    **i)** cyclobutane and chlorine.
  **b)** Explain why more than one organic product is formed in each reaction. (11)

**14 a)** Name the following alkenes:

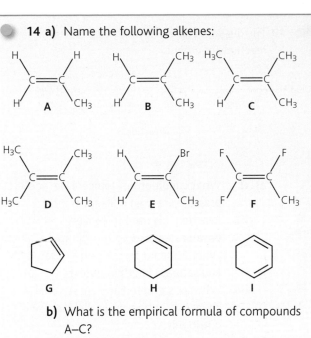

  A    B    C

  D    E    F

  G    H    I

  **b)** What is the empirical formula of compounds A–C?
  **c)** What is the molecular formula of compound D?
  **d)** What is the molecular formula of compound I?
  **e)** What is the empirical formula of all alkenes with one double bond? (13)

**15 a)** Explain why the disposal of polymers made from alkenes is problematic.
  **b)** Explain why polymers made from alkenes are good insulators. (2)

**16** Explain how waste polymers are likely to be treated in the future in order to reduce damage to the environment. (4)

**17 a)** What is meant by electronegativity?
  **b)** Explain why alkanes are unreactive. (2)

**18** Explain what is meant by the term 'electrophile'. Write a balanced equation and the mechanism for the reaction between bromine and cyclohexene, $C_6H_{10}$. Use curly arrows to show the movement of electrons. Show any relevant dipoles and lone pairs of electrons. (6)

**19** The diagram below shows three important reactions of ethene.

$$CH_3CH_3 \xleftarrow[\text{Reaction 1}]{H_2/Ni} CH_2=CH_2 \xrightarrow[\text{Reaction 2}]{} CH_3CH_2OH$$

ethene

↓ Reaction 3

poly(ethene)

a) i) What conditions of temperature and pressure are used in Reaction 1?
   ii) Reaction 1 is used to convert unsaturated alkenes to saturated alkanes. What is meant by the terms unsaturated and saturated in this context?
   iii) Why are saturated and unsaturated chemicals important to dieticians and nutritionists?

b) i) What chemicals are used to produce $CH_3CH_2OH$ in Reaction 2?
   ii) A common name for $CH_3CH_2OH$ is alcohol. What is its systematic name?
   iii) What is the major use of $CH_3CH_2OH$ from Reaction 2?

c) Draw a section of the poly(ethene) structure that consists of three monomer units.

d) State four properties of poly(ethene) that make it particularly suitable for making plastic bags. (12)

**20** Calculate the formulae of alkanes A, B, and C.

a) When $100\,cm^3$ of alkane A is burned in excess oxygen, $300\,cm^3$ carbon dioxide gas and $400\,cm^3$ water (steam) are produced. Calculate the formula of alkane A.

b) When burned, 0.1 mol of alkane B produces $10.8\,g$ of water. Calculate the formula of alkane B.

c) When $60\,cm^3$ of alkane C reacts with $700\,cm^3$ of oxygen (an excess) and is cooled to room temperature, a gaseous mixture with a volume of $460\,cm^3$ is produced. When this gaseous mixture is passed through limewater, the volume is reduced to $40\,cm^3$.
   i) Deduce the volume of carbon dioxide gas in the mixture.
   ii) Deduce the volume of oxygen gas used in the reaction.
   iii) Calculate the formula of alkane C. Write a balanced equation for the reaction. (12)

**21** Alkenes can undergo polymerisation. Write an equation to show the polymerisation of cyclohexene. Draw two repeat units of the polymer. (3)

**22** The unsaturated compound cyclohex-1,4-diene, $C_6H_8$, reacts with an excess of HBr to produce a saturated compound with molecular formula $C_6H_{10}Br_2$. Draw and name all possible isomers of $C_6H_{10}Br_2$. (6)

**23** The action of the bacterium *Clostridium acetobutylicum* on biomass, such as sugar beet, wheat or straw, is used to produce compound X, which is a fuel. The fuel, X, can produce almost as much power as petrol, although it cannot be used in a pure form because it exists as a gel below 25.5 °C.

a) A sample of X contained the following elements by mass: carbon 64.9%, hydrogen 13.5%, oxygen 21.6%. Calculate the empirical formula of X.

b) When X is vaporised, it is found that $0.2\,g$ of X has a volume of $65\,cm^3$ measured at room temperature and pressure. Calculate the molecular formula of X.

c) Compound X reacts with concentrated sulfuric acid and a hydrocarbon, Y, is formed that contains the same number of carbon atoms as X. When bubbled into aqueous bromine, the bromine is decolorised. Compound Y could be one of four isomers.
   i) Draw the displayed formulae of these isomers.
   ii) Choose one of the isomers and write a balanced equation for its reaction with bromine.

d) Compound Y reacts with hydrogen to give an unbranched hydrocarbon, Z.
   i) Give the conditions for the reaction of compound Y with hydrogen.
   ii) Name hydrocarbon Z.

e) When compound Y reacts with steam, two products are obtained, one of which is X.
   i) Give the structural formulae of the two products.
   ii) The functional group of X is at the end of its carbon chain. Name compound X. (14)

## Challenge

**24** Pent-1,4-diene, $CH_2CHCH_2CHCH_2$ reacts at high temperature and pressure with steam in the presence of suitable acid catalyst. A mixture of alcohols is formed, each alcohol having molecular formula $C_5H_{12}O_2$.

a) Draw and name all possible diols that could be formed.

b) Suggest which is the major product. Justify your answer. (10)

# Chapter 14

# Alcohols and haloalkanes

## Prior knowledge

*In this chapter it is assumed that you are familiar with:*

- electronegativity and dipoles
- intermolecular bonds such as hydrogen bonds and induced dipole–dipole interactions

In organic chemistry the bonds that hold the carbon-based molecules together are covalent bonds. These bonds are strong bonds and are usually only broken when a chemical reaction occurs. If two different atoms with different electronegativity are covalently bonded the bond becomes polarised and dipoles are formed. As well as the bonds within a molecule there are also bonds that form between molecules. These are called intermolecular forces and are relatively weak and can be broken by physical processes such as melting or boiling.

## Test yourself on prior knowledge

1 Define electronegativity.
2 The electronegativity of H = 2.1 and of Cl = 3.0. Explain, with the aid of a diagram, the bonding in a hydrogen chloride, HCl, molecule.
3 Explain, with the aid of a diagram, the intermolecular forces in water.
4 Draw two molecules of ammonia, $NH_3$, and show how hydrogen bonds are formed between the two ammonia molecules.
5 a) Write equations for the manufacture of ethanol by:
   i) fermentation
   ii) hydration.
   b) Calculate the atom economy for each manufacturing process.
   c) Suggest which manufacturing process is more sustainable. Explain your answer.

This chapter covers the reactions of two more homologous series: alcohols and haloalkanes.

# Alcohols

**Figure 14.1** Menthol is a naturally occurring alcohol that is extracted from mint oils. It has local anaesthetic and anti-irritant properties and is used to relieve minor throat irritation.

A molecule of an alcohol contains the hydroxy group, −OH. The names of all alcohols end with -ol. Molecules of many naturally occurring compounds, such as carbohydrates, pheromones, vitamins and steroids, contain an −OH group.

## Classification of alcohols

All alcohols contain a carbon atom bonded to a hydroxy group, C−OH. They are classified as primary, secondary or tertiary alcohols:

**Table 14.1** Alcohol classification.

| Primary alcohol | Secondary alcohol | Tertiary alcohol |
|---|---|---|
| $H_3C-\overset{\displaystyle H}{\underset{\displaystyle H}{C}}-OH$ | $H_3C-\overset{\displaystyle CH_3}{\underset{\displaystyle H}{C}}-OH$ | $H_3C-\overset{\displaystyle CH_3}{\underset{\displaystyle CH_3}{C}}-OH$ |
| The carbon atom bonded to the –OH group is bonded to only one other carbon atom | The carbon atom bonded to the –OH group is bonded to two carbon atoms | The carbon atom bonded to the –OH group is bonded to three carbon atoms |

**Tip**

R, R′ and R″ are used to represent groups other than –H.

**Test yourself**

1 Draw, name and classify as primary, secondary or tertiary the alcohols with molecular formula $C_4H_9OH$.
2 Draw, name and classify as primary, secondary or tertiary the alcohols with molecular formula $C_5H_{11}OH$.
3 Classify each of the following as either primary, secondary or tertiary. Give the systematic name of each alcohol.

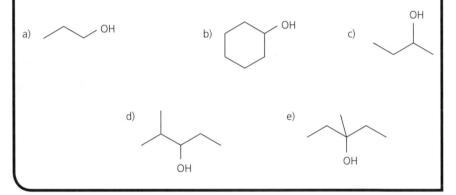

## Physical properties of alcohols

The simplest alcohols are liquids at room temperature and are miscible with water. Alcohols have relatively high boiling and melting points. This can be explained in terms of hydrogen bonding (page 108). A hydrogen bond forms between the lone pair of electrons on the oxygen in the –OH group in one alcohol molecule and the hydrogen in the –OH group in an adjacent alcohol molecule:

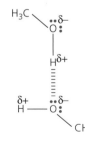

## Boiling point and melting point

When alcohols are boiled (or melted), energy is required to break the intermolecular forces. All substances, including alcohols, have induced dipole–dipole interactions that help bind the molecules together.

In addition, alcohol molecules have hydrogen bonds. Hydrogen bonds are the strongest of the intermolecular forces, so more energy is required to break them than to overcome induced dipole–dipole interactions. Hydrogen bonding decreases the volatility and, therefore, results in an increase in boiling point.

### Miscibility with water

Methanol and ethanol are freely miscible (mix together to make a single layer) with water. When mixed, some of the hydrogen bonds in the individual liquids are broken, but they are then replaced by new hydrogen bonds between the alcohol and water:

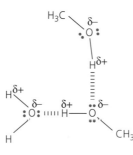

As the relative molecular mass of the alcohol increases, the miscibility with water decreases.

## Reactions of alcohols

Alcohols are more reactive than alkanes because the C–O and O–H bonds are both polar bonds.

$$\overset{\delta+}{C}\!-\!\overset{\delta-}{O} \qquad \overset{\delta-}{O}\!-\!\overset{\delta+}{H}$$

### Combustion

Alcohols burn to produce carbon dioxide and water. The reactions are exothermic, releasing large amounts of energy:

$$CH_3OH + 1\tfrac{1}{2}O_2 \rightarrow CO_2 + 2H_2O$$
$$C_2H_5OH + 3O_2 \rightarrow 2CO_2 + 3H_2O$$

### Oxidation

Primary and secondary alcohols can be oxidised to produce either an aldehyde, a ketone or a carboxylic acid. Tertiary alcohols do not readily undergo oxidation.

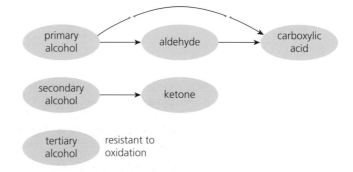

Oxidation of an alcohol can only occur if the C bonded to the O–H is also bonded to at least one H atom.

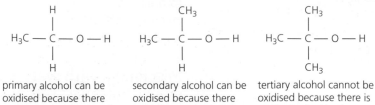

primary alcohol can be oxidised because there are two H attached to the C bonded to the OH

secondary alcohol can be oxidised because there is one H attached to the C bonded to the OH

tertiary alcohol cannot be oxidised because there is no H attached to the C bonded to the OH

The most common oxidising agent is an acidified solution of potassium dichromate, $K_2Cr_2O_7$. The acidified dichromate(VI), $H^+/Cr_2O_7^{2-}$, is a bright orange solution which turns dark green when an alcohol is oxidised.

Balanced equations for the oxidation of alcohols can be written using [O] to represent the oxidising agent. The equations for the oxidation of a **primary alcohol** to an **aldehyde** or a **secondary alcohol** to a **ketone** are similar in that water is always formed:

- primary alcohol + [O] → aldehyde + water
- secondary alcohol + [O] → ketone + water

**Aldehydes** and **ketones** both contain the carbonyl, C=O, group:

**Figure 14.2** Result of warming two alcohols, one primary and the other tertiary with an acidic solution of potassium dichromate(VI). The orange dichromate(VI) is reduced to green chromium(II) ions when there is a reaction. Tertiary alcohols do not react with potassium dichromate(VI).

- aldehydes – the carbonyl group is on the end of the carbon chain
- ketones – the carbonyl group is **not** on the end of the carbon chain.

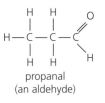

propanal
(an aldehyde)

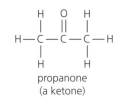

propanone
(a ketone)

**Oxidation of a primary alcohol to an aldehyde**
Methanol is oxidised to methanal:

$$CH_3OH + [O] \rightarrow HCHO + H_2O$$

methanol          methanal

Ethanol is oxidised to ethanal:

$$CH_3CH_2OH + [O] \rightarrow CH_3CHO + H_2O$$

ethanol          ethanal

**Oxidation of a secondary alcohol to a ketone**
Propan-2-ol is oxidised to propanone:

$$CH_3CH(OH)CH_3 + [O] \rightarrow CH_3COCH_3 + H_2O$$

propan-2-ol          propanone

**Tip**

When balancing oxidation reactions, count the [O] which has to be balanced alongside any other oxygens in the equation.

Pentan-2-ol is oxidised to pentan-2-one:

$$CH_3CH_2CH_2CH(OH)CH_3 + [O] \rightarrow CH_3CH_2CH_2COCH_3 + H_2O$$

pentan-2-ol　　　　　　pentan-2-one

### Further oxidation of a primary alcohol to a carboxylic acid

After an aldehyde has been formed by the oxidation of a primary alcohol, a further oxidation step can take place. The product is a **carboxylic acid** which contains the COOH functional group. Balanced equations for the oxidation of a primary alcohol to a carboxylic acid can be represented as:

primary alcohol + 2[O] → carboxylic acid + water

Methanol can be oxidised to methanoic acid:

$$CH_3OH + 2[O] \rightarrow HCOOH + H_2O$$

methanol　　　methanoic acid

**Tip**

In an oxidation reaction water will **always** be formed along with an aldehyde or a ketone or a carboxylic acid. Hydrogen, $H_2$, will **never** be formed.

Ethanol can be oxidised to ethanoic acid:

$$CH_3CH_2OH + 2[O] \rightarrow CH_3COOH + H_2O$$

ethanol　　　ethanoic acid

### Apparatus used in the oxidation of alcohols

Organic compounds are volatile and reactions are often carried out under reflux or by distillation. A primary alcohol can be oxidised to form either an aldehyde or a carboxylic acid and the product depends on the apparatus used.

to fume cupboard or sink
heat
propan-1-ol + sodium dichromate(VI) + dilute sulfuric acid
propanal

**Figure 14.3** Distillation allows the aldehyde to distil off as it is formed which prevents formation of the carboxylic acid by further oxidising the aldehyde.

reflux condenser
propan-1-ol with excess sodium dichromate(VI) and sulfuric acid
heat

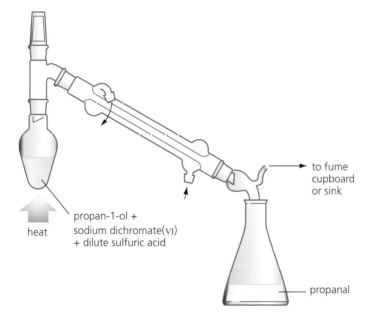

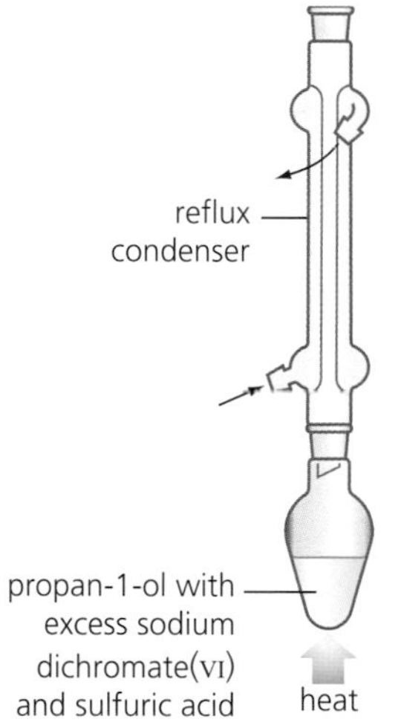

**Figure 14.4** Reflux ensures that any volatile aldehyde condenses and flows back into the flask, where excess oxidising agent ensures complete oxidation to form the carboxylic acid.

If ethanol is oxidised, either ethanal or ethanoic acid can be produced. The boiling point of:

- ethanol, $CH_3CH_2OH$, is 78 °C
- ethanal, $CH_3CHO$, is 21 °C
- ethanoic acid, $CH_3COOH$, is 118 °C.

The product depends on the apparatus used.

## Oxidation of ethanol to ethanal

The distillation process involves evaporation followed by condensation, which allows the most volatile component to be separated. When ethanol is heated with a mixture of potassium dichromate(VI) and sulfuric acid in a distillation apparatus (Figure 14.3), the volatile component evaporates first. Ethanal has the lowest boiling point and, therefore, vaporises most readily. The condenser has cold water circulating in the outer sleeve, so when the ethanal reaches the condenser, it condenses and is separated from the reaction mixture and can be collected.

## Oxidation of ethanol to ethanoic acid

Reflux involves a process of continuous evaporation and condensation, which prevents volatile components from escaping.

As with the distillation, when the reaction mixture is heated the most volatile component vaporises first. Ethanal has the lowest boiling point and is vaporised most readily. The vertical reflux condenser has cold water circulating in the outer sleeve. When the ethanal reaches the condenser, it condenses and falls back into the oxidising mixture. Therefore, it is not separated from the reaction mixture. The ethanal is oxidised further to form ethanoic acid.

## Elimination or dehydration

When reacted with hot, concentrated sulfuric acid or hot pumice/$Al_2O_3$, an alcohol is dehydrated to form an alkene.

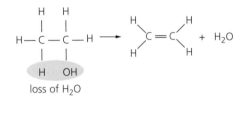

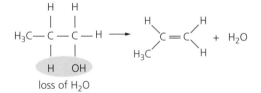

**Figure 14.5** Elimination reaction: ethanol loses water to become ethene, and propan-1-ol is dehydrated to form propene.

For alcohols such as butan-2-ol, water can be lost in two ways:

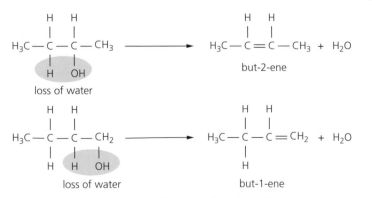

**Figure 14.6** Losing water from butan-2-ol produces but-2-ene or but-1-ene.

Both the *E*- and *Z*-versions of but-2-ene can be formed.

The C=C bond is formed by an **elimination reaction**. In this case, the water is eliminated and the reaction could also be described as a **dehydration reaction**.

## Substitution with halide ions

Primary, secondary and tertiary alcohols all react with a halide ion (Br⁻) in the presence of an acid (H⁺). The haloalkane is produced by reacting the alcohol with $NaBr/H_2SO_4$.

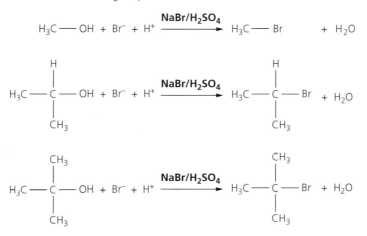

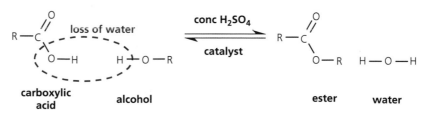

## Test yourself

4 Explain, with the aid of equations, why the hydration of propene produces a mixture of alcohols but the hydration of but-2-ene produces a single alcohol.

5 Write equations for the reaction of pentan-1-ol when:
   a) burnt with excess $O_2(g)$
   b) refluxed with acidified potassium dichromate
   c) heated with concentrated sulfuric acid
   d) distilled with acidified potassium dichromate
   e) reacts with NaBr in the presence of concentrated sulfuric acid.

## Ester formation

All alcohols, primary, secondary and tertiary, can react with carboxylic acids to form esters. This is covered fully in Module 6, *Carboxylic acids and derivatives* of the A level book. The equation for the formation of an ester is given below.

A haloalkane is a compound in which one or more hydrogen atoms of an alkane is replaced by a halogen atom. If one hydrogen is replaced, the general formula is $C_nH_{2n+1}X$ (where X = F, Cl, Br or I).

Like alcohols, haloalkanes are subdivided into primary, secondary and tertiary (Table 14.2). The rules for classification are the same. If the carbon atom that is bonded to the halogen (X) is bonded to one carbon atom only, then the compound is a primary haloalkane. If the carbon atom in the C–X bond is bonded to two other carbon atoms, the compound is a secondary haloalkane; if it is bonded to three other carbon atoms, the compound is a tertiary haloalkane.

**Table 14.2** Classification of haloalkanes.

| Primary haloalkane | Secondary haloalkane | Tertiary haloalkane |
|---|---|---|
| $H_3C-\overset{\displaystyle H}{\underset{\displaystyle H}{C}}-Cl$ | $H_3C-\overset{\displaystyle CH_3}{\underset{\displaystyle H}{C}}-Cl$ | $H_3C-\overset{\displaystyle CH_3}{\underset{\displaystyle CH_3}{C}}-Cl$ |

There are four structural isomers of $C_4H_9Cl$: two are primary, one secondary and one tertiary.

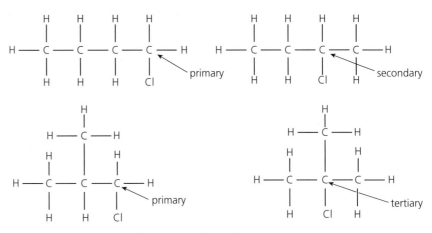

**Figure 14.7** Structures of the isomers of $C_4H_9Cl$.

Primary, secondary and tertiary haloalkanes behave similarly. They react with the same reagents, but the rate at which they react differs. The rate of reaction depends on the strength of the C–X bond. Tertiary haloalkanes react faster than secondary haloalkanes; primary haloalkanes react most slowly. This indicates that the C–X bond in a tertiary haloalkane is the weakest and the C–X bond in a primary haloalkane is the strongest.

## Substitution reactions of haloalkanes

### Hydrolysis
The substitution reactions of haloalkanes with water are examples of hydrolysis.

> **Key term**
>
> Hydrolysis is a reaction that involves water and results in the organic compound being split and two new products being formed.

The word 'hydrolysis' comes from: 'hydro', which is related to water, and 'lysis', which means splitting. So the term hydrolysis is used to describe any reaction in which water causes another molecule to split apart. Hydrolysis reactions are often catalysed by acids or alkalis. They are also often substitution reactions in which the chemical attack is by water molecules (or hydroxide ions).

Hydrolysis of haloalkanes using water is very slow and results in the replacement of the halogen with −OH groups to form alcohols.

$$CH_3CH_2I(l) + H_2O(l) \rightarrow CH_3CH_2OH(l) + H^+(aq) + I^-(aq)$$

The hydrolysis reaction is much quicker in the presence of an alkali catalyst, :ÖH⁻. When a primary haloalkane is heated under reflux with an aqueous solution of an alkali (e.g. sodium hydroxide or potassium hydroxide), the haloalkane is hydrolysed to a primary alcohol.

**Reagent:** sodium (or potassium) hydroxide (an alkali is required)

**Conditions:** the solvent **must** be water and the reaction mixture **must** be heated under reflux

**Equation:** $CH_3CH_2Br + NaOH \rightarrow CH_3CH_2OH + NaBr$

## Nucleophilic substitution of primary halogenoalkanes

The carbon–halogen bond is polar, $C^{\delta+}–Hal^{\delta-}$. The charge separation leaves the carbon atom open to attack by a nucleophile.

Common nucleophiles include $OH^-$, $:NH_3$, $^-CN$ and $H_2O$.

The reaction is a substitution reaction. The general equation for the reaction is:

$$R—X + :Nu^- \rightarrow R—Nu + :X^-$$

Nucleophilic substitution of a **primary** haloalkane is described by the following general mechanism:

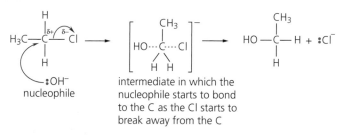

The nucleophile approaches the $C^{\delta+}$ in the $C^{\delta+}—X^{\delta-}$ from the opposite side to the halogen. This is sometimes referred to as 'backside attack'. An intermediate is formed that involves both the haloalkane and the nucleophile. The carbon–halogen bond is partially broken and the carbon–nucleophile bond is partially formed. The carbon–halogen bond then breaks to form the product and a halide ion.

The mechanism for the nucleophilic substitution of a primary amine can be simplified to:

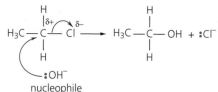

233

## Trend in rates of hydrolysis of primary haloalkanes

The rate of hydrolysis is different for different haloalkanes. The rate of reaction can be compared by carrying out the reaction in the presence of silver ions. The halogen atoms in haloalkanes are covalently bonded to carbon and give no precipitate of a silver halide. Hydrolysis releases halide ions, which immediately precipitate as the silver halide.

The rate of reaction can be monitored in one of two ways:

### Method 1

The haloalkane is mixed with water and a small amount of aqueous silver nitrate solution containing some ethanol in a water bath at 60 °C. The water can act as a nucleophile (using a lone pair of electrons on its oxygen atom), but the reaction is slow. As the hydrolysis reaction occurs, the halide ion is displaced and reacts with the $Ag^+$ ions to give a precipitate. By monitoring the rate at which the precipitate appears, it is possible to deduce the rate of hydrolysis. The ethanol is added to aid better mixing of the reagents since, although haloalkanes and water are immiscible (i.e. they form separate layers when mixed), they both dissolve in ethanol. The ethanol may also be involved as a nucleophile. If a comparison has to be made between haloalkanes, the conditions have to be carefully controlled.

### Method 2

The haloalkane is reacted with excess sodium hydroxide for a fixed period of time. The excess sodium hydroxide is then neutralised by the addition of nitric acid, $HNO_3$. Aqueous silver nitrate, $AgNO_3$, is added and the density of the precipitate is monitored (Figure 14.9). The neutralisation by nitric acid is essential to avoid a precipitate of silver oxide being formed. The experiment is repeated for other fixed periods of time.

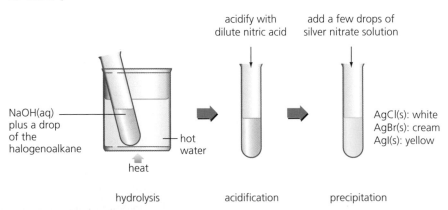

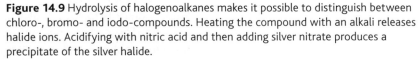

**Figure 14.9** Hydrolysis of halogenoalkanes makes it possible to distinguish between chloro-, bromo- and iodo-compounds. Heating the compound with an alkali releases halide ions. Acidifying with nitric acid and then adding silver nitrate produces a precipitate of the silver halide.

## Reactivity of haloalkanes

When 1-chlorobutane, 1-bromobutane and 1-iodobutane are reacted under identical conditions, 1-iodobutane reacts the fastest and 1-chlorobutane reacts the slowest. This may seem surprising, since the dipole is greatest for C–Cl and least for C–I such that the nucleophile has the strongest attraction to the carbon atom in a chloroalkane.

**Table 14.3** Bond enthalpies of carbon–halogen bonds.

| Bond | Bond enthalpy/kJ mol$^{-1}$ |
|------|------|
| C–F | 467 |
| C–Cl | 340 |
| C–Br | 280 |
| C–I | 240 |

> **Tip**
>
> The comparison of primary, secondary and tertiary is not in the specification and you will not be tested on it.

However, the other factor that determines the rate of the reaction is the strength of the carbon–halogen bond. Bond strengths (enthalpies) are listed in Table 14.3.

The C–I bond is the weakest and the least energy is required to break it. The C–F bond is so strong that it rarely undergoes hydrolysis.

The strength of the carbon–halogen bond is the dominant factor in determining rates of reaction of haloalkanes. For example, when 1-chlorobutane, 2-chlorobutane and 2-chloro-2-methylpropane are reacted under identical conditions, 2-chloro-2-methylpropane reacts most quickly and 1-chlorobutane reacts most slowly. This can be explained by comparing the relevant carbon–chlorine bond enthalpies (Table 14.4).

The C–Cl bond in the tertiary chloroalkane is the weakest and, therefore, requires the least amount of energy to break it. The C–Cl bond in the primary chloroalkane is the strongest, so the primary chloroalkane is the most difficult to hydrolyse.

**Table 14.4** Carbon chlorine bond enthalpies in primary, secondary and tertiary chloroalkanes.

| Bond | Bond enthalpy/kJ mol$^{-1}$ |
|------|------|
| C–Cl (primary) in 1-chlorobutane | 340 |
| C–Cl (secondary) in 2-chlorobutane | 331 |
| C–Cl (tertiary) in 2-chloro-2-methylpropane | 289 |

## Environmental concerns from use of organohalogen compounds

Haloalkanes are used in the preparation of a wide range of products, including pharmaceuticals (such as ibuprofen) and polymers (such as PVC and PTFE). Haloalkanes are also used in industry as solvents and lubricants.

Table 14.5 shows three complex halogenated compounds and their uses, other than in the manufacture of the plastics PVC and PTFE.

**Table 14.5** Complex halogenated compounds and their uses.

| Compound | Structure | Use |
|----------|-----------|-----|
| Halothane | | Anaesthetic |
| Thyroxine | | Thyroid hormone |
| Dichlorodiphenyltrichloromethane | | Insecticide used only as a last resort because it is harmful to wildlife |

## CFCs

Haloalkanes were used to produce chlorofluorocarbons (CFCs), such as dichlorodifluoromethane, $CF_2Cl_2$, and trichlorofluoromethane, $CFCl_3$. Their role in the destruction of the ozone layer is discussed below.

The use of CFCs has been phased out. However, it is worth appreciating why they were originally chosen for use in air conditioning, refrigeration units and aerosols.

These applications require a liquid of suitable volatility that can be readily evaporated and re-condensed. In addition, the liquid must be unreactive, non-flammable and non-toxic. CFCs fit these criteria and they were, therefore, considered to be safe and environmentally friendly. However, it is these properties that make them so persistent in the atmosphere. At the time of their introduction, the dangerous effect they would have in the stratosphere was not understood. The manufacture of CFCs was banned by the Montreal Protocol in 1987 and chemists worked to find alternatives.

Much effort was put into the development of hydrochlorofluorocarbons, HCFCs, which introduced hydrogen, making HCFCs less stable and more likely to break down before reaching the ozone layer. However, when HCFCs do break down chlorine radicals are still produced. Current thinking is to use hydrocarbons, but chemists are also developing hydrofluorocarbons, HFCs, which contain a C–H bond and therefore degrade in the atmosphere but contain no chlorine so they do not directly affect stratospheric ozone. The exact nature of the products of HFC degradation is not fully understood and they may have other harmful environmental effects.

## The ozone layer

Oxygen normally exists as an $O_2$ molecule, but oxygen atoms can also combine to form ozone, $O_3$. The structure of ozone can be represented as follows:

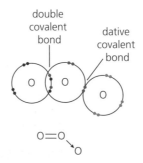

Ozone is a reactive gas that does not occur normally in the atmosphere close to the Earth's surface. However, some 25 km higher up, in the stratosphere, there is a layer of ozone approximately 15 km deep. This may sound substantial, but it would have a depth of no more than 3 mm if it existed at lower altitudes, where the atmospheric pressure is much greater. The ozone layer is important because it absorbs much of the harmful ultraviolet radiation emitted by the Sun. If this were to reach the surface of the Earth, it would cause an increase in sunburn and skin cancers in humans.

**Figure 14.10** CFCs were commonly used as the propellant in aerosols.

**Figure 14.11** CFCs were commonly used as a refrigerant and still present a problem in the disposal of old fridges.

In the stratosphere, ozone is in equilibrium with oxygen:

$$O_3 \rightleftharpoons O_2 + O$$

ozone      oxygen molecule      oxygen radical

At high altitude the oxygen radical, O, is formed from oxygen molecules:

$$O_2 \xrightarrow{\text{uv}} O + O$$

An oxygen radical then reacts with an oxygen molecule to produce a molecule of ozone:

$$O_2 + O \rightarrow O_3$$

## Causes of the hole in the ozone layer

It was a shock when it was discovered, during a polar spring season, that the ozone layer above the Antarctic had thinned to the point where a hole had developed. This discovery showed how important the management of the atmosphere is.

### Chlorine radicals

The main cause of the hole in the ozone layer is almost certainly chlorine radicals, created by the homolytic fission of a covalent bond. One source of chlorine-containing chemicals in the atmosphere is CFCs. Chlorine radicals attack the ozone layer, as shown below.

### Mechanism of action of chlorine radicals

First, there is an initiation step resulting in the formation of a chlorine radical, Cl·. This is most likely to be the result of the breaking of a C—Cl bond in a CFC such as $C_2F_2Cl_2$.

Initiation:          $C_2F_2Cl_2 \rightarrow \dot{C}_2F_2Cl + Cl\cdot$

The chlorine radical is then involved in the propagation steps:

Propagation:        $Cl\cdot + O_3 \rightarrow \cdot ClO + O_2$

Propagation:        $\cdot ClO + O \rightarrow Cl\cdot + O_2$

The net reaction of the two propagation steps can be written as $O_3 + O \rightarrow 2O_2$ and the chlorine radical, Cl·, used in the first propagation step but regenerated in the second, can be regarded as a catalyst. It has been estimated that approximately 1000 ozone molecules could be destroyed as the result of the production of a single chlorine radical.

The sequence is terminated only if the chlorine radical combines with another radical to remove it from the propagation sequence.

### Nitrogen(II) oxide (NO)

Compounds that contain a C—Cl bond are not alone in supplying radicals that initiate reactions responsible for the destruction of the ozone layer. Of growing concern is the role of nitrogen monoxide. Thunderstorms are a major source of this gas, as lightning causes nitrogen and oxygen to combine:

$$N_2(g) + O_s(g) \rightarrow 2NO(g)$$

> **Tip**
>
> You should appreciate the similarity between this mechanism and the reactions of alkanes with free radicals.

Nitrogen monoxide is also produced by the high-temperature combustion of nitrogen and oxygen that occurs to a small extent in internal combustion engines. More significantly, it is present in the exhaust gases of aircraft.

Complex reactions in the atmosphere may result in most of this gas being removed. For example, some may be oxidised and combine with moisture to create nitric acid, which is a contributor to acid rain. However, a small amount of nitrogen monoxide migrates to the stratosphere, where it attacks the ozone layer in a series of reactions analogous to those of the chlorine radical. There is no initiation step, because nitrogen monoxide already possesses an 'unpaired' single electron.

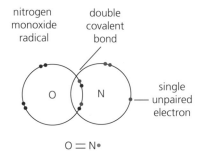

The propagation steps are as follows:

$$\cdot NO + O_3 \rightarrow \cdot NO_2 + O_2$$

$$\cdot NO_2 + O \rightarrow \cdot NO + O_2$$

The net reaction of the two propagation steps can again be written as $O_3 + O \rightarrow 2O_2$ and the nitrogen monoxide radical, NO, which used in the first propagation step but regenerated in the second can be regarded as a catalyst.

## Activity

**Alternatives to CFCs**

**Table 14.6** Alternatives to CFCs.

| Formula | Type of compound | Boiling point/°C | Flammable | Ozone depleting potential | Global warming potential over 100 years |
|---|---|---|---|---|---|
| $CCl_3F$ | CFC-11 | 24 | No | 1.0 | 4600 |
| $CCl_2F_2$ | CFC-12 | −30 | No | 1.0 | 7300 |
| $CHClF_2$ | HCFC-22 | −41 | No | 0.05 | 1700 |
| $CF_3CCl_2H$ | HCFC-123 | +29 | No | 0.02 | 100 |
| $CH_2F_2$ | HFC-32 | −51.6 | No | 0 | 650 |
| $CF_3CH_2F$ | HFC-134a | −26.6 | No | 0 | 1300 |
| $CH_4$ | Alkane | – | Yes | 0 | 23 |
| $C_3H_8$ | R-290 | −42 | Yes | 0 | |
| $CO_2$ | R-744 | Liquid under pressure | No | 0 | 1 |
| $NH_3$ | R-717 | −33 | Yes | 0 | 0 |

▶▶▶

1 What is the chemical name for HFC-134a?
2 Draw 'dot-and-cross' diagrams for these two compounds:
   a) CFC-12
   b) ammonia.
3 Suggest reasons why CFC-12 and ammonia have similar boiling points.
4 Suggest reasons why ammonia is much more chemically reactive than CFC-12.
5 What properties made CFC-12 suitable as a replacement for ammonia as a refrigerant?
6 Refrigerators and air-conditioning units keep the working fluid in a sealed system. Even so, the refrigerants continue to cause environmental problems. Suggest reasons why.
7 Suggest reasons why ammonia is now being reintroduced as a refrigerant.
8 The HCFCs and HFCs have been introduced as replacements for CFCs. They are now being phased out. Suggest why.
9 Carbon dioxide from burning fuels is an environmental problem, yet when it is used as a blowing agent or propellant, it is classified as having minimal global warming potential. How do you account for this?
10 Suggest disadvantages of using propane as an aerosol propellant for spray paints.

## Test yourself

6 Draw, name and classify the four structural isomers of $C_4H_9Cl$.
7 Which of the following are non-polar: $CH_3Br$, $CH_2Br_2$, $CHBr_3$ and $CBr_4$. Explain your answer.
8 Describe a chemical test for a haloalkane.
9 The manufacture of PVC is a three-stage process:
   • ethene reacts with chlorine
   • 1,2-dichloroethane is heated strongly to chloroethene and hydrogen chloride
   • chloroethene is polymerised.
   a) Write equations for the manufacture of PVC.
   b) Explain the problems of disposal of PVC.
10 Iodopropane is hydrolysed by aqueous sodium hydroxide.
   a) Write an equation for the reaction.
   b) Write the mechanism for the reaction. Show curly arrows, relevant dipoles and lone pairs of electrons.

# Practice questions

## Multiple choice questions 1–10

1 Haloalkanes such as $CH_3CH_2Cl$, $CH_3CH_2Br$ and $CH_3CH_2I$ can be hydrolysed but the rate of hydrolysis varies.

Which of the following statements is true?

A $CH_3CH_2I$ would be the fastest because the C–halogen bond has the biggest dipole.

B $CH_3CH_2Cl$ would be the fastest because the C–halogen bond has the biggest dipole.

C $CH_3CH_2I$ would be the fastest because it has the weakest C–halogen bond.

D $CH_3CH_2Cl$ would be the fastest because it has the weakest C–halogen bond.

2 Which one of the following is **not** readily oxidised?

A pentan-2-ol

B pentan-3-ol

C 2-methylbutan-2-ol

D 3-methylbutan-2-ol

3 When pentan-2-ol is dehydrated, which one of the following is the number of isomers formed?

A 1        B 2

C 3        D 4

4 The correct systematic name of this compound is:

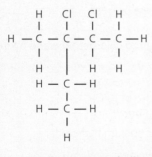

A 2,3-dichloro-2-ethylbutane

B 2,3-dichloro-3-ethylbutane

C 3,4-dichloro-2-methylpentane

D 2,3-dichloro-3-methylpentane.

5 A compound with the formula $C_nH_{2n+2}O$ when fully oxidised forms a compound $C_nH_{2n+1}O$.

$C_nH_{2n+2}O$ is likely to be

A a primary alcohol

B a secondary alcohol

C an aldehyde

D a ketone

6 Which of the following is a propagation step when chlorine radicals react with ozone?

A $Cl\cdot + O_3 \rightarrow ClO_2 + O$

B $Cl\cdot + O_2 \rightarrow ClO + O$

C $\cdot ClO + O \rightarrow Cl\cdot + O_2$

D $\cdot ClO_2 + O \rightarrow Cl\cdot + O_3$

Use the key below to answer Questions 7 to 10.

| A | B | C | D |
|---|---|---|---|
| 1, 2 & 3 correct | 1, 2 correct | 2, 3 correct | 1 only correct |

7 Which of the following when fully oxidised will form a molecule containing three atoms of oxygen?

1 2-methylbutan-2,3-diol

2 2-methylbutan-1,2-diol

3 3-methylbutan-1,2-diol

8 Which of the following can be dehydrated to give 2-methylpropene as one of the products of the reaction?

1 2-methylpropan-1-ol

2 2-methylpropan-2-ol

3 2,2 dimethylpropan-1-ol

9 Which of the following statements is correct?

1 If $AgNO_3(aq)$ is added to $CH_3CH_2Br$ a cream precipitate is formed that dissolves in dilute aqueous ammonia.

2 When $AgNO_3(aq)$ is added to $CH_3CH_2Br$ the nitrate ion, $NO_3^-(aq)$ behaves as a nucleophile.

3 If $AgNO_3(aq)$ is added to $CH_3CH_2Br$ a cream precipitate is formed only after hydrolysis has occurred.

10 Butane-1,3-diol could be oxidised to form:

1 an aldehyde

2 a ketone

3 a carboxylic acid.

**11 a)** Name the alcohols A–F:

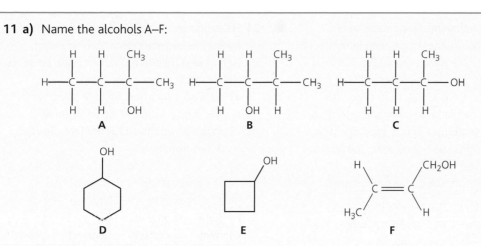

A    B    C

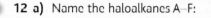

D    E    F

**b)** Answer the following questions about compounds A–F in part (a).
  **i)** What is the molecular formula of compound D?
  **ii)** Identify any secondary alcohols.
  **iii)** What do compounds E and F have in common?  (9)

**12 a)** Name the haloalkanes A–F:

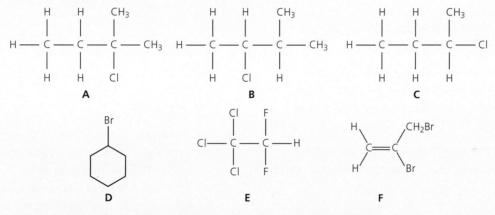

A    B    C

D    E    F

**b)** Classify compounds A, B and C as either primary, secondary or tertiary.
**c)** Write a balanced equation for the reaction of compound D with OH⁻.
**d)** When compound E is exposed to ultraviolet light it forms radicals.
Explain what is meant by the term 'radical'. Identify the radicals that are most likely to be formed.  (14)

**13** The structure of chlordane is shown below. What are the molecular and empirical formulae of chlordane?  (1)

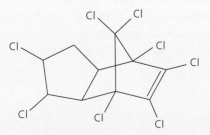

**14** Haloalkanes undergo nucleophilic substitution reactions.
  **a)** Define a nucleophile.
  **b)** Explain fully the mechanism when 1-bromopropane reacts with potassium hydroxide, KOH, to produce propan-1-ol.  (8)

**15** A few drops of a halogenoalkane were added to 2 cm³ of ethanol in a test tube, and 5 cm³ of aqueous silver nitrate were then added. Finally, the test tube was placed in a water bath for a few minutes. A cream precipitate formed. This precipitate was soluble in concentrated ammonia.
  a) Why was ethanol used in this experiment?
  b) Why was the test tube containing the mixture warmed in a hot water bath?
  c) What was the formula of the precipitate? Explain your answer.
  d) What can you conclude about the halogenoalkane? (4)

**16** Explain what is meant by the terms 'reflux' and 'distillation'. (2)

**17** Write a balanced equation for each of the following reactions:
  a) the complete combustion of propan-1-ol
  b) the dehydration of pentan-3-ol
  c) the oxidation of butan-2-ol (use [O] to represent the oxidising agent). (4)

**18** Explain, with the aid of equations, why the dehydration of pentan-3-ol gives two alkenes, whereas the dehydration of pentan-2-ol gives a mixture of three alkenes. (4)

**19** Identify the organic products formed when 2-methylcyclohexanol undergoes dehydration. (3)

**20** Classify the following conversions as: addition, elimination, substitution, oxidation, reduction, hydrolysis or polymerisation reactions. (Note that a reaction may belong to more than one category.)
  a) butan-2-ol to but-2-ene
  b) butane to 1-bromobutane
  c) but-1-ene to 1,2-dichlorobutane
  d) butanal to butanoic acid
  e) 1-bromobutane to butan-1-ol
  f) buta-1,3-diene to synthetic rubber (6)

**21** Haloalkanes, such as 1-bromobutane, are hydrolysed by reaction with NaOH(aq). $AgNO_3$(aq) is added to confirm the presence of the bromide.
  a) Write an equation, include state symbols, for the precipitation of silver bromide.
  b) Write an equation for the hydrolysis of 1-bromobutane.
  c) Describe, with the aid of curly arrows, the mechanism for the hydrolysis of 1-bromobutane. Show relevant dipoles and lone pairs of electrons.
  d) Why is hydrolysis necessary before testing with $AgNO_3$(aq)?
  e) Why must nitric acid, $HNO_3$(aq), be added before the $AgNO_3$(aq)?
  f) Explain why hydrochloric acid could not be used in place of the nitric acid. (10)

**22** The boiling points of ethane, ethanol and ethane-1,2-diol are shown below.

| Compound | Boiling point/°C |
|---|---|
| Ethane | −89 |
| Ethanol | 78 |
| Ethane-1,2-diol | 198 |

  a) Explain the variation in boiling points.
  b) Which would you expect to be most soluble in water? Explain your answer.
  c) Ethane-1,2-diol can be used as an anti-freeze to prevent the freezing of water. State three reasons why ethane-1,2-diol is suitable to use as an anti-freeze.
  d) Write a balanced equation for the complete combustion of ethane-1,2-diol.
  e) Using [O] to represent the oxidising agent, write a balanced equation for the complete oxidation of ethane-1,2-diol.
  f) When propane-1,2,3-triol is oxidised a mixture of organic products is formed. Identify as many of the oxidation products as possible. (14)

**23** Haloalkanes react by nucleophilic substitution but under certain conditions an elimination reaction can occur. The mechanism for the elimination reaction is shown below, but the curly arrows have been omitted.

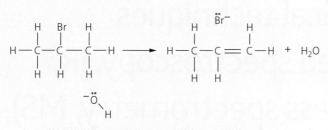

Copy the mechanism above and add curly arrows to show the movement of the electron pairs. (3)

## Challenge

**24** There are four isomers, L, M, N and P, of molecular formula $C_4H_9X$, where X is a halogen.

**a)** Draw skeletal formulae for the four isomers.

**b)** Following hydrolysis, each isomer reacts with aqueous silver nitrate to produce a cream precipitate that dissolves in concentrated ammonia. Identify X. Explain your answer.

**c)** Write an ionic equation for the reaction with silver nitrate.

**d)** Isomer L can be hydrolysed to produce a compound, $C_4H_{10}O$, that does *not* react when warmed with a mixture of potassium dichromate(VI) and sulfuric acid. Identify isomer L. Explain your answer.

**e)** Isomer M can also be hydrolysed. The compound obtained from this reaction *does* react with a mixture of potassium dichromate(VI) and sulfuric acid. The organic product obtained from the reaction with acidified dichromate under reflux is neutral. Identify isomer M. Explain your answer and write equations for the reactions described.

**f)** The isomers N and P react in a similar way. Each produces an alkane when the following reactions are carried out:

**i)** Compounds N and P are hydrolysed to produce $C_4H_{10}O$.

**ii)** $C_4H_{10}O$ is then dehydrated to $C_4H_8$.

**iii)** $C_4H_8$ is reacted with hydrogen using a nickel catalyst.

The alkane formed from compound N has a lower boiling point than that formed from compound P. Identify N and P. Explain your answers.

**g)** Consider reactions (i)–(iii) in part (f) above. Place them in order of decreasing atom economy. Calculations are not necessary, but you must justify your answer.

**h)** The mass spectrum of isomer P is shown below.

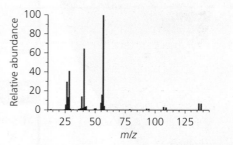

**i)** Calculate the relative molecular mass of the isomers.

**ii)** There are two peaks of equal height at $m/z$ 136 and 138. Suggest why these two peaks occur.

**iii)** The mass spectrum also has a prominent line at $m/z$ 57. Suggest the source of this peak.

**iv)** There is also a peak at $m/z$ 29. Explain how this peak would enable isomer P to be distinguished from isomer L. (27)

# Chapter 15

# Organic synthesis and analytical techniques (infrared spectroscopy, IR, and mass spectrometry, MS)

## Prior knowledge

*In this chapter it is assumed that you are familiar with:*
- the terms reagents, products and conditions
- empirical and molecular formula calculations
- percentage yield calculations
- mass spectrometry.

## Test yourself on prior knowledge

1 Give the empirical formula of the following organic molecules:
   a) $C_4H_{10}$
   b) $C_6H_6$
   c) $C_6H_{12}O_6$
   d) $CH_3COOCH_2CH_3$.

2 Compound A is an alcohol and has the following composition by mass: C, 48.6%; H, 8.1%; O, 43.3%.
   a) Calculate the empirical formula.
   b) The molar mass, $M$, of compound A is 74. Deduce molecular formula.
   c) Draw and name all possible isomers of compound A.

3 In the reaction below 4.6 g ethanol was reacted with an excess of ethanoic acid. 4.4 g of the ester ethyl ethanoate was produced.

$$H_3C-\underset{\underset{\displaystyle OH}{|}}{\overset{\overset{\displaystyle O}{\parallel}}{C}} + H_3C-CH_2-OH \underset{\substack{conc.\ H_2SO_4 \\ catalyst}}{\overset{heat}{\rightleftharpoons}} H_3C-\underset{\underset{\displaystyle O-CH_2-CH_3}{|}}{\overset{\overset{\displaystyle O}{\parallel}}{C}} + H_2O$$

   a) Why was an excess of ethanoic acid used?
   b) Calculate the percentage yield.
   c) Suggest why the percentage yield was so low.

## Practical skills

Chemists have developed a range of practical techniques for the synthesis of solid and liquid organic compounds. These methods allow for the fact that reactions involving molecules with covalent bonds are often slow and that it is difficult to avoid side reactions which produce by-products. There are several stages in the preparation of an organic compound. These are summarised in Figure 15.1.

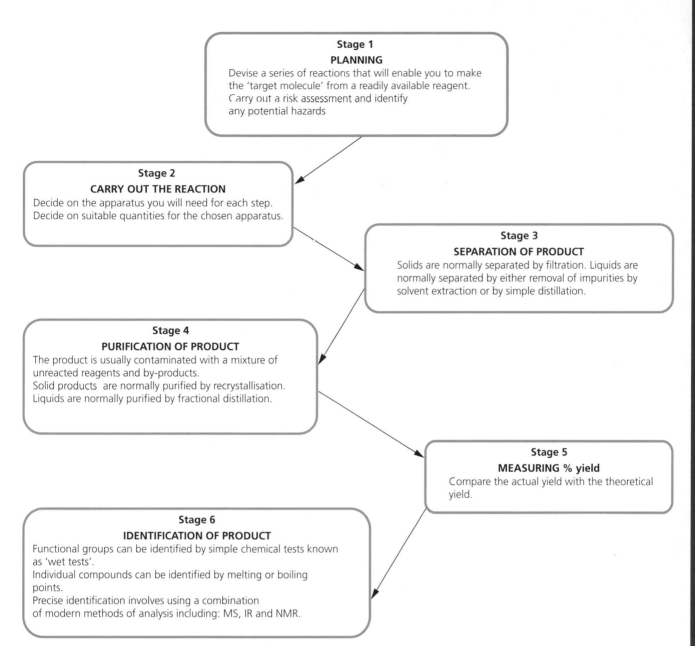

**Stage 1**
**PLANNING**
Devise a series of reactions that will enable you to make the 'target molecule' from a readily available reagent. Carry out a risk assessment and identify any potential hazards

**Stage 2**
**CARRY OUT THE REACTION**
Decide on the apparatus you will need for each step.
Decide on suitable quantities for the chosen apparatus.

**Stage 3**
**SEPARATION OF PRODUCT**
Solids are normally separated by filtration. Liquids are normally separated by either removal of impurities by solvent extraction or by simple distillation.

**Stage 4**
**PURIFICATION OF PRODUCT**
The product is usually contaminated with a mixture of unreacted reagents and by-products.
Solid products are normally purified by recrystallisation.
Liquids are normally purified by fractional distillation.

**Stage 5**
**MEASURING % yield**
Compare the actual yield with the theoretical yield.

**Stage 6**
**IDENTIFICATION OF PRODUCT**
Functional groups can be identified by simple chemical tests known as 'wet tests'.
Individual compounds can be identified by melting or boiling points.
Precise identification involves using a combination of modern methods of analysis including: MS, IR and NMR.

**Figure 15.1** Preparation and identification of an organic compound.

**Tip**

IR - infrared spectroscopy
MS - mass spectrometry
NMR - nuclear magnetic resonance spectroscopy

## Stage 1: Planning

In very many experiments it is not possible to convert a reagent into the desired product in a single reaction and often an intermediate step is required. This is dealt with in the section 'Synthetic routes', which follows (page 249).

## Stage 2: Carry out the reaction

When carrying out the reaction, one of the commonest techniques is to heat the reaction mixture in a flask fitted with a reflux condenser (see Chapter 14, page 229).

## Stage 3: Separation of products

When separating a solid product from the reaction mixture, solids are removed by filtration using a Buchner or Hirsch funnel with suction from a water pump (Figure 15.2).

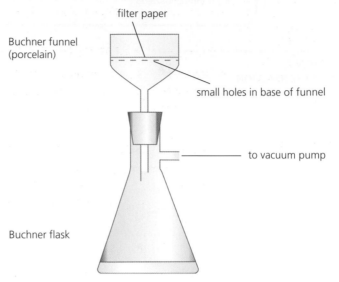

**Figure 15.2** Separation of solid product by vacuum filtration.

Liquids can often be separated by simple distillation, as shown in Figure 15.3.

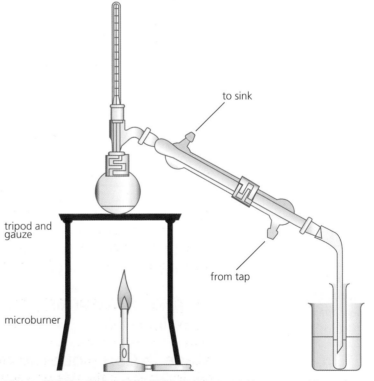

**Figure 15.3** Distillation. Volatile products are quickly separated from the reagents. Their boiling point can be measured as they distil over.

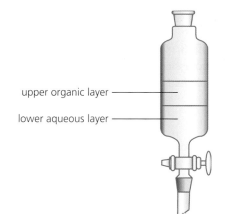

upper organic layer ——————

lower aqueous layer ——————

**Figure 15.4** Separating funnel for immiscible liquids.

Another method of separating liquids is using a separating funnel. (Figure 15.4). When separating a liquid product from the reaction mixture, the organic liquid is usually immiscible with water and will form a separate layer. If the organic liquid is less dense than water, the lower aqueous layer can be run off and discarded. The organic layer can then be dried by shaking with an anhydrous salt such as $MgSO_4$ or $CaCl_2$.

## Stage 4: Purification of products

The 'crude' product is usually contaminated with unreacted reagents or with by-products. The method of purifying this 'crude' product depends on whether it is a solid or a liquid.

### Purifying solids by recrystallisation

Solids are normally purified by recrystallisation. Recrystallisation is based on using a solvent which dissolves the product readily when hot, but only sparingly when cold. The choice of solvent is usually made by trial and error.

Recrystallisation can be broken down into several steps:

1 Dissolve the impure solid in the minimum volume of hot solvent.

2 If the hot solution is not clear, filter the hot mixture through a heated funnel to remove insoluble impurities leaving the product dissolved in the hot solution.

3 Cool the filtrate so that the product recrystallises, leaving the soluble impurities in solution.

4 Filter the cold solution to recover the purified product.

5 Wash the purified solid with small amounts of pure cold solvent to wash away any solution containing impurities.

6 Allow the solvent to evaporate from the purified solid in the air.

### Purifying organic liquids

Chemists often purify organic liquids which are insoluble in water by shaking with aqueous reagents in a separating funnel to extract impurities. This is followed by washing with distilled water, drying with an anhydrous salt and finally fractional distillation. Fractional distillation separates mixtures of liquids with different boiling points.

## Stage 5: Measuring percentage yield

Comparing the actual yield with the yield expected from the chemical equation is a good measure of the efficiency of a process. The yield expected from the equation, assuming that the reaction is 100% efficient, is called the **theoretical yield**.

The efficiency of a synthesis, like that of other reactions, is normally calculated as a **percentage yield**. This is given by the relationship:

$$\text{percentage yield} = \frac{\text{actual yield of product}}{\text{theoretical yield of product}} \times 100$$

### Example 1

a) What is the theoretical yield of ethanal if 13.14 g of ethanol is heated, under distillation, with excess acidified potassium dichromate?

b) What is the percentage yield if the actual yield of ethanal is 9.43 g?

**Answer**

Start by writing an equation for the reaction.

You need to know the mole ratio of reagent : product.

$$CH_3CH_2OH + [O] \rightarrow CH_3CHO + H_2O$$

| mole ratio | 1 mol | : | 1 mol |
|---|---|---|---|
| molar mass | $46\,g\,mol^{-1}$ | | $44\,g\,mol^{-1}$ |

a) From the equation 1 mol of ethanol will produce 1 mol of ethanal, so: 46 g of ethanol will produce 44 g ethanal

hence 13.14 g of ethanol will produce $\frac{44}{46} \times 13.14\,g = 12.57\,g$ ethanal

The theoretical yield is 12.57 g ethanal.

b) $$\text{percentage yield} = \frac{\text{actual yield of product}}{\text{theoretical yield of product}} \times 100$$

$$= \frac{9.43}{12.57} \times 100\% = 75\%$$

# Stage 6: Identifying products and checking their purity

Pure solids have sharp melting points – databases now include the melting points of all known compounds, which makes it possible to check the identity and purity of a product by checking its melting point. If the solid is impure the melting point will be lower than expected; the solid will not have a sharp melting point but will soften and melt over a range of temperature.

Boiling points can be used to check the purity and identity of liquids. If a liquid is pure, it should all distil over a narrow range, at the expected boiling point. The boiling point can be measured as the liquid distils over during fractional distillation. Impurities increase the boiling point and the range over which it boils.

## Qualitative tests for functional groups

Functional groups can be identified by simple qualitative tests. These 'wet tests' are summarised in Table 15.1.

**Table 15.1** Chemical tests to identify functional groups.

| Test | Observation | Conclusions |
|---|---|---|
| pH of solution (add litmus) | Red | Carboxylic acid |
| $Br_2$ | Decolorises | Alkene |
| $Na_2CO_3$ | Gas ($CO_2$) given off, bubbles, fizzes | Carboxylic acid |
| $AgNO_3$(aq) in water bath at about 60 °C | White precipitate<br>Cream precipitate<br>Yellow precipitate | Chloroalkane<br>Bromoalkane<br>Iodoalkane |
| Heat with $H^+/Cr_2O_7{}^{2-}$ | Orange to green | Primary alcohol<br>Secondary alcohol<br>Aldehyde |

**Test yourself**

1 **a)** Explain what is meant by:
   **i)** reflux
   **ii)** distillation.
   **b)** Ethanal can be prepared by oxidising ethanol. Would you use reflux or distillation? Explain your answer.
   **c)** Ethanoic acid can be prepared by oxidising ethanol. Would you use reflux or distillation? Explain your answer.

2 State the practical techniques you would use to obtain a pure sample each of the following:
   **a)** butanal from a mixture of butanal and butan-1-ol (both are liquids)
   **b)** octane from a mixture of octane and water (both are liquids)
   **c)** benzoic acid and phenyl ethanone (both are solids).
   Explain your reasoning for each.

3 Explain what happens to the melting point of pure ice if the ice is contaminated with salt (NaCl).

4 A sample of benzoic acid was contaminated with mixture of potassium dichromate and carbon. Use the information in the table to explain how you could obtain a pure sample of benzoic acid. Describe a chemical test that you could use to confirm that the product was a carboxylic acid.

| Chemical | Appearance | Solubility in cold water | Solubility in hot water |
|---|---|---|---|
| Benzoic acid | White crystalline solid | Insoluble | Soluble |
| Potassium dichromate | Orange crystalline solid | Soluble | Soluble |
| Carbon | Black solid | Insoluble | Insoluble |

5 Propan-1-ol was heated under distillation with a acidified potassium dichromate. The percentage yield of propanol was 75%.
   **a)** Suggest why the percentage yield was significantly below the theoretical yield.
   **b)** Suggest the identity of any likely organic impurities.
   **c)** State a simple chemical test that would confirm the presence of the organic impurities.

# Synthetic routes

Functional groups provide the key to organic molecules. Knowledge of the properties and reactions of a limited number of functional groups enables the preparation of a wide variety of organic compounds.

Chemists often think of an organic molecule as a relatively unreactive hydrocarbon skeleton with one or more functional groups in place of one or more hydrogen atoms. The functional group in a molecule is responsible for most of its reactions. In contrast, the carbon–carbon bonds and carbon–hydrogen bonds are relatively unreactive, partly because they are both strong and have very little polarity.

Your knowledge of functional groups should now cover the reactions of alkanes, alkenes, alcohols and haloalkanes. Table 15.2 summarises the reactions of these groups.

**Table 15.2** Reactions of functional groups

| Functional group | | Type of reactions | Reagents that react |
|---|---|---|---|
| Name | Group | | |
| Alkane | $C_nH_{2n+2}$ | Radical substitution | $Cl_2$, $Br_2$ |
| Alkene | $\overset{\diagdown}{\diagup}C{=}C\overset{\diagup}{\diagdown}$ | Electrophilic addition | $H_2$, HBr, $Br_2$ $H_2O(g)$ |
| Alcohol | R—OH | Oxidation<br>Esterification*<br>Elimination<br>Halogenation | $H^+/Cr_2O_7^{2-}$<br>RCOOH (carboxylic acids)<br>$H_2SO_4$<br>$NaBr/H_2SO_4$ |
| Haloalkane | R—Cl | Nucleophilic substitution<br>Hydrolysis | Common nucleophiles include: $:OH^-$, $:NH_3^*$, $:CN^{-*}$ |

*Only likely to be tested after the second year of the A Level course.

Organic chemists synthesise new molecules using their knowledge of functional groups, reaction mechanisms and molecular shapes – as well as the factors which control the rate and extent of chemical change. They often start by examining the 'target molecule'. Then, they work backwards through a series of steps to find suitable starting chemicals that are available and cheap enough. In recent years, chemists have developed computer programs to help with the process of working back from the target molecule to a range of possible starting molecules. Table 15.3 outlines some reactions of functional groups.

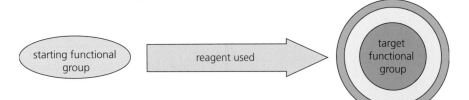

**Table 15.3** Target functional groups.

| Functional group | Reagent | Target functional group |
|---|---|---|
| Alkane | Halogen | Haloalkane |
| Alkene | Hydrogen halides | Haloalkanes |
| | Halogens | Di-haloalkanes |
| | Steam | Alcohol |
| | Hydrogen | Alkanes |
| Alcohols | Carboxylic acids* | Esters |
| | $H^+/Cr_2O_7^{2-}$ | Aldehyde, ketone or carboxylic acid |
| | Hot concentrated $H_2SO_4$ | Alkene |
| | NaBr in presence of $H_2SO_4$ | Haloalkane |
| Haloalkane | NaOH(aq) | Alcohol |
| | $NH_3$(ethanol)* | Amine |
| | Cyanide, $^-C{\equiv}N^*$ | Nitrile |

*Only likely to be tested after the second year of the A Level course.

A simple example is shown below.

## Example 2

Explain how you could prepare propanoic acid starting from chloropropane.

**Answer**

In any two-stage synthesis:

● start with the **target molecule** and identify the functional groups that can be used to make the **target** functional group

● secondly look at the **starting molecule** and identify the functional groups that can be made from the **starting** functional group.

In this case:

1-chloropropane $\rightarrow$ ? $\rightarrow$ propanoic acid

**Target molecule** – Using your knowledge so far, the carboxylic acid functional group can be made from a primary alcohol or an aldehyde.

**Starting molecule** – The haloalkane acid functional group can be used to form a primary alcohol.

Clearly the connecting intermediate molecule is a primary alcohol such that the two-step synthesis is:

1-chloropropane $\rightarrow$ propan-1-ol $\rightarrow$ propanoic acid

**Step 1** 1-chloropropane $\rightarrow$ propan-1-ol

- ● Reagents: $NaOH(aq)$
- ● Conditions: warm
- ● Equations: $CH_3CH_2CH_2Cl + NaOH \rightarrow CH_3CH_2CH_2OH + NaCl$
- ● Type of reaction: nucleophilic substitution

**Step 2** propan-1-ol $\rightarrow$ propanoic acid

- ● Reagents: acidified dichromate, $H^+/Cr_2O_7^{2-}$
- ● Conditions: heat (under reflux) with excess $H^+/Cr_2O_7^{2-}$
- ● Equations: $CH_3CH_2CH_2OH + 2[O] \rightarrow CH_3CH_2COOH + H_2O$
- ● Type of reaction: oxidation

Example 3

Explain how you could prepare butanone starting from but-2-ene.

**Answer**

In this case, the 'target molecule' is butanone and the 'starting molecule' is an alkene, but-2-ene.

**Step 1** Start with the target molecule and identify the compounds that could readily be converted directly into the target – concentrate on the functional group.

Butanone is a ketone which can be made from the oxidation of a secondary alcohol, butan-2-ol.

**Step 2** Look at your starting molecule, but-2-ene. What reactions of alkenes do you know?

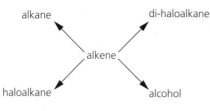

You should now see a possible two-stage synthetic route from your starting molecule to the target molecule. In this case, the route can go via the alcohol.

$$H_3C - CH = CH - CH_3 \longrightarrow H_3C - CH_2 - CH(OH) - CH_3 \longrightarrow H_3C - CH_2 - \overset{\overset{\displaystyle O}{\|}}{C} - CH_3$$
starting molecule   intermediate molecule   target molecule

You will need to know the reagents and conditions for each step:

$$H_3C - CH = CH - CH_3 \xrightarrow[\substack{\text{acid catalyst} \\ \text{high temp./pressure}}]{\text{steam}} H_3C - CH_2 - CH(OH) - CH_3 \xrightarrow[\substack{\text{heat under} \\ \text{reflux}}]{H^+/Cr_2O_7^{2-}} H_3C - CH_2 - \overset{\overset{\displaystyle O}{\|}}{C} - CH_3$$
starting molecule   intermediate molecule   target molecule

You may have to write equations for each step:

Step 1    $CH_3CH{=}CHCH_3 + H_2O \rightarrow CH_3CH_2CH(OH)CH_3$

Step 2    $CH_3CH_2CH(OH)CH_3 + [O] \rightarrow CH_3CH_2CHOCH_3 + H_2O$

Chemists normally seek a synthetic route that has the least number of stages and which, therefore, produces a higher yield of the product. It is rare for any one reaction to be 100% efficient; normally the percentage yield is significantly below the theoretical yield.

**Test yourself**

6 Compound A decolorises bromine and when heated with acidified dichromate the dichromate turns from orange to green. Which of the three compounds below is compound A most likely to be?
$CH_3CH_2CH_2OH$   $CH_3CH{=}CHCOOH$   $CH_3CHCHCH_2OH$
Explain your answer.
7 Devise a two-stage synthesis for converting:
  **a)** methane to methanol
  **b)** propene into propanone.
State the reagents and conditions needed for each conversion.

# Analytical techniques

**Figure 15.5** Scientists using an infrared spectrometer. The instrument covers a range of infrared wavelengths, and a detector records how strongly the sample absorbs at each wavelength. Wherever the sample absorbs, there is a dip in the intensity of the radiation transmitted, which shows up as a dip in the plot of the spectrum.

## Infrared spectroscopy

Substances used to be identified by means of distinctive chemical tests. In simple cases, such tests are still useful. However, more powerful and flexible methods have been developed that use instruments to detect molecules by their structural characteristics. One such method is **infrared spectroscopy**. Infrared radiation causes covalent bonds to vibrate; particular bonds respond at different frequencies (energies).

Infrared radiation is part of the electromagnetic spectrum, which extends from low-energy waves, such as radio waves, through to high-energy waves, such as X-rays and gamma rays. Infrared is the region of the spectrum that has energy just below that of red light.

### Absorptions

Covalent bonds respond to infrared radiation in a number of ways. The frequencies at which the bonds vibrate are called the **absorptions** of that bond. Some absorptions occur over a small range of frequencies (such as those of the $C=O$ bond); others have a wider spread (such as those of an O–H bond). The absorptions are affected by neighbouring bonds. For example, the O–H bond of an alcohol responds differently from the O–H bond in a carboxylic acid group, –COOH. The complicated patterns that are produced can make the interpretation of these spectra difficult.

There is no need to consider here how an infrared spectrometer works, but Table 15.4 below gives some key absorptions. Absorptions are identified by **wavenumber** ($cm^{-1}$).

**Table 15.4** Key infrared absorptions.

| Wavenumber range/$cm^{-1}$ | Bond | Functional groups |
|---|---|---|
| 750–1100 | C–C* | Alkanes, alkyl chains |
| 500–800 | C–X* | Haloalkanes (X = Cl, Br or I) |
| 1000–1350 | C–F* | Fluoroalkanes |
| 1000–1300 | C–O* | Alcohol, ester, carboxylic acid |
| 1620–1680 | C=C | Alkenes |
| 1630–1820 | C=O | Aldehyde, ketone, carboxylic acid, ester |
| 2850–3100 | C–H | Any organic compound with a C–H bond |
| 2500–3300 (very broad) | O–H | Carboxylic acid |
| 3200–3600 | O–H | Alcohol |

*These absorptions are difficult to identify as they appears in the 'fingerprint region' where there are usually very many peaks.

Data sheets are supplied when required in exams, so there is no need to learn the wavenumbers in Table 15.4. The absorptions can be roughly divided into three distinct sections, as shown in Figure 15.6.

> **Tip**
>
> The **fingerprint region** is the part of the spectrum from about 500 to 1500 $cm^{-1}$, which usually contains a very complicated series of absorptions.

Wavenumber range
4000
2500
1900
1500
500 cm⁻¹
C — H
O — H
N — H
C = O
C = C
fingerprint region
contains many peaks often unique to a particular compound
e.g. C — O

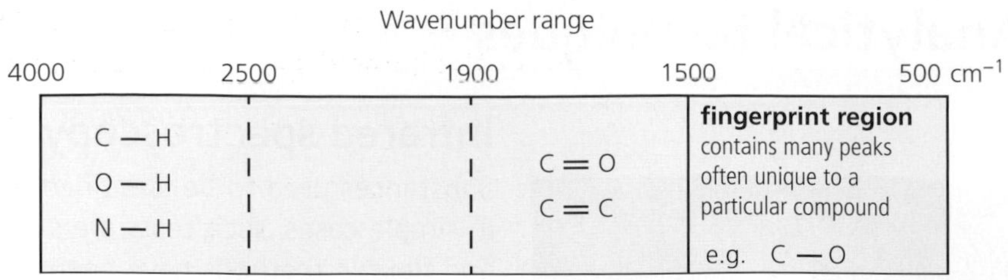

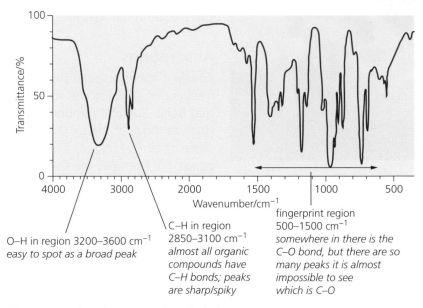

**Figure 15.6** Wavenumber ranges.

### Infrared spectrum of an alcohol

The identifying absorptions are indicated on the spectrum in Figure 15.7. Notice the characteristic broad absorption due to the O–H bond in the region 3200–3600 cm$^{-1}$.

100
50
0
Transmittance/%
4000
3000
2000
1500
1000
500
Wavenumber/cm⁻¹
O–H in region 3200–3600 cm⁻¹
*easy to spot as a broad peak*
C–H in region 2850–3100 cm⁻¹
*almost all organic compounds have C–H bonds; peaks are sharp/spiky*
fingerprint region 500–1500 cm⁻¹
*somewhere in there is the C–O bond, but there are so many peaks it is almost impossible to see which is C–O*

**Figure 15.7** Infrared spectrum of an alcohol.

### Infrared spectrum of a carbonyl compound

The infrared spectrum of a carbonyl compound (Figure 15.8) has an absorption due to the C=O bond in the region 1630–1820 cm$^{-1}$.

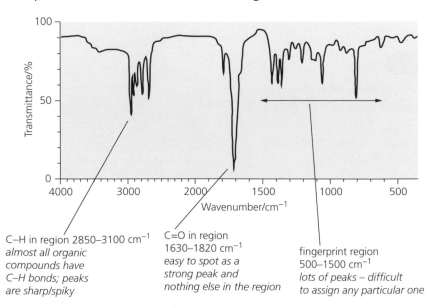

100
50
0
Transmittance/%
4000
3000
2000
1500
1000
500
Wavenumber/cm⁻¹
C–H in region 2850–3100 cm⁻¹
*almost all organic compounds have C–H bonds; peaks are sharp/spiky*
C=O in region 1630–1820 cm⁻¹
*easy to spot as a strong peak and nothing else in the region*
fingerprint region 500–1500 cm⁻¹
*lots of peaks – difficult to assign any particular one*

**Figure 15.8** Spectrum of a carbonyl compound.

### Infrared spectrum of a carboxylic acid

The infrared spectrum of a carboxylic acid has absorptions due to:

- the C=O bond (in the region 1630–1820 cm$^{-1}$)

- the O–H bond, which is a very broad absorption (in the region 2500–3300 cm$^{-1}$).

These can be seen in Figure 15.9.

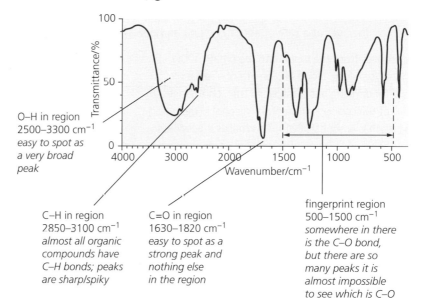

O–H in region 2500–3300 cm$^{-1}$ *easy to spot as a very broad peak*

C–H in region 2850–3100 cm$^{-1}$ *almost all organic compounds have C–H bonds; peaks are sharp/spiky*

C=O in region 1630–1820 cm$^{-1}$ *easy to spot as a strong peak and nothing else in the region*

fingerprint region 500–1500 cm$^{-1}$ *somewhere in there is the C–O bond, but there are so many peaks it is almost impossible to see which is C–O*

**Figure 15.9** Spectrum of a carboxylic acid.

The infrared spectra for alcohols, carbonyls and carboxylic acids make it easy to identify compounds that contain the O–H and C=O functional groups and they also illustrate the use of infrared spectroscopy. Modern breathalysers employ this technique quantitatively to measure the amount of ethanol in exhaled breath. The detection of alcohol is explored more fully in the activity 'Detecting alcohol' on page 256, which is best used as a class-based discussion exercise.

### Infrared spectrum of an alkene

The infrared spectrum of an alkene has absorptions due to the C=C bond (in the region 1620–1680 cm$^{-1}$). These can be seen in Figure 15.10.

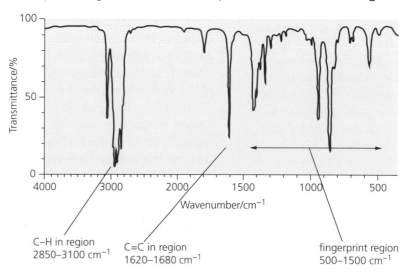

C–H in region 2850–3100 cm$^{-1}$

C=C in region 1620–1680 cm$^{-1}$

fingerprint region 500–1500 cm$^{-1}$

**Figure 15.10** Spectrum of an alkene.

### Detecting alcohol

Infrared spectroscopy is one of the methods used to analyse breath samples from drivers suspected of drinking. Alcohol from drinks is absorbed into the bloodstream through the walls of the stomach and intestines. The alcohol circulates with the blood through all parts of the body, including the lungs. It is slowly removed by the liver at a rate of about one unit an hour.

Alcohol moves from the blood into the breath in the lungs. Analysis of the blood and breath of a range of people after drinking has established that, on average, the ratio of the concentration of the alcohol in the blood to the breath is around 2300:1. This makes it possible to estimate someone's blood alcohol level by measuring the concentration of alcohol in their breath. The current legal limits for drivers in the United Kingdom excluding Scotland correspond to 35 mg of alcohol per 100 cm³ breath or 80 mg of alcohol per 100 cm³ of blood. In Scotland the limit is 50 mg of alcohol per 100 cm³ of blood.

The police use test instruments that contain fuel cells for their roadside tests. In these devices, any alcohol in a driver's breath acts as fuel to produce an electrical voltage from the cell. The voltage is automatically converted by the instrument to a measure of the concentration of alcohol in the blood. These instruments are not used as evidence in court, but they allow the police to decide whether to take a driver to the police station for a further test. For many years, the accurate breath tests at police stations relied solely on infrared instruments. The newer machines combine infrared and fuel cell technologies.

The infrared spectrum of ethanol has strong peaks corresponding to the O–H and C–H bond vibrations (see Figure 15.7). The absorption at 2950 cm⁻¹ corresponds to the C–H bond vibration and is used for analysis. The test instrument passes infrared radiation at 2950 cm⁻¹ through a standard cell containing a sample of the driver's breath. The strength of the absorption is a measure of the alcohol concentration. A built-in computer processes the signal from the instrument, and the machine prints out the result.

1 A unit of alcohol in the UK is 10 cm³ ethanol. Estimate the number of units of alcohol in:
a) half a pint (284 cm³) of lager that contains 5% by volume of alcohol
b) a small glass (125 cm³) of wine that contains 14% by volume of alcohol.
2 Show that the values of 80 mg of alcohol per 100 cm³ of blood and 35 mg of alcohol per 100 cm³ breath correspond to a concentration ratio of 2300:1.
3 Suggest reasons why:
a) prosecution does not follow a breath test unless the breath alcohol concentration is at least 40 mg per 100 cm³
b) drivers are given the option of having a blood or urine test if the value is between 40 and 50 mg per 100 cm³ but not if it is higher than this range.
4 Suggest reasons why the C–H peak in the infrared spectrum and not the O–H peak is used to analyse alcohol in breath.
5 Suggest a procedure for checking the accuracy of a breath test machine in a police station.

Modern breathalysers (like the one described in the activity above) are used to measure the ethanol in breath. In addition, infrared spectroscopy is used to monitor gases such as carbon monoxide and oxides of nitrogen that cause air pollution.

# Absorption of infrared radiation by atmospheric gases

The Earth is warmed mostly by energy transmitted from the Sun. This energy consists largely of visible light, but there is also some ultraviolet and infrared radiation. Most ultraviolet radiation is removed in the upper parts of the atmosphere (the stratosphere) by the ozone layer. Of the remaining radiation, some is reflected and some is absorbed by the atmosphere. Carbon dioxide and water molecules remove part of the incoming infrared radiation. Once at the Earth's surface, chemical reactions absorb and transmit this energy. Over time, the surface temperature of the Earth has remained approximately constant because an equilibrium has been established between arriving and departing energy. The departing energy – which is almost wholly infrared radiation – does not pass unhindered into outer space. If it did, the Earth would be a very cold place. Many gases in the atmosphere absorb some of this energy and reflect it back to Earth. Carbon dioxide and water molecules are not the only molecules to behave in this way. Gases that have bonds that vibrate in a way that alters the electrical balance of the molecule also absorb infrared radiation. In fact the only molecules that do not absorb infrared radiation are diatomic molecules made up of the same atoms (e.g. $O_2$, $N_2$).

## Greenhouse gases

Gases that do absorb infrared are collectively known as **greenhouse gases**. There is a wide range of greenhouse gases, since most molecules contain polar bonds.

### Contribution of a gas to the greenhouse effect

The overall contribution of a gas to the greenhouse effect depends on:

- its ability to absorb infrared radiation
- its atmospheric concentration
- its residence time (how long it stays in the atmosphere).

These factors can be treated quantitatively. The relative greenhouse effect of four gases is shown in Table 15.5 below.

Table 15.5 Contribution of gases to the greenhouse effect.

| Gas | Formula | Approximate relative greenhouse effect | Approximate atmospheric concentration (%) |
|---|---|---|---|
| Carbon dioxide | $CO_2$ | 1 | 0.035 |
| Methane | $CH_4$ | 25 | 0.00017 |
| Dinitrogen oxide | $N_2O$ | 250 | 0.00003 |
| CFC-12 | $CCl_2F_2$ | 25000 | $\sim 4 \times 10^{-8}$ |

Table 15.5 shows that there are large differences in the contributions to the greenhouse effect made by individual molecules. However, this must be set against the amount of each gas present in the atmosphere. Carbon dioxide has the highest concentration, whereas that of CFC-12 is low and, as CFCs are phased out, will become lower. A major source of atmospheric methane is the reduction of carbon-containing compounds under anaerobic conditions. It may cause amusement that this occurs in the digestive tract of cows, but it is also a hazard of the decomposition of rubbish in compacted landfill sites. Dinitrogen monoxide is released as a result of the reduction of nitrates on agricultural land.

Each of the bonds in the greenhouse-gas molecules (C=O in carbon dioxide, C–H in methane, O–H in water and so on) absorbs infrared radiation of a particular wavelength and vibrates with increased energy. This energy is then randomly dispersed, with much of it returning to the Earth's surface.

### Effect of greenhouse gases

As the concentration of greenhouse gases rises, the average temperature at the Earth's surface will increase. It is this increase that is usually referred to as the greenhouse effect. However, it is more accurate to call it the 'enhanced greenhouse effect' because it is the *extra* heating that is the cause for concern.

The role of carbon dioxide as a contributor to **global warming** is the focus of much attention. This is not because it is the most effective greenhouse gas or that it is a major contributor, but because it is produced in huge quantities by burning fuels such as wood, coal and oil products. (In fact, water vapour is the biggest contributor to the greenhouse effect.) Many attempts have been made to predict the result of continued emissions of carbon dioxide into the atmosphere at the current rate. There is no universal agreement, but there is serious concern that it could result in the melting of the polar ice-caps, causing extensive flooding of low-lying land, and that changing temperature patterns could lead to severe droughts in some parts of the world.

### Test yourself

8 The three spectra below are the infrared spectra of propan-1-ol, propanone and propanoic acid.

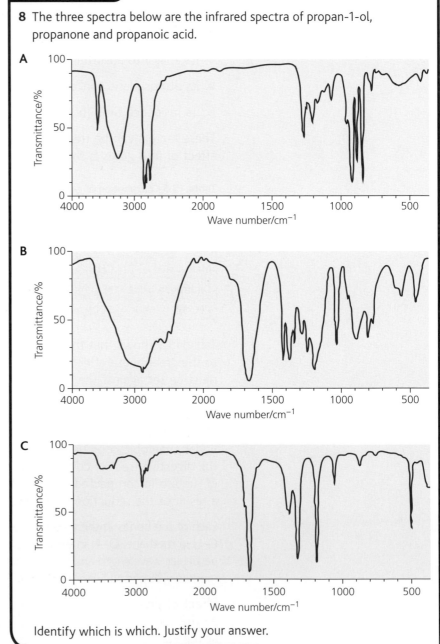

Identify which is which. Justify your answer.

# Mass spectrometry

The use of the mass spectrometer was discussed in Chapter 2. The mass spectrometer can be used to analyse both elements and molecules.

The mass spectrum of an element shows a peak for each isotope of that element. The mass corresponding to each peak, together with their relative abundance, can be used to calculate the relative atomic mass of that element.

The mass spectrum of a molecule shows multiple peaks. The peak with the largest $m/z$ value, the molecular ion (or the parent ion), shows the relative molecular mass of the molecule. The rest of the peaks show fragment ions formed when the molecule breaks up. Molecules break up more readily at weak bonds or at bonds that give rise to fragments that are more stable. It turns out that ions with a positive charge on a tertiary carbon atom are more stable than ions with a charge on a primary carbon atom.

Some important fragmentation ions are listed in Table 15.6.

**Table 15.6** Common fragment ions.

| $m/z$ value | Ion responsible |
|---|---|
| 15 | $[CH_3]^+(g)$ |
| 29 | $[CH_3CH_2]^+(g)$ |
| 43 | $[CH_3CH_2CH_2]^+(g)$ |
| Alkyl chains extend by $CH_2$ so it is possible that you will get peaks at 57, 71 etc. | |
| 31 | $[CH_2OH]^+(g)$ (primary alcohol) |

Figure 15.11 shows the mass spectrum of ethanol. There is a molecular ion peak (M peak) at $m/z = 46$ but the spectrum also shows a small peak at $m/z = 47$, which is known as the M+1 peak and is due to the presence of the carbon-13 isotope that is found in all organic compounds.

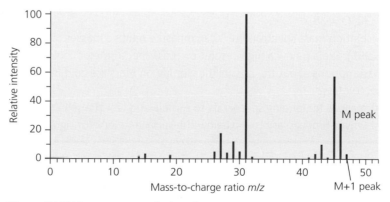

**Figure 15.11** Mass spectrum of ethanol.

## Activity

### Mass spectrometry in space research

In December 1978, space probes landed on Venus for the first time. Weight limitations meant that the low-resolution mass spectrometers on board could give measurements of relative masses to only 1 decimal place. A molecule of relative molecular mass 64.0 was identified, but the analysis could not show whether it was $SO_2$ or $S_2$.

Since then, mass spectrometers with higher resolutions have been developed. One reason for this is that space scientists are keen to explore whether or not there is, or ever has been, life on Mars. Scientists want to explore the north pole of the planet because evidence from surveys by orbiting spacecraft suggests that the pole is rich in ice just below the surface.

Beagle was a British-led effort to land a spacecraft on Mars in 2003 to look for signs of life. Like half of all missions to Mars, the project failed. In this case, it failed because the spacecraft crashed on landing. However, the mass spectrometers developed for the mission have been developed further and have been used in later projects, such as the Phoenix lander, which landed on Mars in May 2008.

**Figure 15.12** Artist's impression of the Phoenix lander on Mars just as it is beginning to dig a trench and gather samples for analysis by mass spectrometry.

Instruments on spacecraft sent to Mars include a combination of high-temperature ovens and a mass spectrometer. These are designed to study ice and soil samples. After the spacecraft lands, a robotic arm digs a trench. It then scoops up samples and drops them into a hopper that feeds them into ovens no bigger than a ballpoint pen. As the oven temperatures increase up to 1000 °C, any gases can be mixed with oxygen and then carried into the mass spectrometer by a stream of gas. The mass spectrometer can detect and measure the amount of carbon dioxide formed by organic material burning or from decomposing minerals or released from gases trapped in rocks.

Data from the mass spectrometer also allow scientists to determine ratios of various isotopes of hydrogen, oxygen, carbon and nitrogen. The ability to measure the ratios of carbon isotopes is key to the search for life, because photosynthesis is known to bring about a slight separation of these isotopes. During photosynthesis on Earth, plants show a slight preference for carbon-12 rather than carbon-13. This means that the proportion of the carbon-13 isotope is slightly lower than average in the chemicals in plants.

1 Why could the mass spectrometers used in 1978 not show whether the molecule found on Venus was $SO_2$ or $S_2$? ($A_r$: S = 32, O = 16)

2 High-resolution mass spectrometers can measure relative masses to 3, 4 or even 5 decimal places ($A_r$: S = 31.972, O = 15.995). Explain how a high-resolution mass spectrometer could determine whether a molecule is $SO_2$ or $S_2$.

3 What assumptions about the possibilities for life on Mars are built into the design of the methods of analysis for missions to Mars?

4 Suggest reasons for landing spacecraft to sample and test the soil at the north pole of Mars.

5 Suggest why it is necessary to dig below the surface to look for signs of life on Mars.

**Test yourself**

9 Use this mass spectrum for lithium to calculate the relative atomic mass of lithium.

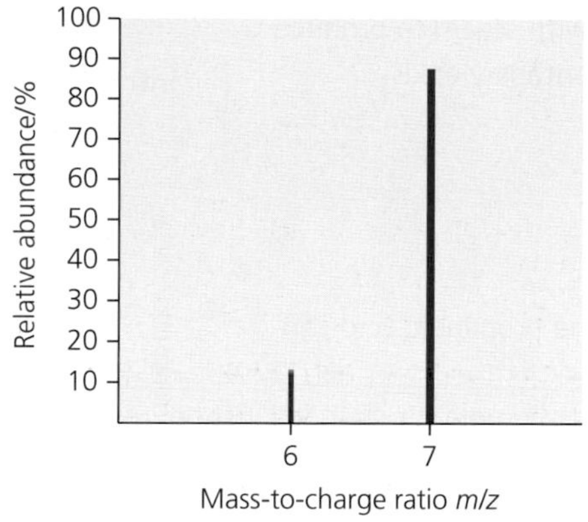

Mass-to-charge ratio *m/z*

10 The graph shows the mass spectrum of ethanol. Match the numbered peaks with the formulae of these positive ions formed: $[C_2H_5]^+$, $[CH_2OH]^+$, $[C_2H_5O]^+$, $[C_2H_5OH]^+$ and $[C_2H_3]^+$.

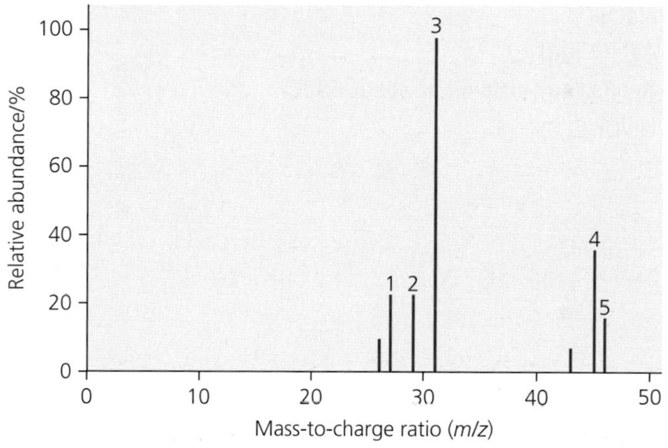

Mass-to-charge ratio (*m/z*)

# Practice questions

## Multiple choice questions 1–10

**1** $12.0\,dm^3$ ethene reacts with steam to produce $18.4\,g$ ethanol. The percentage yield is

A 65.2

B 80.0

C 95.2

D 100.

**2** Propan-1-ol, propanal and propanoic acid can best be separated by

A filtration

B separating funnel

C distillation

D reflux.

**3** Haloalkanes, such as $C_2H_5Cl$, can be detected by reacting it with

A NaOH

B $AgNO_3(aq)$

C $AgNO_3(aq)$, ethanol at about 60 °C

D $H^+/Cr_2O_7^{2-}$.

**4** The infrared spectrum of compound A is shown below.

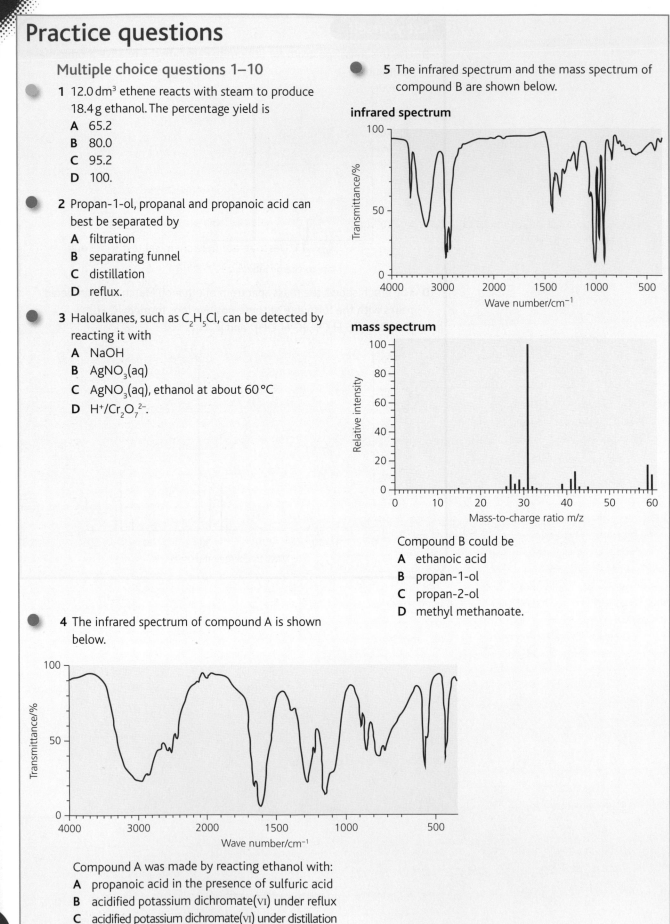

Compound A was made by reacting ethanol with:

A propanoic acid in the presence of sulfuric acid

B acidified potassium dichromate(VI) under reflux

C acidified potassium dichromate(VI) under distillation

D concentrated sulfuric acid at about 170 °C.

**5** The infrared spectrum and the mass spectrum of compound B are shown below.

**infrared spectrum**

**mass spectrum**

Compound B could be

A ethanoic acid

B propan-1-ol

C propan-2-ol

D methyl methanoate.

**6** The mass spectrum of 1-chlorobutane shows a peak at *m/z*

   **A** 92

   **B** 92.5

   **C** 93

   **D** 93.5.

**7** Ethene can be converted into compound Y via a two-stage synthesis

ethene $\xrightarrow{\text{reaction 1}}$ compound X $\xrightarrow{\text{reaction 2}}$ compound Y

Compound Y can be oxidised to form a compound HOOC–COOH

Compounds X and Y could be:

|   | X | Y |
|---|---|---|
| A | ethanol | ethanal |
| B | bromoethane | ethanol |
| C | dibromoethane | dihydroxyethane |
| D | ethanol | dihydroxyethane |

**8** HOCH$_2$CHCHCH$_2$CHO contains three different functional groups. They are:

   **A** alcohol, ketone, alkene

   **B** alcohol, ketone, carboxylic acid

   **C** alcohol, alkene, ketone

   **D** alcohol, alkene, aldehyde.

Use the key below to answer Questions 9 and 10.

| A | B | C | D |
|---|---|---|---|
| 1, 2 & 3 correct | 1, 2 correct | 2, 3 correct | 1 only correct |

**9** The compound with the infrared spectrum shown below could be:

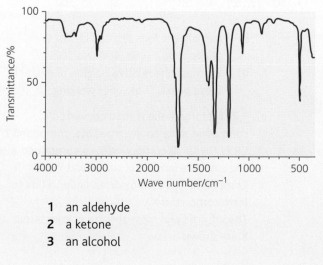

   **1** an aldehyde

   **2** a ketone

   **3** an alcohol

**10** Compound X is shown below.

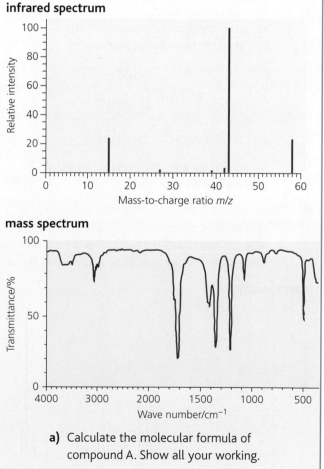

Compound X will:

   **1** decolorise bromine.

   **2** react with H$^+$/Cr$_2$O$_7$$^{2-}$ and the colour will change from orange to green

   **3** react with H$^+$/Cr$_2$O$_7$$^{2-}$ and the colour will change from green to orange

**11** Describe a simple chemical test that could distinguish between:

   **a)** cyclohexane and cyclohexene

   **b)** 1-chloropropane and propan-1-ol

   **c)** chlorobutane and iodobutane

   **d)** ethanol and ethanoic acid.

For each part, state the reagent(s), observations and write an equation. (16)

**12** Compound A contains by mass: 62.1% C; 10.3% H and 27.6% O. The mass spectrum and the infrared spectrum of compound A are shown below.

**infrared spectrum**

**mass spectrum**

   **a)** Calculate the molecular formula of compound A. Show all your working.

**b)** Determine the functional group, draw and name possible isomers of compound A. Show all your working.

**c)** Use the fragmentation pattern in the mass spectrum to suggest an identity for compound A. Explain your reasoning. (11)

**13** The mass spectrum of 2-chloropropane is shown below.

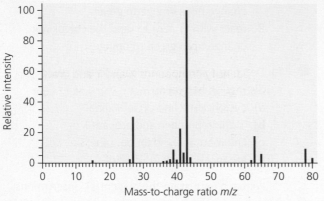

Mass-to-charge ratio *m/z*

The relative molecular mass of 2-chloropropane is 78.5. Explain why the molecular ion peak in the mass spectrum is not at *m/z* = 78.5. (5)

**14** Suggest a two-stage synthesis for each of the following conversions:

**a)** 1-bromopropane → propanoic acid

**b)** cyclohexene → cyclohexanone

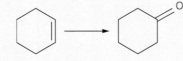

**c)** $CH_3CH_2CH_2OH \rightarrow CH_3CHBrCH_3$.

For each stage give the reagents, conditions, if any, and write an equation. (18)

**15** Compound W decolorises bromine and also reacts with acidified potassium dichromate to produce an acidic compound. The mass spectrum and the infrared spectrum of compound W are shown below.

**mass spectrum**

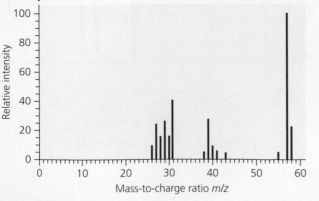

Mass-to-charge ratio *m/z*

**infrared spectrum**

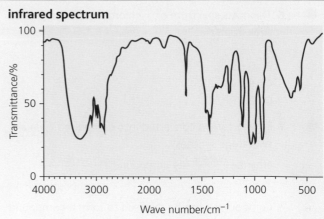

Wave number/cm$^{-1}$

Use all of the information to identify compound W. Show all of your working. (6)

## Challenge

**16** Ethene can be converted into compound X via a three-stage synthesis.

**a)** In stage 1 ethene is converted into 1,2-dichloroethane.
Write an equation for the reaction and calculate the relative molecular mass of 1,2-dichloroethane.

**b)** The mass spectrum of 1,2-dichloroethane is shown below.

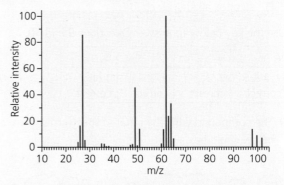

m/z

**i)** Identify the ions responsible for the peaks at *m/z* = 98, 100 and 102.

**ii)** Estimate the relative heights of each of these peaks. Show your working.

**c)** 1,2-dichloroethene is then converted compound X via an intermediate, compound Y. $ClCH_2CH_2Cl \rightarrow$ compound Y $\rightarrow$ compound X The empirical formula of compound Y is $CH_3O$. Compound Y is then oxidised under reflux to form compound X.
The infrared and mass spectra of compound X are shown below.

**infrared spectrum**

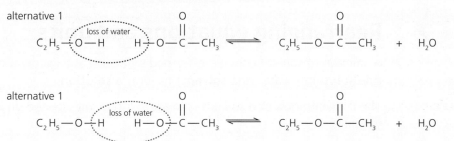

**mass spectrum**

Identify compounds X and Y and draw their displayed formula. Show all of your working. (19)

**17** Ethyl ethanoate can be prepared by the reaction between ethanol and ethanoic acid. There are two possible alternative ways of producing the ester:

alternative 1

$$C_2H_5-O-H \quad H-O-C-CH_3 \rightleftharpoons C_2H_5-O-C-CH_3 + H_2O$$ (loss of water)

alternative 1

$$C_2H_5-O-H \quad H-O-C-CH_3 \rightleftharpoons C_2H_5-O-C-CH_3 + H_2O$$ (loss of water)

The reaction was studied by using ethanol in which the oxygen was labelled with the isotope $^{18}O$:

$$C_2H_5-^{18}O-H$$

The mass spectrum of the ester, $CH_3COOC_2H_5$, is shown below.

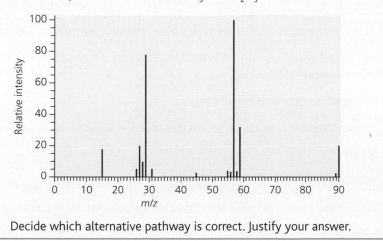

Decide which alternative pathway is correct. Justify your answer. (4)

# Chapter 16

# Maths in chemistry

> ## Prior knowledge
>
> *In this chapter it is assumed that you are familiar with routine mathematics. Calculations will always be part of your chemistry. There are two key requirements for solving numerical calculations:*
>
> - You need to understand the chemistry.
> - You need to be able to process the mathematics.
>
> Calculations in chemistry are straightforward and logical – you will be given some numerical data and asked to use it to calculate some other numerical value. The connection between the data you are given and the values you have to calculate is the chemical relationship. You will need to know your chemistry to recognise what the chemical relationship is.

## Rearranging equations and units

A solution of sodium chloride is formed by adding solid sodium chloride (the **solute**) to water (the **solvent**) to form a **solution**.

The concentration of a sodium chloride solution will depend on two variables:

- the amount of sodium chloride dissolved
- the volume of the solution formed.

If either of these is changed then the concentration will change.

If the mass of sodium chloride is doubled, the concentration will double.

If the volume of the solution is doubled, the concentration will halve.

The related variables can be linked together by an equation which is their chemical relationship:

The equation can also be written as:

concentration = mass of solute dissolved ÷ volume of solution

and as:

$$\text{concentration} = \frac{\text{mass of solute dissolved}}{\text{volume of solution}}$$

If you know two quantities it is possible to calculate the third.

If 10 g of NaCl(s) is dissolved to produce 500 cm³ of solution, calculate the concentration.

$$\text{concentration} = \frac{\text{mass}}{\text{volume}} = \frac{10}{500} = 0.020$$

The numerical value is correct but it has little value unless it also has units.

The units can be worked out from the equation:

which means that the units of concentration are:

$$\text{concentration} = \frac{g}{cm^3}$$

The units of concentration are grams divided by $cm^3$ or grams per $cm^3$, which can be written as $g/cm^3$. This is more correctly written as $g\,cm^{-3}$.

Concentration (conc.) can be calculated if we know the mass (of the solute) and the volume (of the solution):

You might want to use the above equation to calculate mass. For instance you might be asked what mass of sodium chloride would you need to make a $400\,cm^3$ of a solution with a concentration of $0.05\,g\,cm^{-3}$.

In the equation above mass is divided by volume, so to get mass by itself the right-hand side has to be multiplied by volume BUT whatever you do to one side of an equation you must also do to the other side.

which gives      conc. × volume = mass

$$0.05 \times 400 = 20\,g$$

It is always worth checking that the units balance:   $g\,cm^{-3} \times cm^3 = g$

We can also use this relationship to find the volume if both concentration and mass are known:

We need to move volume across to the left-hand side and transfer concentration (conc.) to the right-hand side. This can be done in two ways:

- in two separate steps

  Step 1 – multiply each side by volume

  $$\text{conc.} \times \text{volume} = \frac{mass}{\cancel{volume}} \times \cancel{volume} \qquad \text{which gives}$$
  $$\text{conc.} \times \text{volume} = \text{mass}$$

  Step 2 – divide each side by conc.

  $$\frac{\cancel{conc.} \times volume}{\cancel{conc.}} = \frac{mass}{conc.} \qquad \text{which gives} \qquad \text{volume} = \frac{mass}{conc.}$$

- by cross-multiplying

  $$\text{conc.} = \frac{mass}{volume} \qquad \text{which gives} \qquad \text{volume} = \frac{mass}{conc.}$$

A triangle can be used for rearranging equations.

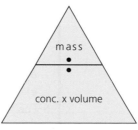

Cover up the quantity you want to find and what is left gives you the equation you have to use.

To find the relationship for mass place your finger over 'mass' and the triangle shows that:

$$\text{mass} = \text{conc.} \times \text{volume}$$

## Test yourself

1 Draw a triangle to show the equation:
   amount in moles = conc. × volume
   Rearrange to give the equation for:
   **a)** concentration
   **b)** volume.

2 Draw a triangle to show the equation:
   mass = density × volume
   Rearrange to give the equation for:
   **a)** density
   **b)** volume.

3 The relationship between voltage ($V$), resistance ($R$) and current ($I$) is
   voltage = resistance × current    or    $V = R \times I$
   (This equation could be written as $V = RI$ – you do not need to include the multiplication sign.)
   Rearrange to give the equation for:
   **a)** $R$
   **b)** $I$.

# Cross-multiplying

When you are sure that you can rearrange a simple equation you should be able to use cross-multiplying:

$$\frac{a}{c} \diagdown \frac{b}{d}$$     $c$ and $d$ can be cross-multiplied to give     $ad = bc$

To obtain an equation for $d$ use $ad = bc$ and divide each side by $a$ to get $d = bc/a$. You should also be able to obtain equations for each of the variables.

$$a = \frac{bc}{d} \quad b = \frac{ad}{c} \quad c = \frac{ad}{b} \quad d = \frac{bc}{a}$$

## Test yourself

4 The relationship between pressure, volume and temperature of a gas are related to the gas constant $R$ by:

$$\frac{P}{T} = \frac{nR}{V}$$

Cross-multiply the equation and then rearrange as necessary to give the equation for:
   **a)** $P$
   **b)** $V$
   **c)** $T$
   **d)** $R$
   **e)** $n$

5 Energy, $E$, can be related to mass, $m$, and speed of light, $c$, by the equation $E = mc^2$.
   Rearrange to give an equation for:
   **a)** $m$
   **b)** $c^2$
   **c)** $c$.

▶▶▶

6 Kinetic energy, $E$, is related to mass, $m$, and velocity, $v$, by the equation $E = \frac{1}{2}mv^2$.

   a) Rearrange this to give the equation for:

      i) $m$

      i) $v^2$

      ii) $v$

   b) Calculate the energy of an object with mass 2.0 kg and a velocity of 6.0 m s$^{-1}$.

   c) If an object of mass 0.50 kg has an energy of 6.25 kg m$^2$ s$^{-2}$, calculate its velocity.

7 Rearrange the equation:

   $$\frac{abc}{f} - \frac{de}{gh}$$

   to give an equation for:

   a) $h$

   b) $f$

   c) $b$.

# Ratio calculations

Many of the calculations in chemistry relate to the equation for a reaction and to the mole ratio given by the equation.

You will have met ratios in your mathematics. The principles are exactly the same for chemistry as they are for mathematics.

## Example 1

Malcolm buys two tickets for a concert for £28 pounds.

How much does his friend Mavis pay for 5 tickets?

**Answer**

You can work out the answer by first working out the cost of **one** ticket. This is sometimes referred to as the **unitary** method.

2 tickets cost £28

1 ticket therefore cost £28/2 = £14 (**unit** cost = £14)

So 5 tickets will cost 5 × £14 = £70

## Example 2

If 5.0 g of $CaCO_3(s)$ is heated strongly 1.2 dm$^3$ of $CO_2(g)$ is produced.

Calculate how much $CO_2(g)$ could be produced from 8.0 g of $CaCO_3(s)$.

**Answer**

5.0 g $CaCO_3(s)$ produce 1.2 dm$^3$ of $CO_2(g)$

1.0 g $CaCO_3(s)$ produce $\frac{1.2}{5}$ = 0.24 dm$^3$ of $CO_2(g)$ (**unit** mass of $CaCO_3(s)$ )

So 8.0 g $CaCO_3(s)$ produce 8.0 × 0.24 = 1.92 dm$^3$ of $CO_2(g)$

**Tip**

Temperature and pressure have to be the same as any change in either pressure or temperature would result in a change in volume of the gas

**Test yourself**

8  2.43 g of Mg(s) reacts with exactly 1.20 dm³ of oxygen.
   Calculate how the volume of oxygen required to react exactly with 121.5 g Mg(s) (at the same temperature and pressure).

9  If 4.00 g NaOH(s) is dissolved in 100 cm³ water a solution with a concentration of 1.00 mol dm⁻³ is formed.
   Calculate the concentration if 5.00 g NaOH(s) is dissolved in 100 cm³ water.

10  A mining company obtained 40 tonnes of CaCO₃(s) from 50 tonnes of limestone. What mass of limestone must be mined to obtain 100 tonnes of CaCO₃(s)?

# Using numbers in standard (index) form

**Key term**

Standard notation means writing the number in the form $A \times 10^n$ where:

- $A$ is a number between 1 and 10
- $n$ is an integer.
This is also called **standard form**.

**Tip**

It is always useful to estimate the answer to a calculation before doing the calculation on a calculator. It makes it easy to spot whether or not you have input the data into the calculator correctly.

Numbers can be written in different formats. A common way to write numbers is to use the decimal notation, for example 123 456.78 and 0.000 3456. When working with very large numbers (123 456.78) or very small numbers (0.000 3456) it is convenient to write these number in standard notation.

For example:

| 3124 | can be written as | $3.124 \times 10^3$ |
|---|---|---|
| 312.4 | can be written as | $3.124 \times 10^2$ |
| 31.24 | can be written as | $3.124 \times 10^1$ |
| 3.124 | can be written as | $3.124 \times 10^0$ (since $10^0 = 1$, the factor $10^0$ is normally omitted) |
| 0.3124 | can be written as | $3.124 \times 10^{-1}$ |
| 0.03124 | can be written as | $3.124 \times 10^{-2}$ |
| 0.003 124 | can be written as | $3.124 \times 10^{-3}$ |

One advantage of using numbers in standard form is that estimating the answer to a calculation is easier.

For example, the calculation $\dfrac{2473 \times 79}{651}$

is almost impossible to do in your head, but if you change the numbers into standard form it is easier to estimate the answer.

$$\frac{\left(2.473 \times 10^3\right) \times \left(7.9 \times 10^1\right)}{\left(6.51 \times 10^2\right)} \quad \text{which is} \quad \frac{\left(2.473 \times 7.9\right) \times \left(10^3 \times 10^1\right)}{\left(6.51 \times 10^2\right)}$$

$$\text{which approximately gives} \quad \frac{\left(2.5 \times 8\right) \times 10^4}{6.5 \times 10^2} = \frac{20 \times 10^2}{6.5} = 3 \times 10^2$$
$$= 300$$

The calculator value is 300.1029186 so 300 is a very good approximation and reassures you that you haven't made any slips whilst using your calculator. (But note that you wouldn't normally quote an answer to 10 significant figures when data you have been given is much less accurate than that. See the section on significant figures on page 271.)

When multiplying powers of 10 you add the powers, and when dividing powers of 10 you subtract the powers. For example:

$$10^3 \times 10^5 = 10^8$$

$$10^7 \div 10^4 = \frac{10^7}{10^4} = 10^3$$

$$10^5 \times 10^{-2} = 10^3$$

$$\frac{10^4}{10^{-3}} = 10^7$$

---

**Test yourself**

11 Convert the following numbers into standard (index) form.
   a) 734.8
   b) 69 845.6
   c) 0.003 45
   d) 333¼
   e) 0.6745
   f) 276 545

12 **Estimate** the value of each of the following calculations (DO NOT use your calculator before you have estimated the answers).

   a) $\dfrac{(6.5 \times 10^8)}{(4.2 \times 10^2)\,(3.9 \times 10^{-4})}$

   b) $\dfrac{349 \times 6578}{174.8}$

   c) $\dfrac{2576 \times 698}{219 \times 4.98}$

---

# Significant figures

In simple cases the number of significant figures is simply the number of digits in the answer. For example, 31.21 has 4 significant figures, while 31.2 has 3 significant figures and 31 has just 2.

In other cases numbers may need rounding up or down before quoting the answer to a particular number of significant figures. The number 17.87 has 4 significant figures but to 3 significant figures this is 17.9 (as 17.87 is nearer to 17.9 than 17.8). To 2 significant figures it would be 18 (17.87 is nearer to 18 than 17).

A number ending in a '5' is rather arbitrarily raised to the number above. So 17.5 must be written as 18 when quoted to 2 significant figures or 25.15 to 3 significant is 25.2.

When a number has 0s **after** the decimal point these are not considered as 'significant'. So 0.004 has only one significant figure because the 0s don't count. 0.0526 has 3 significant figures, which to 2 significant figures is 0.053.

A number like 1800 is ambiguous with regards to significant figures. This presents a problem if the intention is to quote 1800 to 2 significant figures. The way round it is to write the number in standard (index) form, i.e. $1.8 \times 10^2$. Written like this it would be taken to mean that

2 significant figures is intended. If 4 significant figures were intended then it should be written as $1.800 \times 10^2$.

Often your calculator will display an answer containing more digits than you were given in the data. Suppose you were asked to calculate the concentration of a NaOH(aq) in which $16.8\,cm^3$ of the NaOH(aq) was neutralised by $25.0\,cm^3$ of $0.500\,mol\,dm^{-3}$ HCl(aq). If you did this calculation correctly your calculator would show the concentration to be $0.74404761904\,mol\,dm^{-3}$. The concentration of the solution isn't known to this degree of precision. The accuracy will be limited to the precision of the data or in an experiment the accuracy of the apparatus. In the example above the data is given to 3 significant figures and the answer should also be limited to 3 significant figures. The figures after the third are dropped and the number is **rounded**. When rounded to 3 significant figures, 0.74404761904 is 0.744.

In numbers like 0.74404761904, discount the 0 at the start of the number such that '7' is the first significant figure.

The list below shows a few simple examples of rounding to a stated number of significant figures.

| 4567.46 | to 5 significant figures is | 4567.5 |
| 4567.46 | to 4 significant figures is | 4567 |
| 0.000 3462 | to 3 significant figures is | 0.000 346 |
| 0.000 3462 | to 2 significant figures is | 0.000 35 |

Remember in the last two examples we do not count the 0s before the number starts. If in doubt put the number into standard form and the 0s will disappear: 0.000 3462 in standard form is $3.462 \times 10^{-3}$, which clearly now has 4 significant figures.

When carrying out a calculation always quote your answer to the same number of significant figures given in the data. If the number of significant figures in the data varies, the least accurate should be used.

### Example 3

Two students measured a temperature rise and recorded their results as shown in the table below.

| | Student 1 | Student 2 |
|---|---|---|
| Initial temperature / °C | 21.0 | 21.5 |
| Final temperature / °C | 29.5 | 29.0 |
| Rise in temperature / °C | 8.5 | 7.50 |

The initial and final temperatures are recorded to 3 significant figures and at first glance it looks like student 2 has recorded their results correctly. However, this is incorrect because the accuracy of the thermometer limits the answer to 1 decimal place. The final reading cannot be more accurate than individual readings.

When adding or subtracting numbers the number of figures after the decimal point indicate the accuracy and may limit the number of significant figures in the answer. Student 2's value of 7.50 would not be accepted in an exam. The correct value for student 2 should have been 7.5 (to 1 dp) and not 7.50 (to 3 sf).

**Tip**

If the number after the significant number of figures is 4 or below the number is rounded down.
If the number after the significant number of figures is 5 or above the number is rounded up.

**Tip**

When adding or subtracting, the number of decimal places is maintained.

**Tip**

When carrying out calculations it is essential that you do not round until the end of the calculation. If necessary you must use the 'memory' function on your calculator.

# Drawing graphs

The sketch below shows a simple relationship between two variables.

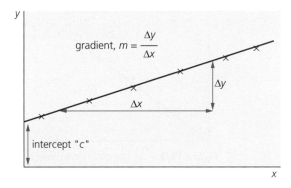

**Figure 16.1** Straight-line graph $y = mx + c$.

The relationship between $x$ and $y$ is $y = mx + c$ (where $m$ is the gradient and $c$ is the intercept).

$\Delta x$ is the change in $x$ and $\Delta y$ is the change in $y$.

There are a number of points to note when drawing graphs.

1   Choose a scale that will allow the graph to cover as much of the graph paper as possible. It is helpful to start both axes at zero. If all the points on one axis are between 90 and 100, to start at zero on that axis would cramp your graph into a small section of the paper (see Figure 16.2a). It is much better to truncate the axis so that the graph fills as much of the paper as possible (see Figure 16.2b).

**Tip**

The symbol $\Delta$ is used to represent 'change in ...' such that :
$\Delta T$ is change in temperature
$\Delta P$ is change in pressure
$\Delta V$ is change in volume.

> **Tip**
>
> The choice of scale for the axes depends on whether or not the origin is a valid point. In a rates graph the origin (0, 0) is a valid point and should always be included.

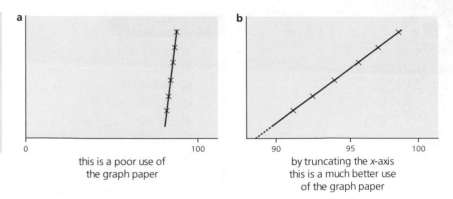

this is a poor use of the graph paper

by truncating the *x*-axis this is a much better use of the graph paper

**Figure 16.2** Choosing the graph scale.

2   Label the axes with the dimensions and the units, for example:

Volume/cm$^3$

Concentration/mol dm$^{-3}$

3   After plotting all the points on a graph, often due to experimental error you may not get a perfect straight line or a curve that goes through all of the points. You have to draw a line of best fit for the points (Figure 16.3).

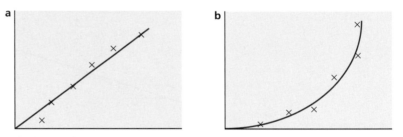

**Figure 16.3** a) All the points are close to the straight line with some slightly above balanced by some slightly below. b) It isn't possible to draw a straight line that is close to all the points but a curve can be drawn that is close to all the points with some points above and some below the curve.

4   Drawing tangents to a curve.

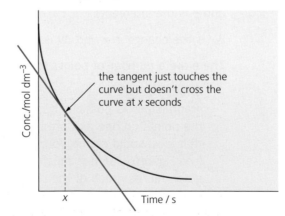

the tangent just touches the curve but doesn't cross the curve at *x* seconds

**Figure 16.4** Drawing the tangent to a curve.

By calculating the gradient of the tangent it is possible to work out how the concentration changes with respect to time after *x* seconds. This enables you to calculate the rate of reaction after *x* seconds. The units of rate are the units of *y*/*x* which are mol dm$^{-3}$/s; this is written as mol dm$^{-3}$ s$^{-1}$.

# Chapter 17

# Preparing for the exam

## Revising for the exam

Revising for exams is one of those boring tasks that no one enjoys and there is no one way of approaching it that will suit everyone. Nevertheless, an organised programme that covers all the work by the date of the exam is essential. At the end of this chapter there is a suggested timetable for revising Module 2 but it is really better if you generate your own scheme based on what you feel needs your greatest attention and the timescale available. When organising your time, make sure that you plan carefully, allowing enough time to cover each of the modules. It sounds easy, but it is a difficult thing to do. The most important point is that any plan at least enables you to see what you should be doing and when you should be doing it. Don't try to be too ambitious – *little and often is by far the best way*. It would of course be sensible to put together a longer rolling programme to cover all your AS subjects. Do *not* leave it too late. Start sooner rather than later.

For many students there are likely to be topics in chemistry that need practice over an extended period of time and it is no good leaving these to the last minute. For example, calculations are a stumbling block that can only be solved by tackling sufficient examples until the routines and procedures become second nature. Some students may find they need practice with the structures and naming of organic molecules, or perhaps with the steps in the mechanisms of reactions. Everyone will have their own list, but topics that depend on understanding should not be left until just before the exam. They are all tasks for the April holidays and not the summer term. However, there are also many topics in chemistry that simply require careful learning and these can be tackled nearer to the time that the exam takes place. Examples are definitions, atomic structures, the reactions of organic compounds, Group 2 and the halogens.

There is also considerable evidence to show that revising for 2–3 hours at a time is counter-productive and that it is much better to work in short, sharp bursts of between 30 minutes and an hour. Whatever your style, you must be organised. Sitting down the night before the examination with a file full of notes and a textbook does not constitute a revision plan – it is just desperation – and you must not expect a great deal from it.

Whatever your personal style, there are a number of things you *must* do and a number of other things you *could* do.

### Things you *must* do
- Leave yourself enough time to cover *all* the material.

- Make sure that you actually *have* all the material to hand (use this book or your notes as a basis) and the specification.

- Identify weaknesses early in your preparation so that you have time to do something about them.

- Familiarise yourself with the terminology used in examination questions (see the section below).

- Practise examination questions on the topic.

**Things you *could* do to help you learn**
- Copy selected portions of your notes.

- Write a summary of your notes, which includes all the key points.

- Write key points on postcards (carry them round with you for quick revision during a coffee break!).

- Discuss a topic with a friend also studying the same course.

- Try to explain a topic to someone *not* on the course.

## Some general advice

There are various bits of advice that may be of help. Firstly, it is no good just staring at a page of chemistry in the hope that by some magic its contents will enter your head. It is all too easy to waste time turning the pages of a textbook and achieving nothing. It is very possible to read a page of text while your mind wanders to something else. You really should force yourself to become engaged in the revision process. For example, arm yourself with the specification, find a topic and on a blank sheet of paper scribble down all the main points you can think of that are relevant. Maybe you won't think of much, but if you then refer to your notes or a textbook your attention will be much more directed to what is written there.

Secondly, do practise some questions against the clock. In the exam you have to respond quickly to the topic being tested and you only have a limited time to think. Many students can provide an answer when given a generous amount of time to collect their thoughts but this won't be possible in an exam. Once you feel you have revised sufficiently try doing some questions as fast as possible. Then check the answers to see what mistakes you have made. Under pressure perhaps you have missed a crucial point in a definition or maybe you are prone to making errors when using your calculator. It is helpful to be aware of where you tend to go wrong so that you are particularly careful on the day of the exam.

Thirdly, do make sure you are familiar with the data sheet that will be provided in the exam. It is surprising how many students seem unaware that this contains a lot of useful information to help them when answering a question.

Chemistry students often accumulate a large quantity of notes so it is useful to keep these in a well-ordered and logical manner. You should review your notes regularly, maybe rewriting those taken during lessons so that they are clear and concise, with key points highlighted. You should check what you have written using textbooks and fill in any gaps. Make sure that you go back and ask your teacher if you are unsure about anything, especially if you find conflicting information in your class notes and textbook. It is a good idea to file everything in specification order using a consistent series of headings. Before starting your revision you may also find it helpful to mark your notes to show which topics you feel you can approach with confidence, which ones require some more work and which require more intensive revision. This will allow you to construct a revision plan that focuses on the right areas.

# The examination

## Terms used in examination questions

You will be asked precise questions in the examinations, so you can save a lot of valuable time, as well as ensuring you score as many marks as possible, by knowing what is expected. Examiners spend a lot of time making sure the questions set specify exactly what they want in the answer. Terms most commonly used are explained below.

- **Define** ... is intended literally. Only a clear, accurate statement is required.

  *Define the standard enthalpy change of formation.* This simply requires a statement quoting the definition which is on page 153.

- **Explain** ... usually requires reference to some aspect of theory but may expect an answer to involve reasoning. The number of marks given for the question will indicate the amount of detail expected.

  *Explain the trend in boiling points of the halogens* This would require a statement of the trend and an explanation of intermolecular forces in the context of the halogens and how they arise.

- **State** ... implies a concise answer usually with no explanation required.

  *State one reason why chlorine is added to drinking water.* This would require a statement that it makes it safe to drink by killing the bacteria.

- **Describe** ... requires candidates to provide the key points of a topic. Diagrams may be required although usually if these are expected it will be indicated in the question. It can be used with reference either to trends or patterns or to experiments. The number of marks given for the question will indicate the amount of description required.

  **Describe and explain** are sometimes linked together in a question.

  *Describe what you would see when excess dilute hydrochloric acid is added to calcium carbonate. Explain your observations.* This requires the observation that bubbling would be seen and the calcium carbonate would dissolve to form a colourless solution. The explanation is that carbon dioxide is given off and calcium chloride solution is formed.

- **Deduce/Predict** ... implies that candidates are not expected to know the answer but that it can be worked out from other pieces of information provided in the question or by using the answers to previous parts of the question. Predict also implies a concise answer with little or no supporting statement required.

*CaO reacts with $H_3PO_4$, deduce/predict the formula of calcium phosphate.* This requires using the formula of $H_3PO_4$ to deduce the valency of the phosphate ion and then to use that to work out the formula of calcium phosphate to be $Ca_3(PO_4)_2$.

- **Suggest** ... can be used in two different ways. It may mean that there is more than one acceptable answer to the question or alternatively that candidates should be able to use their knowledge and imagination to provide a likely answer.

  *Suggest why $SiO_2$ has a very high melting point and $SiCl_4$ has a very low melting point.* Neither of these chemicals is directly mentioned on the specification so you are expected to use your knowledge that substances with high melting points have a giant lattices and that bonds between non-metals are covalent.

- **Calculate** ... is used when a numerical answer is required. Working should always be shown if the answer involves more than one step.

  *Calculate the amount, in mol, of HCl used* ... would require using the data given in the question and processing it using the correct mathematics.

- **Sketch/draw** ... when applied to diagrams, implies that a simple, freehand drawing is acceptable. Nevertheless, care should be taken over proportions and important details should be clearly labelled. If a graph is expected essential points might be implied. For example the position with respect to a line already drawn or the passing of a line through the origin.

  *Sketch, with suitable labels, an enthalpy profile diagram for an exothermic reaction.* This requires a simple sketch with the axes labelled and which clearly shows $\Delta H$ and the activation energy.

- **Compare** ... means that candidates should include both the similarities and differences between substances or concepts.

  *Compare the relative reactivities of magnesium and barium in their reaction with water* ... would require noting that although the reaction is similar, barium reacts more readily and forms a hydroxide while magnesium reacts very slowly. The amount of detail expected would be implied by the marks given for the question.

## On the day of the exam

When you finally open the test paper, it can be quite a stressful moment and you need to be certain of your strategy.

Time will be very tight; you must be calm and you must have a practised approach that includes:

- do *not* begin writing as soon as you open the paper

- briefly scan *all* the questions before you begin to answer any

- identify those questions about which you feel most confident

- *read the question carefully* – if you are asked to explain, then explain, do *not* just describe

- take notice of the mark allocation and do not supply the examiner with all your knowledge of any topic if there is only 1 mark allocated – similarly, you have to come up with *four* ideas if 4 marks are allocated

- try to stick to the point in your answer – it is easy to stray into related areas that will not score marks and will use up valuable time

- do not get stuck on a particular question which you are finding difficult – it is better to move on to make sure that you complete the paper

- try to answer *all* the questions.

### Multiple choice questions

Clearly these are questions with just one correct answer from those suggested. You may find you are under pressure for time so don't spend too long on any question trying to deduce the correct answer. It is better to press on and return to questions that you are not sure about once you have worked through the whole paper. Don't leave any question unanswered but if you are forced to guess at least make sure that you have rejected any responses that are clearly incorrect.

### Structured questions

These are questions that may require a single-word answer, a short sentence or a response amounting to several sentences. The setter for the paper will have thought carefully about the amount of space required for the answer and the marks allocated, so the space provided usually gives a good indication of the amount of detail required. Remember though that this space is always generous so if you find yourself filling it completely you are probably writing too much.

### Free-response questions

These questions enable you to demonstrate the depth and breadth of your knowledge as well as your ability to communicate chemical ideas in a concise way. The questions may include marks for the quality of written communication. You are expected to use appropriate scientific terminology and to write in continuous prose, paying particular attention to spelling, punctuation and grammar.

# Module 1

The first module of the specification covers the development of practical skills leading to the award of a 'Practical Endorsement' at A Level. It is not part of the assessment at AS chemistry but a minimum of 15% of the marks available in the exams will be for the assessment of practical skills. Practical work is an essential part of A Level and can only be fully appreciated as a result of extensive laboratory experience. This book provides advice to support this work. For example, explanations of the use of pipettes and burettes and the procedures used in a titration are covered as well as the principles of procedures such as refluxing and distillation. The 'Activities' are mostly based on practical work and the interpretation and evaluation of the results. Chapter 1 is also specifically devoted to providing a consideration of the reliability of different types of apparatus.

# Sample revision plan

The plan that follows suggests a possible way you might approach your revision of Module 2. The main work is carried out 3 weeks (or more) before the exam. It does pre-suppose that you are already broadly familiar with the topics. In the second and third weeks it focuses on doing past questions and it is assumed that during these two weeks you will also be revising the other two modules (and of course the other subjects that you are studying). It is by no means the only way of going about things, but it does emphasise the need for a clear plan of what you are going to do.

| Day | Week 1 | Week 2 | Week 3 |
|---|---|---|---|
| | Each revision session should be approximately **30 minutes** | | |
| Mon | **30 min** Atomic structure and isotopes, formulae and equations | Re-read all your summary notes at least twice. Check that you can write formulae correctly. Write a few equations showing the reactions of acids. | You should now have revised all of Module 2 and have attempted questions relating to each topic. Make a list of your weaknesses and ask your teacher for help. Re-read all your summary notes at least twice. Ask someone to test you. |
| Tue | **20 min** Acids, oxidation numbers and redox reactions<br>**10 min** Moles – determination of empirical formulae and calculating masses from equations | Using past papers or other question sources, try a couple of questions on the determination of empirical formulae and masses from equations.<br>Mark it and list anything you do not understand.<br>**Allow about 30 minutes** | |
| Wed | **10 min** Electron structure – orbitals<br>**10 min** Moles – gas volumes<br>**10 min** Atom economy and percentage yields | Using past papers or other question sources, try structured questions on electron structure and volumes of gases from equations.<br>Mark it and list anything you do not understand.<br>**Allow about 30 minutes** | Re-read all your summary notes at least twice<br>Concentrate on the weaknesses you identified on Monday (by now you should have talked to your teacher about them). Ask someone to test you. |
| Thu | **15 min** Bonding and structure<br>**10 min** Bond polarity and Intermolecular forces<br>**5 min** Moles – masses and volumes from equations | Using past papers or other question sources, try a structured question on bonding and structure.<br>Mark it and list anything you do not understand.<br>**Allow about 30 minutes** | |
| Fri | **10 min** Shapes of molecules<br>**10 min** Bonding and structure<br>**10 min** Electron structure, acids and redox reactions | Using past papers or other question sources, try a structured questions on shapes of molecules.<br>Mark it and list anything you do not understand.<br>**Allow about 30 minutes** | Collect together about five structured questions and one extended answer question covering all six topics and try them under exam conditions.<br>**Allow 60 minutes**<br>Mark them and list anything you do not understand. |
| Sat | **15 min** Moles – solution volumes and concentrations<br>**10 min** Checking definitions<br>**5 min** Bond polarity<br>**5 min** Electron structure, acids and redox reactions, moles and equations | Using past papers or other question sources, try a structured question involving solution volumes and concentrations and one on any other section of Module 2.<br>Mark it and list anything you do not understand.<br>**Allow about 30 minutes** | |
| Sun | **10 min** Moles – calculations from equations<br>**20 min** Electron structure and shapes of molecules, bond polarity, types of bonding, | Using past papers or other question sources, try tackling a range of multiple choice questions.<br>Mark it and list anything you do not understand.<br>**Allow about 30 minutes** | When other modules have been revised, try a complete past exam paper under exam conditions.<br>Check your answers against the mark schemes. |

# Glossary

**acid** – a proton donor.

**activation energy** – the minimum energy needed for a reaction to take place.

**addition** – a reaction in which two molecules combine together to form a single product.

**alkali** – a soluble base; it can accept protons in solution.

**anion** – a negatively charged ion.

**atomic number** – the number of protons in an atom of an element.

**base** – a proton acceptor.

**bond enthalpy** – the enthalpy change required to break and separate 1 mol of bonds in the molecules of a gas so that the resulting gaseous (neutral) particles exert no forces upon each other. Expressed as an equation this is: X–Y(g) → X(g) + Y(g)

**cation** – a positively charged ion.

**covalent bond** – a bond formed by the sharing of two electrons between two adjacent atoms. Each atom provides one electron. The electrostatic attraction between the shared pair of electrons and the nuclei of the two bonded atoms constitutes the covalent bond.

**dative covalent bond** – a bond in which two atoms share two electrons but one atom provides both electrons. The electrostatic attraction between the shared pair of electrons and the nuclei of the two bonded atoms constitutes the dative covalent bond.

**delocalised electrons** – bonding electrons that are not fixed between two atoms in a bond. They are mobile and are shared by several or many atoms.

**disproportionation** – when the same element both increases and decreases its oxidation numbers so that the element is simultaneously oxidised and reduced.

**dynamic equilibrium** – equilibrium reached when the rate of the forward reaction equals the rate of the reverse reaction. The concentrations of the reagents and products remain constant; the reactants and the products react continuously.

**electron pair repulsion theory** – states that the electron pairs around a bonded atom repel each other and the overall shape of the molecule depends on the number and type of electron pairs around the central atom. Lone pairs (non-bonded pairs) repel more than bonded pairs of electrons.

**electronegativity** – the ability of an atom of an element to attract the shared pair of electrons in a covalent bond.

**electrophile** – an electron-pair acceptor that forms a covalent bond.

**empirical formula** – gives the simplest ratio of the elements in a compound. These can be calculated from the amounts of each element.

**enthalpy change, $\Delta H$** – the difference between the enthalpy of the reactants and the enthalpy of the products.
$\Delta H$ = enthalpy of products − enthalpy of reactants

**first ionisation energy** – the energy required to remove one electron from the ground state of each atom in a mole of gaseous atoms of that element, to form a mole of gaseous ions of charge 1+.

**functional group** – either a structural feature (e.g. a carbon-to-carbon double bond, C=C), a group of atoms (e.g. a hydroxyl group, O–H) or a single atom (e.g. Cl). It is the functional group that determines much of the chemistry of a compound.

**ground state** – state of an atom that shows how it naturally exists with its electrons in their lowest energy position.

**Hess' law** – states that, if a reaction can take place by more than one route, the enthalpy change for the reaction is the same irrespective of the route taken, provided that the initial and final conditions are the same.

**homologous series** – a group of organic compounds that have the same general formula; contain the same functional group; and in which each member of the homologous series differs from the next by $CH_2$.

**homolysis** or **homolytic fission** – occurs when a covalent bond is broken so that the atoms joined by the covalent bond each take one of the shared electrons.

**hydrocarbon** – a compound that contains hydrogen and carbon only.

**hydrogen bonds** – the relatively strong electrostatic attractions between polar molecules that contain hydrogen covalently bonded to elements with high electronegativity such as fluorine, oxygen and nitrogen.

**hydrolysis** – a reaction that involves water and results in the organic compound being split and two new products being formed.

**indicator** – a substance that changes colour with a change in pH.

**induced dipole–dipole interaction** – attractive force between atoms/molecules as a result of electrostatic attraction between neighbouring dipoles: 1) the movement of electrons generates an instantaneous dipole; 2) the instantaneous dipole induces other dipoles in neighbouring atoms/molecules; 3) two dipoles generate a weak temporary force of attraction between the atoms/molecules.

**intermolecular forces** – forces of attraction that occur between molecules.

**ion** – an electrically charged particle formed by the loss or gain of one or more electrons from an atom or a group of atoms.

**ionic bond** – the electrostatic attraction between oppositely charged ions, the attraction between positive and negative ions.

**isotopes** – atoms of the same element that have different masses are isotopes of that element. The isotopes of an element all have the same number of protons (and electrons), but different numbers of neutrons.

**le Chatelier's principle** – states that if a closed system at equilibrium is subject to a change, the system will move to minimise the effect of that change.

**mass number** – the number of protons + neutrons in the nucleus of an atom.

**metallic bond** – the electrostatic attraction between the delocalised electrons and the positive ions held within the lattice.

**molecular formula** – represents the actual number of atoms of each element in the molecule. It does not provide any detail of the arrangement of the atoms. For example, the molecular formula of ethanol is $C_2H_6O$.

**molecular ion peak** – the peak on the mass spectrum corresponding to the molar mass of the compound. (It is sometimes referred to as the parent ion peak.)

***n*th ionisation energy** – the energy required to remove 1 electron from each $(n-1)+$ ion in 1 mole of gaseous $(n-1)+$ ions to form 1 mole of gaseous $n+$ ions.

**nucleophile** – an ion or molecule that is an electron-pair donor and can form a new covalent bond.

**orbital** – a region around the nucleus of an atom that can hold up to a maximum of two electrons.

**percentage error** – $\dfrac{\text{the maximum error}}{\text{actual value}} \times 100$.

**periodicity** – a repeating pattern, in either physical or chemical properties, across different periods.

**permanent dipole–dipole interactions** – the weak electrostatic attractions between polar molecules that are essentially covalent but have some ionic character.

**radical** – a particle that has at least one unpaired electron.

**salt** – formed when an acid has one or more of its hydrogen ions replaced by either a metal ion or an ammonium ion.

**saturated** – a saturated compound is a molecule in which all carbon – carbon bonds are linked together by C–C single ($\sigma$) bonds only.

**second ionisation energy** – the energy required to remove 1 electron from each 1+ ion in 1 mole of gaseous 1+ ions to form 1 mole of gaseous 2+ ions.

**specific heat capacity** – for a substance, the specific heat capacity is the energy required to increase the temperature of 1.0 g of the substance by 1 °C (1 K).

**standard enthalpy change of combustion** ($\Delta_c H^\ominus$) – the enthalpy change when 1 mol of a substance is burned completely, in an excess of oxygen, under standard conditions.

**standard enthalpy change of formation** ($\Delta_f H^\ominus$) – the enthalpy change when 1 mol of a substance is formed from its elements, in their standard states, under standard conditions.

**standard enthalpy change of neutralisation** ($\Delta_{neut}H^{\ominus}$) – the enthalpy change when 1 mol of water is formed in a reaction between an acid and a base.

**standard enthalpy change of reaction** ($\Delta_r H^{\ominus}$) – the enthalpy change when the amount in moles of the substances in the equation *as* written react under standard conditions.

**standard notation** – writing a number in the form $A \times 10^n$ where $A$ is a number between 1 and 10 and $n$ is an integer. This is also called standard form.

**standard solution** – a solution with a precisely known concentration.

**stereoisomers** – compounds with the same molecular and structural formulae but a different three-dimensional spatial arrangement.

**strong acid** – a proton donor that completely dissociates into its ions.

**structural isomers** – compounds that have the same molecular formula but different structural formulae.

**substitution** – a reaction in which an atom or a group of atoms is replaced by another atom or group of atoms.

**unsaturated** – molecules that contain one or more C=C double (or triple) bond.

**weak acid** – a proton donor that only partially dissociates into its ions.

# Index

# Free online resources

Answers for the following features found in this book are available online:

- Test yourself questions
- Activities

You'll also find Practical skills sheets and Data sheets. Additionally there is an Extended glossary to help you learn the key terms and formulae you'll need in your exam.

Scan the QR codes below for each chapter.

Alternatively, you can browse through all chapters at:
www.hoddereducation.co.uk/OCRChemistry1

## How to use the QR codes

To use the QR codes you will need a QR code reader for your smartphone/tablet. There are many free readers available, depending on the smartphone/tablet you are using. We have supplied some suggestions below, but this is not an exhaustive list and you should only download software compatible with your device and operating system. We do not endorse any of the third-party products listed below and downloading them is at your own risk.

- for iPhone/iPad, search the App store for Qrafter
- for Android, search the Play store for QR Droid
- for Blackberry, search Blackberry World for QR Scanner Pro
- for Windows/Symbian, search the store for Upcode

Once you have downloaded a QR code reader, simply open the reader app and use it to take a photo of the code. You will then see a menu of the free resources available for that topic.

## 1 Practical skills

## 3 Compounds, formulae and equations

## 2 Atoms and electrons

## 4 Amount of substance – moles in solids and gases

5 Amount of substance – moles in solution

12 Basic concepts in organic chemistry

6 Types of reaction – precipitation, acid–base and redox

13 Hydrocarbons

7 Bonding and structure

14 Alcohols and haloalkanes

8 The periodic table and periodicity

15 Organic synthesis and analytical techniques (IR and MS)

9 Group 2 and the halogens, qualitative analysis

16 Maths in chemistry

10 Enthalpy changes

17 Preparing for the exam

11 Rates and equilibria

# The Periodic Table

**Key**
atomic number
symbol
relative atomic mass

| 1 | 2 | | | | | | | | | | | 13 | 14 | 15 | 16 | 17 | 18 |
|---|---|---|---|---|---|---|---|---|---|---|---|---|---|---|---|---|---|
| 1<br>**H**<br>1.0 | | | | | | | | | | | | | | | | | 2<br>**He**<br>4.0 |
| 3<br>**Li**<br>6.9 | 4<br>**Be**<br>9.0 | | | | | | | | | | | 5<br>**B**<br>10.8 | 6<br>**C**<br>12.0 | 7<br>**N**<br>14.0 | 8<br>**O**<br>16.0 | 9<br>**F**<br>19.0 | 10<br>**Ne**<br>20.2 |
| 11<br>**Na**<br>23.0 | 12<br>**Mg**<br>24.3 | 3 | 4 | 5 | 6 | 7 | 8 | 9 | 10 | 11 | 12 | 13<br>**Al**<br>27.0 | 14<br>**Si**<br>28.1 | 15<br>**P**<br>31.0 | 16<br>**S**<br>32.1 | 17<br>**Cl**<br>35.5 | 18<br>**Ar**<br>39.9 |
| 19<br>**K**<br>39.1 | 20<br>**Ca**<br>40.1 | 21<br>**Sc**<br>45.0 | 22<br>**Ti**<br>47.9 | 23<br>**V**<br>50.9 | 24<br>**Cr**<br>52.0 | 25<br>**Mn**<br>54.9 | 26<br>**Fe**<br>55.8 | 27<br>**Co**<br>58.9 | 28<br>**Ni**<br>58.7 | 29<br>**Cu**<br>63.5 | 30<br>**Zn**<br>65.4 | 31<br>**Ga**<br>69.7 | 32<br>**Ge**<br>72.6 | 33<br>**As**<br>74.9 | 34<br>**Se**<br>79.0 | 35<br>**Br**<br>79.9 | 36<br>**Kr**<br>83.8 |
| 37<br>**Rb**<br>85.5 | 38<br>**Sr**<br>87.6 | 39<br>**Y**<br>88.9 | 40<br>**Zr**<br>91.2 | 41<br>**Nb**<br>92.9 | 42<br>**Mo**<br>95.9 | 43<br>**Tc** | 44<br>**Ru**<br>101.1 | 45<br>**Rh**<br>102.9 | 46<br>**Pd**<br>106.4 | 47<br>**Ag**<br>107.9 | 48<br>**Cd**<br>112.4 | 49<br>**In**<br>114.8 | 50<br>**Sn**<br>118.7 | 51<br>**Sb**<br>121.8 | 52<br>**Te**<br>127.6 | 53<br>**I**<br>126.9 | 54<br>**Xe**<br>131.3 |
| 55<br>**Cs**<br>132.9 | 56<br>**Ba**<br>137.3 | 57–71 | 72<br>**Hf**<br>178.5 | 73<br>**Ta**<br>180.9 | 74<br>**W**<br>183.8 | 75<br>**Re**<br>186.2 | 76<br>**Os**<br>190.2 | 77<br>**Ir**<br>192.2 | 78<br>**Pt**<br>195.1 | 79<br>**Au**<br>197.0 | 80<br>**Hg**<br>200.6 | 81<br>**Tl**<br>204.4 | 82<br>**Pb**<br>207.2 | 83<br>**Bi**<br>209.0 | 84<br>**Po** | 85<br>**At** | 86<br>**Rn** |
| 87<br>**Fr** | 88<br>**Ra** | 89–103 | 104<br>**Rf** | 105<br>**Db** | 106<br>**Sg** | 107<br>**Bh** | 108<br>**Hs** | 109<br>**Mt** | 110<br>**Ds** | 111<br>**Rg** | 112<br>**Cn** | | 114<br>**Fl** | | 116<br>**Lv** | | |

| 57<br>**La**<br>138.9 | 58<br>**Ce**<br>140.1 | 59<br>**Pr**<br>140.9 | 60<br>**Nd**<br>144.2 | 61<br>**Pm**<br>144.9 | 62<br>**Sm**<br>150.4 | 63<br>**Eu**<br>152.0 | 64<br>**Gd**<br>157.2 | 65<br>**Tb**<br>158.9 | 66<br>**Dy**<br>162.5 | 67<br>**Ho**<br>164.9 | 68<br>**Er**<br>167.3 | 69<br>**Tm**<br>168.9 | 70<br>**Yb**<br>173.0 | 71<br>**Lu**<br>175.0 |
|---|---|---|---|---|---|---|---|---|---|---|---|---|---|---|
| 89<br>**Ac** | 90<br>**Th**<br>232.0 | 91<br>**Pa** | 92<br>**U**<br>238.1 | 93<br>**Np** | 94<br>**Pu** | 95<br>**Am** | 96<br>**Cm** | 97<br>**Bk** | 98<br>**Cf** | 99<br>**Es** | 100<br>**Fm** | 101<br>**Md** | 102<br>**No** | 103<br>**Lr** |